Power Systems

AF810973

More information about this series at http://www.springer.com/series/4622

Naser Mahdavi Tabatabaei
Sajad Najafi Ravadanegh
Nicu Bizon
Editors

Power Systems Resilience

Modeling, Analysis and Practice

Springer

Editors
Naser Mahdavi Tabatabaei
Department of Electrical Engineering
Seraj Higher Education Institute
Tabriz, Iran

Sajad Najafi Ravadanegh
Department of Electrical Engineering
Azarbaijan Shahid Madani University
Tabriz, Iran

Nicu Bizon
Faculty of Electronics, Communications
and Computers
University of Pitesti
Pitesti, Arges, Romania

ISSN 1612-1287 ISSN 1860-4676 (electronic)
Power Systems
ISBN 978-3-030-06852-3 ISBN 978-3-319-94442-5 (eBook)
https://doi.org/10.1007/978-3-319-94442-5

© Springer Nature Switzerland AG 2019
Softcover re-print of the Hardcover 1st edition 2019
This work is subject to copyright. All rights are reserved by the Publisher, whether the whole or part of the material is concerned, specifically the rights of translation, reprinting, reuse of illustrations, recitation, broadcasting, reproduction on microfilms or in any other physical way, and transmission or information storage and retrieval, electronic adaptation, computer software, or by similar or dissimilar methodology now known or hereafter developed.
The use of general descriptive names, registered names, trademarks, service marks, etc. in this publication does not imply, even in the absence of a specific statement, that such names are exempt from the relevant protective laws and regulations and therefore free for general use.
The publisher, the authors and the editors are safe to assume that the advice and information in this book are believed to be true and accurate at the date of publication. Neither the publisher nor the authors or the editors give a warranty, express or implied, with respect to the material contained herein or for any errors or omissions that may have been made. The publisher remains neutral with regard to jurisdictional claims in published maps and institutional affiliations.

This Springer imprint is published by the registered company Springer Nature Switzerland AG
The registered company address is: Gewerbestrasse 11, 6330 Cham, Switzerland

Dedicated to
all our teachers and colleagues
who enabled us to write this book,
and our family and friends
for supporting us all along

Foreword

Power Systems Resilience has always been a major challenge for the scientists and the engineers. However, in the recent years, the increased public solicitude for the resiliency of power systems and the resilience enhancement of cyber-physical systems used here has powerful motivated activities of the research and development in this field. More than 18,000 technical articles written in last decade and stored in "Science Direct Elsevier" database have "Power Systems Resilience" in their title, abstract, or list of keywords. Other 2000 are already published in 2018, so this interest will grow exponentially in next years. There were less than 1450 until 2000.

The aim behind this enormous progress is quite simple: both scientific and industry are involved in resilience enhancement of cyber-physical systems used for securing the critical infrastructure networks such as power systems.

The book generally explains the fundamentals and contemporary materials in Power Systems Resilience. It will be very efficient for electrical engineers to have the book which containing important subjects in considering modeling, analysis, and practice related to Power Systems Resilience. The book comprises knowledge and theoretical and practical issues as well as up to date contents in these issues and methods for controlling the reliabilities against attacks and risks in AC power systems.

Some textbooks and monographs are previously presented for people want to learn more on Power Systems Resilience. The worth of the present book is that it tries to put forward some practical ways for impact analysis of risk events on power system operation, bringing together the topics such "resiliency", "smart grid or microgrids", and "cybersecurity", which are now more and more organized. The editors wisely designated the topics to be preserved, and the chapters written by well-recognized experts in the field are placed in four sections.

The book introduces the reader to the modeling, analysis, and operation of resilience networks, and optimal operations of microgrids. Then, the main subjects related to planning, attacks, and recovery in resilience systems are presented and explained. The last section presents new research solutions for challenging issues appeared in last years. The book also includes informative case studies and many instances.

The book can be used in the classroom, to teach Power Systems Resilience courses to graduate students, and be suggested as further reading to undergraduate students in engineering sciences. It will also be a valuable information resource for the researchers and engineers concerned by Power Systems Resilience issues or involved in the development of cybersecurity applications.

Baku, Azerbaijan
May 2018

Academician Arif M. Hashimov
Institute of Physics
Azerbaijan National Academy of Sciences

Preface

It is obvious that the electrical infrastructure, which is responsible for providing electricity to all other critical infrastructures, has an important place in the functioning and development of today's world, and this network is to be reliable and at the same time to operate safely. In last decade, the needs for electric energy and the technological developments have increased annually, but the fact that the raw or renewable energy sources cannot be actuated in the same way has made it necessary to optimize their use. The power plants are dispersed in different regions and the optimum operation of the energy systems could be ensured by interconnection of these grid networks. Consequently, different problems arise during the planning and operation of microgrids, and the use of new technologies has become compulsory in order to solve the emerging problems, including enhance of Power Systems Resilience.

Resiliency and cybersecurity are two essential issues for today's power systems, but especially for smart grid or microgrids implemented in many developed countries. In the progress of power system components, one of the key aspects is to enhance their resilience in order to become more safety face to actual complex cyberattacks. The planning and operation of distribution network infrastructures all over the world are based on reliability principles of security and sufficiency. These principles enable distribution network to operate safety against common threats and as a result provide high-quality supply with few interruptions for consumers. However, it is evident that more considerations beyond the classical reliability are required to keep the lights on. Disruptive events, whether man-made or based on natural causes, can lead to component failures, and then breaking down the entire power system. The origin of the resiliency is word "resilio", which mean the ability of an object to return to its original state. Here, the ability of power system to recover quickly after such destructive events is called as power system resilience.

Chapter 1 gives an overview of challenges and possibilities of optimal planning of the conventional electric distribution network based on distribution network resilience enhancement. The optimal size and site of microgrid components and its topology are determined using optimization algorithm. A resilient-based fitness function is proposed to meet the resilient network requirement. To model the effect of the natural disasters on resilient-based distribution network planning, the geographical data for hurricane as a natural disaster is combined to create a spatial risk index map.

In this context, Chap. 2 introduces the main methods of developing a system which is robust to deliberate or involuntary perturbations, open in the matter of incorporating renewable energy sources, integrated, and efficient for the whole two-way energy chain. The connectivity types, the opportunities of developing the connectivity in the chain of the grid-connected or autonomous microgrids, were presented to increase their efficiency and performances. The Internet of Things (IoT) technologies will play an increased role in the design of smart grids, assuring accessibility and easy use, interoperability, and future developing. But the IoT developments must be implemented in high-security conditions, on the basis of Big Data principles, analytic analysis, and foresight.

So, the flexibility of the power system must be increased in order to have an adaptable structure against the various circumstances, balancing the power supply and demand in terms of intervals such as minute or hourly. The flexibility of power system is based on appropriate energy management strategy that fast and safety integrate distributed sources such as hydro, wind, and solar. Chapter 3 analyzes the challenges of the flexibility researches, which are focused on rapid deployment of distributed generation, pricing, standards, policies, and microgrids' integration to power system considering the customer features.

Therefore, the resiliency of power systems requires to be handled in terms of physical and cyber-damages. The resiliency is researched in three main topics as damage prevention, system recovery, and survivability of power system. The necessity of quantifying resilience metrics is an important challenge, which mostly depends on how to define the resiliency. The metrics used for quantifying the resiliency of power system are explained in Chap. 4. The metrics investigated in this chapter are quantitative, being defined based on the topology, hardware used, efficiency of the system, reliability indices, and also type and severity of the threat. The accurate assessment of each of these metrics can help to properly understand the concept of resilience in power systems.

It can be highlighted that abovementioned chapters are included in the first part of the book and provide a worthy background to the reader on the modeling, analysis, and operation of resilience networks. Both experimental and theoretical methods to different problems support to know the innovative aspects and cutting-edge information, world-widely, thereby the readers at various educational levels from undergraduate to the professionals can find interesting research topics in order to apply in their own studies.

Other main topic of the book is the optimal operation of the microgrids to enhance the resiliency of power systems. Currently, the number of microgrids is continuously increased in distribution network. In this view, the future advanced distribution network can be regarded as clusters of microgrids. Hence, the microgrid is the building block of smart distribution networks. The technical issues and economic constraints for microgrids' implementation, including and the social reasons, are analyzed in Chap. 5.

Here, it is proposing a framework for analyzing the optimal microgrid-based resilient distribution networks that are detailed in next two chapters of the second part of this book considering the increasing demand for electricity supply along with higher requirements for power quality and system reliability, and margins to usage the existing fossil fuels and minimization of the environmental pollutants.

Chapter 6 proposes an optimization strategy for energy management system to globally reach the minimum cost operation of microgrids in normal condition while meeting the adequacy in resiliency operation mode. The dispatchable unit status, energy storage, and adjustable loads scheduling and the energy trade status between microgrids in each hour are evaluated with stochastic modeling of load and renewable power generations. The algorithms that contain the heuristic methods used on Optimal Power Flow management are described and explained in Chap. 7.

Planning, attacks, and recovery in resilience systems are approached in the four chapters on the third part of this book. Chapter 8 presents an approach for Resilient Distribution System Expansion Planning (RDSEP) considering gas-fired Non-Utility DGs (NUDGs) and Demand Side Providers (DRPs). The RDSEP method explores the NUDGs and DRPs impacts on the planning paradigm. The RDSEP problem is decomposed into multi subproblems to optimize the investment, operational, and reliability costs. Each problem can be solved using embedded systems. On the other hand, the embedded systems control sensitive data and information depending where these systems are installed to accomplish required tasks.

Due to this aspect, cyber-criminals or hackers are motivated and determined to rob intellectual property of these systems through more and more sophisticated attacks. A huge problem in defending against these massive and various types of attacks is that in the last year's attacks increased their complexity while the knowledge of an attacker decreased significantly because of the tools and devices they can find in the online world and free market. Therefore, Chap. 9 is focused on embedded systems based on Field-Programmable Gate Array (FPGA) technology used for security against malicious and deliberate attacks, highlighting the risks, threats, and vulnerabilities that motivated hackers to perform these attacks in time.

The improvement of the power system resiliency by adopting on one hand preventive measures and on the other hand redesign of the damaged infrastructure is studied in Chap. 10. There are defined two types of extreme weather events (severe natural conditions and disasters), respectively, deliberate attacks on power systems, and the impact analysis of these events on power system operation and the system resiliency are analyzed to achieve a good risk management. The interruption period of the power system in extreme weather conditions can be effectively eliminated by using secondary generator systems based on fuel cells and renewable energy sources. The operation, the response time, and the performance of the extended secondary system are described in details here. The potential terrorist threats ranging from cyberattacks under their multiple aspects, to the direct attacks through physical destruction are also presented here, but a review of method and technologies is performed in Chap. 11 for resilience enhancement of the cyber-physical systems.

The last part of this book is dedicated to new research solutions for challenging issues appeared in last years. Chapter 12 presents main issues in securing critical infrastructure networks for smart grid based on SCADA, and other industrial control and communication systems. Chapter 13 presents an analysis of one of the components of the electricity quality, the continuity in the electricity supply of the consumers, indicating possibilities for improvement of the electricity. A case study is presented as a "self-healing" automation to isolate faults on a medium voltage line through reclosers using a General Packet Radio Service.

It is worth to mention that the book covers educational case studies and many examples in all chapters, so it will be of concern for all current researchers and specialists in that field, and for technicians as well. On the whole, comprehensive and complex world of power infrastructure systems such as smart grid, water distribution, telecommunications, and transportation systems have attracted the attention of many researchers and power engineers to build safe, stable, and resilient network to guarantee reliable power supply. Although there are many studies about prevention and protection, the resilience and vulnerability of power system to attack or natural disasters have found their own direction to move forward in researches.

As a conclusion, the researches and aims to advance concepts of cyber-physical Power Systems Resilience give more chances to reliability, quality demanded network, decreasing power failure level, network recovery, as well as reducing the magnitude and/or duration of disruptive events. Therefore, this book tries to highlight the difficulties of the basic methods on Power Systems Resilience and proposes advanced methods to solve these issues. All proposed methods were validated through simulation, experimental results, and case studies. The mentioned subjects will be of interest, challenge, and hard task for researchers, considering the new energy standards due to energy crisis and the intensive use of IoT in the future.

We hope that this book will be very efficient for students, researchers, and engineers, which learn and wish to work in this field, because the chapters of this book cover all important and challenging subjects related to Power Systems Resilience. The book comprises the knowledgeable and up to date contents that present the state-of-the-art equipment and methods used for the Power Systems Resilience. Finally, the main arguments that may recommend this book to be read are the following: (1) it is a comprehensive book on Power Systems Resilience; (2) covers the operating principles, design methods, and real applications; (3) introduces the metrics used for quantifying the resiliency of power system; (4) enables the FPGA-based embedded systems design for Power Systems Resilience; (5) introduces the cybersecurity concepts; (6) provides a comprehensive overview of cyberattacks and provide some practical solutions; and the last, but not the least, (7) can be used as a course text.

The editors and authors made all efforts to have a good book, and hope that interested readers to enjoy by reading this book and to be satisfied by its content.

Tabriz, Iran — Naser Mahdavi Tabatabaei
Tabriz, Iran — Sajad Najafi Ravadanegh
Pitesti, Romania — Nicu Bizon

Contents

Part I Modeling, Analysis and Operation of Resilience Networks

1 **Modeling and Analysis of Resilience for Distribution Networks** 3
Sajad Najafi Ravadanegh, Masoumeh Karimi and Naser Mahdavi Tabatabaei

2 **Power Systems Connectivity and Resiliency** 45
Horia Andrei, Marian Gaiceanu, Marilena Stanculescu, Iulian Nicusor Arama and Paul Cristian Andrei

3 **Power System Flexibility and Resiliency** 81
Ersan Kabalci

4 **Resilience Metrics Development for Power Systems** 101
Hossein Shayeghi and Abdollah Younesi

Part II Microgrids and Optimal Operations of Resilience Systems

5 **Resilience Thorough Microgrids** 119
Shahram Mojtahedzadeh, Sajad Najafi Ravadanegh, Mahmoudreza Haghifam and Naser Mahdavi Tabatabaei

6 **Optimal Scheduling of Networked-Microgrids to Resiliency Enhancement Under Uncertainty** 139
Pouya Salyani, Sajad Najafi Ravadanegh and Naser Mahdavi Tabatabaei

7 **Resilient Optimal Power Flow with Evolutionary Computation Methods: Short Survey** 163
Basar Baydar, Haluk Gozde, M. Cengiz Taplamacioglu and A. Osman Kucuk

Part III Planning, Attacks and Recovery in Resilience Systems

8 Multi-stage Resilient Distribution System Expansion Planning Considering Non-utility Gas-Fired Distributed Generation 193
Mehrdad Setayesh Nazar and Alireza Heidari

9 Malicious and Deliberate Attacks and Power System Resiliency 223
Fernando Georgel Birleanu, Petre Anghelescu and Nicu Bizon

10 Power Systems Recovery and Restoration Encounter with Natural Disaster and Deliberate Attacks 247
Horia Andrei, Paul Cristian Andrei, Marian Gaiceanu, Marilena Stanculescu, Iulian Nicusor Arama and Ioan Marinescu

11 Resilience Enhancement of Cyber-Physical Systems: A Review 269
Sanda Florentina Mihalache, Emil Pricop and Jaouhar Fattahi

12 Issues in Securing Critical Infrastructure Networks for Smart Grid Based on SCADA, Other Industrial Control and Communication Systems 289
Florentina Magda Enescu, Nicu Bizon and Carmen Maria Moraru

13 Continuity of Electricity Supply and Specific Indicators 325
Doru Ursu and Mariana Iorgulescu

Index 351

Contributors

Horia Andrei Doctoral School of Engineering Sciences, University Valahia of Targoviste, Târgovişte, Romania

Paul Cristian Andrei Department of Electrical Engineering, University Politehnica Bucharest, Bucharest, Romania

Petre Anghelescu Faculty of Electronics, Communications and Computers, University of Pitesti, Pitesti, Romania

Iulian Nicusor Arama Department of Control Systems and Electrical Engineering, University Dunarea de Jos Galati, Galaţi, Romania

Basar Baydar Baskent Electricity Distribution Company, Enerjisa, Ankara, Turkey

Fernando Georgel Birleanu Faculty of Electronics, Communications and Computers, University of Pitesti, Pitesti, Romania

Nicu Bizon Faculty of Electronics, Communications and Computers, University of Pitesti, Pitesti, Romania

Florentina Magda Enescu Faculty of Electronics, Communications and Computers, University of Pitesti, Pitesti, Romania

Jaouhar Fattahi Department of Computer Science and Software Engineering, Laval University, Quebec, Canada

Marian Gaiceanu Department of Control Systems and Electrical Engineering, University Dunarea de Jos Galati, Galaţi, Romania

Haluk Gozde Electronic Engineering Department, Military Academy, National Defense University, Ankara, Turkey

Mahmoudreza Haghifam Electric Transmission and Distribution Research Lab, Faculty of Electrical and Computer Engineering, Tarbiat Modares University, Tehran, Iran

Alireza Heidari School of Electrical Engineering and Telecommunication, University of New South Wales, Sydney, Australia

Mariana Iorgulescu Faculty of Electronics, Communications and Computers, University of Pitesti, Pitesti, Romania

Ersan Kabalci Department of Electrical and Electronics Engineering, Faculty of Engineering and Architecture, Nevsehir Haci Bektas Veli University, Nevsehir, Turkey

Masoumeh Karimi Smart Distribution Grid Research Lab, Azarbaijan Shahid Madani University, Tabriz, Iran

A. Osman Kucuk Electrical and Electronics Engineering Department, Gazi University, Ankara, Turkey

Naser Mahdavi Tabatabaei Department of Electrical Engineering, Seraj Higher Education Institute, Tabriz, Iran

Ioan Marinescu Doctoral School of Engineering Sciences, University Valahia of Targoviste, Targoviste, Romania

Sanda Florentina Mihalache Automatic Control, Computers and Electronics Department, Petroleum-Gas University of Ploiesti, Ploiesti, Romania

Shahram Mojtahedzadeh Department of Electrical Engineering, Azarshahr Branch, Islamic Azad University, Azarshahr, Iran

Carmen Maria Moraru National Institute of Research-Development for Cryogenic and Isotope Technologies, Ramnicu Valcea, Romania

Sajad Najafi Ravadanegh Smart Distribution Grid Research Lab, Azarbaijan Shahid Madani University, Tabriz, Iran

Mehrdad Setayesh Nazar Shahid Beheshti University, A.C., Tehran, Iran

Emil Pricop Automatic Control, Computers and Electronics Department, Petroleum-Gas University of Ploiesti, Ploiesti, Romania

Pouya Salyani Smart Distribution Grid Research Lab, Azarbaijan Shahid Madani University, Tabriz, Iran

Hossein Shayeghi Department of Electrical Engineering, University of Mohaghegh Ardabili, Ardabil, Iran

Marilena Stanculescu Department of Electrical Engineering, University Politehnica Bucharest, Bucharest, Romania

M. Cengiz Taplamacioglu Electrical and Electronics Engineering Department, Gazi University, Ankara, Turkey

Doru Ursu Energy Distribution Oltenia, Craiova, Romania

Abdollah Younesi Department of Electrical Engineering, University of Mohaghegh Ardabili, Ardabil, Iran

Abbreviations and Acronyms

AAR	Automatic Reserve Lockout
AAR	Automatic Alternate Routing
AAR-JT	Automatic Reserve Lockout for JT Distribution Networks
AC	Alternative Current
AC	Assimilation Coefficient
ACSI	Abstract Communication Service Interfaces
ADMS	Advanced Distribution Management System
AGA	Adaptive Genetic Algorithm
AI	Analog Input
AI	Artificial Intelligence
AIS	Artificial Immune Systems
AIT	Average Interruption Time
ANN	Artificial Neural Network
AP	Access Point
ASCII	American Standard Code for Information Interchange
ASIC	Application Specific Integrated Circuit
ATM	Asynchronous Transfer Mode
AV	Antivirus
AvLD	Average Load Factor
BA	Bat Algorithm
BOA	Bayesian Optimization Algorithm
CA	Control Access
CAD	Computer-Aided Design
CAES	Compressed Air Energy Storage
CAGR	Compound Annual Growth Rate
CCITT	Consultative Committee for International Telegraphy and Telephony
CCTV	Closed-Circuit Television
CE	Currency Exchange Factor
CHP	Combined Heat and Power

CI	Computer Intelligence
CL	Critical Load
CLP	Classification, Labeling, and Packaging
CPS	Cyber-Physical System
DAS-MT	Data Acquisition System Multi-Tasking
DC	Direct Current
DCS	Distributed Control System
DDoS	Distributed Denial-of-Service
DER	Distributed Energy Resources
DES	Data Encryption Standard
DFR	Digital Fault Recorders System
DG	Distributed Generators
DI	Digital Input
DisCo	Distribution Company
DNP3	Distributed Network Protocol
DO	Digital Output
DOPF	Dynamic Optimal Power Flow
DoS	Denial-of-Service
DPA	Differential Power Analysis
DR	Demand Response
DRP	Demand Side Providers
DRP	Disaster Recovery Plan
DRRI	Device for Reset Trigger Failure
DSM	Demand Side Management
DSO	Distribution System Operator
DSS	Decision Support System
DT	Distribution Transformer
EC	Evolutionary Computation
EDNS	Expected Demand Not Served
EENS	Expected Energy Not Served
EMA	Electromagnetic Analysis
EMS	Energy Management System
ENS	Energy Not Supplied
ENSC	Energy Not Supplied Cost
EPRI	Electric Power Research Institute
ERCOT	Electric Reliability Council of Texas
ESIC	Energy Storage Integration Council
ESS	Electric Storage System
EU	European Union
EV	Electric Vehicle
FAST2	Flexibility Assessment Tool
FC	Fuel Cell
FCH-JU	Fuel Cells and Hydrogen Joint Undertaking
FIB	Focused Ion Beam
FN	Feasible Network

FO	Optical Fiber
FPGA	Field-Programmable Gate Array
FTP	File Transfer Protocol
FWA	Fireworks Algorithm
G	Graph
G2V	Grid-to-Vehicle
GIS	Geographic Information System
GIVARIII	Grid Integration of Variable Renewables
GLS	Guided Local Search Optimization Algorithm
GPRS	General Packet Radio Service
GPS	Global Positioning System
GSM	Global System for Mobile communications
HAN	Home Area Network
HAZUS-MH	Federal Emergency Management Agency's Multi-Hazard
HILP	High-Impact, Low-Probability
HMI	Human Machine Interface
HTTP	Hypertext Transport Protocol of the Internet
HV	High Voltage
IBM	International Business Machines
IC	Information and Communication
IC	Integrated Circuit
ICA	Independent Component Analysis
ICCP	Inter-Control Center Protocol
ICS	Industrial Control System
ICT	Information and Communication Technology
IEC 61850	Standard Protocol
IED	Intelligent Electronic Devices
IoT	Internet of Things
IP	Intellectual Property
IP	Internet Protocol
IPS	Intrusion Prevention System
IT	Information Technology
JESS	Java Expert System Shell
LAN	Local Area Network
LEA	Overhead line
LOLP	Loss of Load Probability
LP	Linear Programming
LV	Low Voltage
M2M	Machine-to-Machine (communication)
MAIFI	Momentary Average Interruption Frequency
MCP	Marginal Clearing Price
MG	Microgrid
MMCS	Networked-Microgrid Control System
MMG	Multiple Microgrid
MMS	Manufacturing Message Specification

MST	Minimum Spanning Tree
MT	Microturbine
MU	Monetary Unit
MV	Medium Voltage
MVAR	Mega Volt Ampere Reactive
MW	Mega Watt
NC	Normalized Constraint
NERC	North American Electric Reliability Corporation
NIAC	National Infrastructure Advisory Council
NIST	National Institute of Standards and Technology
NLP	Natural Language Processing
NMG	Networked-Microgrid
NUDG	Non-Utility DG
OCP	Operational Control Points
OPF	Optimal Power Flow
OS	Operating System
OSI	Open System Interconnection
OT	Operation Technology
P2P	Peer-to-Peer
PA	Power point or transformer station
PBIL	Population-Based Incremental Learning
PC	Central Point
PC	Personal Computers
PCB	Printed Circuit Board
PCC	Point of Common Coupling
PCWL	Path Combination Without Loop
PDF	Probability Density Function
PEDA	Protection Engineering Diagnostic Agents
PEM	Point Estimation Method
PhVPP	Photovoltaic Power Plants
PLC	Programmable Logic Controller
PMU	Phasor Measurement Unit
PN	Possible Network
PROFIBUS	Process Field Bus
PS	Performance Standard
PS	Power Systems
PUF	Physical Unclonable Function
PV	Photovoltaic
PV	Photovoltaic Cell
QoS	Quality of Service
QPS	Quiet, Paranoid, Skilled Hacker
RAR	Automatic Resumption of Reserve
RATT	Relative Arrival Time Technique
RCR	Remote Control Reclosers
RCS	Remote Control Separators

RCS	Resilience Control System
RDSEP	Resilient Distribution System Expansion Planning
RES	Renewable Energy Sources
RIWL	Route Incorporation Without Loop
RSA	Ron Rivest, Adi Shamir, and Len Adleman
RTOS	Regular Operating Systems
RTU	Remote Terminal Units
SAIDI	System Average Interruption Duration Index
SAIFI	System Average Interruption Frequency Index
SCADA	Supervisory Control and Data Acquisition
SCOPF	Security-Constraint OPF
SG	Smart Grid
SM	Smart Meter
SoC	State of Charge
SoH	State of Health
SOPF	Stochastic Optimal Power Flow
SPA	Simple Power Analysis
TASE 2	Telecontrol Application Service Element 2
TCP/IP	Transmission Control Protocol/Internet Protocol
TSC	Tabu Search Continuous Algorithm
TSO	Transmission System Operator
TV	Television
UC	Unit Commitment
UDG	Utility-Owned DG
UNIX	Linux Operating System
UPS	Uninterruptible Power Supply
V2G	Vehicle-to-Grid
VG	Variable Generation
VoD	Voice and Data
VSC	Voltage Source Converter
VT	Voltage Transformer
WAMS	Wide Area Measurement System
WAN	Wide Area Network
WAP	Wireless Application Protocol
WASRI	Weighted Average System Reliability Index
WDO	Wind-Driven Optimization Algorithm
WF	Wind Farms
Wi-Fi	Wireless Fidelity
WLANs	Technology Makes Wireless LANs
WSAN	Wireless Sensor and Actuator Network
WSN	Wireless Sensor Network
WT	Wind Turbine

Part I
Modeling, Analysis and Operation of Resilience Networks

Chapter 1
Modeling and Analysis of Resilience for Distribution Networks

Sajad Najafi Ravadanegh, Masoumeh Karimi and Naser Mahdavi Tabatabaei

Abstract Electric power distribution networks are constructed and expanded among wide geographical areas. Traditionally the focus of distribution networks planning is mainly on network reliability improvement with minimum cost considering the technical constraints. The aim of such vision is based-on the fact that distribution network outages occurred on the network component because of conventional faults with high frequency and low impact characteristics. Nowadays the power distribution networks encounter with unwanted weather conditions and natural disasters facing the network with high impact low probability events on distribution network that can be damage the network components widely and permanently. It can cause costumer interruption and loss of load with high financial cost. The aim of this chapter is the optimal planning of conventional electric distribution network based-on distribution network resiliency enhancement. In this work the optimal size and site of network component and its topology is determined using optimization algorithm. A resilient-based fitness function is proposed to meet the resilient network requirement. To model the effect of the natural disasters on resilient based distribution network planning, the geographical data for hurricane as a natural disaster is combined to create a spatial risk index map. In this work a new methodology is proposed to establish a rational relation between network component fragility curves, component geographical location and hurricane spatial risk index. The proposed method is tested on a real large scale distribution network.

S. Najafi Ravadanegh (✉) · M. Karimi
Smart Distribution Grid Research Lab, Azarbaijan
Shahid Madani University, Tabriz, Iran
e-mail: s.najafi@azaruniv.ac.ir

M. Karimi
e-mail: m.karimi@azaruniv.edu

N. Mahdavi Tabatabaei
Department of Electrical Engineering, Seraj Higher
Education Institute, Tabriz, Iran
e-mail: n.m.tabatabaei@gmail.com

© Springer Nature Switzerland AG 2019
N. Mahdavi Tabatabaei et al. (eds.), *Power Systems Resilience*, Power Systems,
https://doi.org/10.1007/978-3-319-94442-5_1

Keywords Distribution network · Fitness function · Geographical location Natural disaster · Optimization · Reliability · Resilience · Spatial risk index

Nomenclatures

t_e	Time that event occurred
R_{pe}	Post-event resilience level
t_r	Initial time of restorative state
R_0	Initial resiliency level before event
t_{pr}	Initial time of post-restorative state
t_{pir}	End time of infrastructure recovery
t_{pe}	End time of event progress
t_{ir}	Initial time of infrastructure recovery
λ_{Tr}	Transformer failure rate with respect to hurricane
λ_{Pole}	Pole failure rate with respect to hurricane
λ_{Con}	Conductor failure rate with respect to hurricane
w	Wind speed
h	Number of hurricane per year
P	Poisson probability distribution function
λ	Average number of hurricane
S_{HV}^{hv}	HV substation transformer losses
$\cos\phi_{dt}^{Tr}$	Power factor for HV substation
DT_{dt}	Distribution transformer *dt*
P_{DisTr}^{dt}	Active power for MV distribution transformer
N_{DistTr}	Number of distribution transformer
$Cost_{HV}$	Cost of HV transformer (Currency)
HV^N	Number of HV transformer
CC_{HV}	Capital cost of HV transformer
S	Capacity of HV transformer (MVA)
$CL(S_{HV}^{hv})$	Cost of HV transformer losses
T_P	Planning period
γ_i	Cost of energy (kwh/Currency)
$P_{NLL}(S_{HV}^{hv})$	No load loss for HV transformer
$P_{SCL}(S_{HV}^{hv})$	Ohmic loss for HV transformer
$AvLoss(S_{HV}^{hv})$	Average loss factor for HV transformer
$ELCF$	Energy loss cost factor, Currency/kWh
HV_{Load}	HV transformer load
P_{dt}	Active power of MV transformer *dt*
P_z	Active power of load zone *z*
$N_{z,dt}$	Number of load *z* connected to transformer *dt*
$Load_z^{dt}$	Total load of distribution transformer *dt*
$Cost_{DisTr}$	Cost of MV transformer
CC_{DisTr}^{dt}	Capital cost of MV transformer *dt*

CL^{dt}_{DisTr}	Loss cost of MV transformer *dt*
$P_{NLL}\left(S^{dt}_{DisTr}\right)$	No load cost of MV transformer *dt*
$P_{SCL}\left(S^{dt}_{DisTr}\right)$	Ohmic loss cost of MV transformer *dt*
$TL\left(S^{dt}_{DisTr}\right)$	MV transformer loading level
$ALSF$	Average loss factor for MV transformer
k	Number of nodes in graph G
$Cost_{MVF}$	Cost of MV feeder
f	Feeder counter
N_{MVF}	Total number of MV feeder
$I\left(MVF_f\right)$	Current of MV feeder f
CC^{f}_{MVF}	Capital cost of MV feeder j
d_f	Direct length of MV feeder f
VD_{MVF}	Voltage drop within MV feeder f
$VD_{MV,\max}$	Maximum voltage drops within MV feeder f
Cost_F	Total cost function
$Cost^{hv}_{HVDist}$	Total cost of HV substations set
$Cost^{dist}_{MVDist}$	Total cost of MV substations set
$Cost^{f}_{MVFeeder}$	Total cost of MV feeders set
N_{HV}	Number of HV substations
N_{MV}	Number of MV substations
N_{Feeder}	Number of MV feeders
RI_{Con}	Index for conductors resiliency
RI_{Poles}	Index for poles resiliency
RI_{Trans}	Index for transformers resiliency
ω_{Poles}	Coefficient for poles of feeder f
ω_{Trans}	Coefficient for transformers of feeder f
ω_{Con}	Coefficient for conductors of feeder f
$RI_{Feeder}(f)$	Index for feeder resiliency
$F_{Dist}(s_i, s_j)$	Distance between substations s_i and s_j
$Poles_{Num}$	Pole number
x_{s_i}, y_{s_i}	Coordinate for MV substation s_i
x_{s_j}, y_{s_j}	Coordinate for MV substation s_j
F_{Con}	Number of conductor segments along a feeder
N_{pole}	Total number of poles
$RI_{Network}$	Index for network resiliency
α	Cost functions weight.

1.1 Introduction

The planning and operation of distribution network infrastructures all over the world are based on reliability principles of security and sufficiency. These principles enable distribution network to operate safety against common threats and as a result provide high quality supply with few interruptions for consumers. However, it is evident that more considerations beyond the classical reliability are required to keep the lights on. Disruptive events, whether man-made or natural causes can lead to component failures and then breaking down the entire system. Also, in the following a few of High-Impact, Low-Probability (HILP) events that a power infrastructure can be effected are listed. Different destructive catastrophes occurred during last decade such as: The U.S. northeastern states which were affected by Hurricane Sandy in 2012 and caused several serious destructions and outages.

Another example is about Queensland's huge flood lead to massive damages to substations, poles, transformers and overhead wires and as a result, 150,000 interrupted customers that experienced power outages. It should be noted that the differences between blackouts and disasters must be taken into account by considering HILP events. Due to the possibility of power grid blackouts as the result of unpredictable faults, a reliable power system should be planned to minimize the amount of interruptions. While, a disaster is the consequence of serious calamity usually never occurred before. Also, based on the severity of catastrophe and the extent of the affected area, it can last a long period of time. Therefore, it is concluded that based on aforementioned concepts, power system infrastructures must be reliable and resilient to common faults and disasters, respectfully. The origin of the Resilience (or resiliency) is word "resilio," which mean the ability of an object to return to its original state. Here, the ability of power system to recover quickly after these destructive HILP events is called as power system resiliency. Furthermore, resiliency refers to power system capability to enhance the adaptation of its operation and infrastructure through situation assessment, rapid response and effective recovery strategies (considering failure probabilities and reduced time to recover) in order to mitigate the vulnerability to severe calamities.

On the whole, comprehensive and complex world of power infrastructure systems such as smart grid, water distribution, telecommunications and transportation systems have attracted the attention of many researchers and power engineers to build safe, stable and resilient network to guarantee reliable power supply. Although, there are many studies about prevention and protection, the resilience and vulnerability of power system to attacks or natural disasters find their own direction to move forward in researches.

1.1.1 Influence of Weather and Climate Change on Power System Components [1–3]

The reliability and resilience of whole power infrastructure operation of electrical components have been extremely affected by weather events, such as [1]:

- Transfer capacity of transmission lines can be limited by maximum permissible operating temperature which high temperature can lead to energy losses.
- Overhead transmission and distribution lines can be damaged by hurricanes and wind storms.
- Also, overhead lines and towers can be affected and failed by cold waves, heavy snow. Meanwhile, conducting path been provided by gathered snow on insulators can lead to fault.
- Short-circuit faults, as the result of lightning strikes on overhead conductors can cause disconnection of the lines and are considered as transient which can be rapidly restored to service.
- Unlike aforementioned weather events, rain and floods do not damage overhead transmission lines, but have extreme effect on substation equipment.

All in all, it can be concluded that the impact of severe weather can be direct or indirect destructive events, such as tower collapses due to high winds, or affecting the normal operation of components, like heat and cold waves. Also, the situation of electrical equipment is the other significant option that the severity of damages as the result of weather events can be depend on. It means that aged components are more vulnerable to weather events than newer ones. Furthermore, the exposure of power system to climate changes which can pose great impact on the discussed weather parameters, affect the operation and reliability of power systems [2, 3].

Therefore, due to critical impact of weather events on electrical components, more researches are needed to assess their severity in detail. It should be noted that due to traditionally designed power system, and in order to prevent imposing threats as the results of climate changes and weather events, the improvement rate of power system must be rapid enough to adapt to great and disruptive climate patterns. It is clear that making all electrical equipment robust enough in order to meet entirely power system resiliency metrics, seems impossible or at least much cost effective.

1.1.2 Realization of Power Systems Resilience

After the first explanation of resilience in 1973 by C.S. Holling, different definitions of resilience in various aspects such as safety management, organizational, social-ecological, and economic have been presented. In terms of significant role of critical infrastructures of power systems, the concept of resilience should be clearly understandable. In this regard, some energy organization engineering organizations all over the world like U.K. Energy Research Centre and the U.S. Power Systems

Engineering Center, differentiate the concepts of resilience and reliability. Based on the U.K. organization's definition, resilience includes reliability and resistance, redundancy, response, and recovery.

The other interpretation of resilience explained by National Center for Earthquake Engineering Research, declares that robustness, redundancy, resourcefulness, and rapidity consist the framework of word resilience. Based on discussed definitions, there are numerous number of explanations about electrical components resilience that considered the important feature of networks to high-impact, low-probability events as recovery rate of power system after being degraded.

Consequently, the following explanations are provided in order to help distinguish the concepts of resilience and reliability:

- Reliability has been defined for high probability, low impact events, while resilience have been used for low probability, high impact shocks;
- Reliability is static but resilience can be adaptive, ongoing, short and long term;
- Reliability assesses power system states, while resilience considers transition times between states in addition to power system states;
- The concept of both reliability and resilience deal with customer interruption time, while resilience considers recovery time of power system after damage.

1.1.3 Resilience Curve Associated with an Event

Figure 1.1 illustrates resilience curve versus time considering disturbance level. Also, significant parameters of power system to cope with disruptive weather events are shown in this figure.

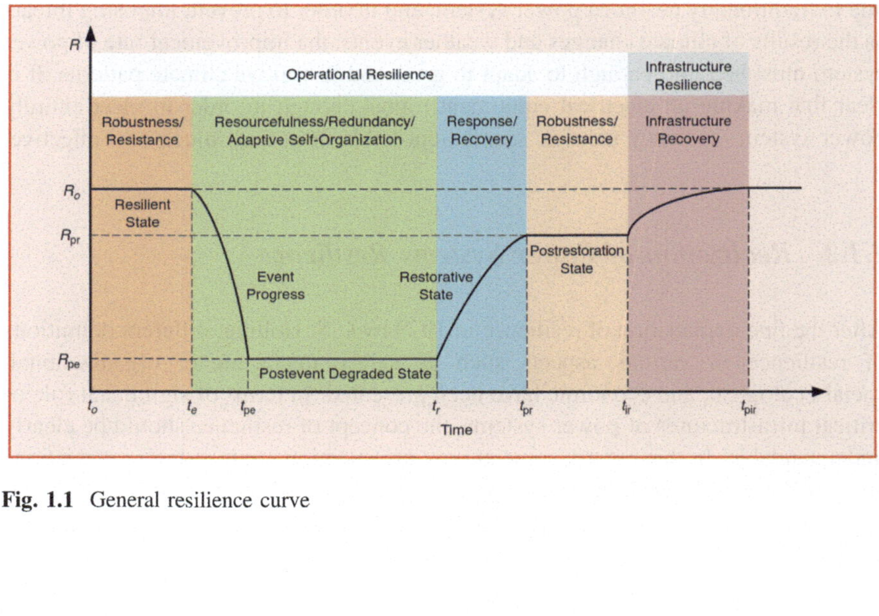

Fig. 1.1 General resilience curve

This curve helps the planners in the context of assessment of power system resilience. Robustness and resistance are the key features of system to make it capable to cope with events before time t_e, i.e., before event occurred. Here, the resilience level of power system is indicated by R, while R_o shows the well planned and operated state of electrical network. The ability of operational flexibility provides efficient conditions for planners to make system more resilient. After event occurred, the system stymie in post-event degraded state and its resilience level reduced remarkably to R_{pe} level. Also, the key features of this state are considered as: resourcefulness, redundancy, and adaptive self-organization which enable the system to cope with the new circumstances that never happened before and as a result minimize the influence of events. It can be concluded that the resilience level at time t_r is decreased to $R_0 - R_{pe}$. The third state is known as restorative stage which its significant options are system's fast response and recovery as quickly as possible. Following restorative state, the system enters to post-restoration stage with resilience level $R_{pr}(R_{pr} < R_0)$.

In the context of provided comparison between operational and infrastructure resilience curves, it can be shown that due to inherent feature of infrastructures, the rate of recovery time to pre-event state for its resilience is more time consuming, i.e., $(t_{pir} - t_{pr}) > (t_{pr} - t_r)$. Also, some resiliency measures may differ from operational and infrastructure point of view. For example, undergrounding overhead transmission and distribution lines enhance the operational resiliency of system but in the case of damaged ones, it takes much longer time and effort to repair. Furthermore, the transition times between the power system states (i.e. $t_{pe} - t_e$, $t_{pr} - t_r$, $t_{pir} - t_{ir}$) take the attention of planners to assess system's resiliency.

1.2 Optimal Distribution Network Planning General Model for Resiliency Enhancement [4]

Boosting resilience of power system due to high impact low probability weather events and climate changes is one of the most vital and essential issues that should be taken into consideration. In this regard, many energy associations all over the world are putting their efforts towards enhancement and performing grid resilience measures. These measures are classified in two categories as short-term and long-term that short-term metrics refer to relative actions before, during and after the weather event, and long-term, refer to the long-term scheduling of power system to make it sufficient resilient and robustness to future weather events and climate change.

1.2.1 Short-Term Resilience Measures [5]

In order to enhance power system resiliency, several efficient solutions for short term resilience measures have been suggested before, during and after event occurrence. These actions can be listed as follows [5].

The most vital preventive actions to mitigate noticeably the influences of natural events before occurrence are:

A. Proper forecast of the weather event's location and severity aims at minimizing generation in the most vulnerable area;
B. High numbers of repair and recovery crews prepositioned effectively to act as quick as possible;
C. Backup equipment provide rapid replacement actions of degraded components; The rest of preventive actions are:
D. Considering near networks;
E. Other actions like: more resilient design of system, reserve planning, and pay attention to smart techniques such as demand side management. Furthermore, most significant corrective actions during natural events are known as:
F. How effectively the system's state can be monitored to improve system's awareness and apply applicable techniques where necessary;
G. Check critical communications failures since communication make it possible to transfer data between system operators and crews;
H. Recovery and repair crews must be coordinated to be able to prevent increasing disturbance;
   Other solutions can be mentioned as followings. Finally, after event occurred these actions seems efficient:
I. Evaluate precisely the degraded level of damaged components to make efficient decisions and start restoration stage to reconnect customers.

1.2.2 Long-Term Resilience Measures

This section focuses on long term resilience measures to reduce disruptive effect of unpredictable weather events and climate changes and so make the system more resilient. These long term measure are described in the following:

- After event occurrence, in terms of operational procedure, defects and imperfections of system can be diagnosed which leads to the enhancement of risk metrics' assessment, emergency schemes and vegetation management [6–8].
- Furthermore, hardening techniques enable system to be much more robust to new severe emergencies. Some efficient actions such as undergrounding the distribution or transmission lines, building more transmission facilities and re-location transmission lines in the less-effected areas would help mitigate impacts and manage the situation [9–12].

- Recently, smart methods have been attracted researchers attention as much more efficient actions to make system sufficient resilience. There are various resilience-boosting actions in the context of smart techniques like, distributed energy resources (DER), including energy storage, distributed generation and demand side management. Also, the emergence of the concept of microgrids are based on the capability of DER's fast response and advanced control schemes which enable microgrids to be operated in two on-grid and off-grid modes [13–15, 20]. So, since the ability of microgrids in self-operating state, it can be separated from network during weather events and as the result increase system's robustness and resilience. It should be noted that noticeable reduction in need of transmission components can be considered as other prons of self-governing mode of microgrids in boosting system's resilience. All together provide effective deployment and control of power system to meet the changes and deal with climates [16–19].

1.2.3 Power Grid Resilience Enhancement

Recently, the vital role of actions about boosting grid resilience to high impact low probability events have been taking into consideration. These actions enable power system to maintain more stable and adequate during and after the events. Resilience measures, system's response following the event and risk management result to enhance efficient grid resilience (Fig. 1.2). Satisfactory of hardening and operational criteria actualize aforementioned resilience aims. Actually, hardening actions are referred to infrastructure reinforcement and operational measures are assumed as control based efforts to provide system more resilient to face destructive future weather events. From cost perspective hardening measures are more cost effective rather than operational actions while, enhancing infrastructure reinforcement make system much more resilient to future disasters. Finally, in order to achieve significant enhancement of power system resilience, system must be efficiently stronger and also smarter developed.

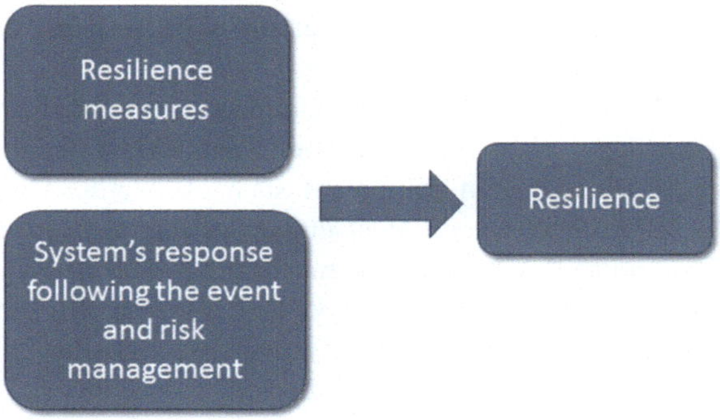

Fig. 1.2 Boosting power system resilience

1.3 Distribution Network Component Damage and Fragility Modelling

The behavior of electric power components facing weather events such as hurricane are described by fragility curves. Although, several states are supposed for modelling power grid components, mostly two states (fail or survive) are considered. A power grid is divided into three main sections known as: generation, transmission, and distribution. It should be noted that since of high reliability of generation side, it is not investigated in components' failure assessment this time [21, 23]. Key electrical components of transmission and distribution system can be mentioned as: electrical substations, transmission lines, towers, transformers, conductors and protective devices and etc.

Also, in order to analysis and assess outages, it is essential to consider a proper demand model based on specific hurricane. So, as an effective solution, Federal Emergency Management Agency's Multi-Hazard (HAZUS-MH) assessment tool FEMA 2008) can be used to predict destructive events of hurricanes, floods, and earthquakes on key electrical components [22]. Also, for example, the probability of a substation's vulnerability to natural events can be taken using fragility functions, obtained from HAZUS-MH 4 internal files (FEMA 2008).

The fragility functions illustrate electric power equipment strength against winds and flood from the hurricane event. Moreover, vulnerability or failure probability of electrical components and their collapse limit are described by fragility curves. Also, it should be noted that while one or more number of support structures (poles or towers) of a transmission line fails, it can be considered as an entire transmission line's failure. Distribution lines transfer electric power from transmission substations to local distribution load points, and then customers are fed through service drops. In the lack of adequate data or experimentally obtained fragility functions, following method approximates failures for transmission and distribution system components.

The proposed exponential damage model relates the failure rates of these components to wind speed. Poisson distribution considering failure rates as λ_{Tr}, λ_{Pole} and λ_{Con} are incorporated to model failures of distribution equipment. Related Eqs. (1.1)–(1.3) equations are described as follows.

For example, a typical fragility curve of an electric power component associated wind speed considering, critical and collapse limits of component are illustrated in Fig. 1.3.

The power system component damage model estimate components failure rates regarding certain hurricane conditions. While there are different types of power distribution system equipment such as overhead and underground components, therefore there are different classes of damage models. Distribution poles, spans, Pad-mount device such as transformers and conductors' damages are the main equipment that should be modelled from fragility point of view. Considering the above infrastructure damage model it is possible to compute total infrastructure damage given the hurricane intensity.

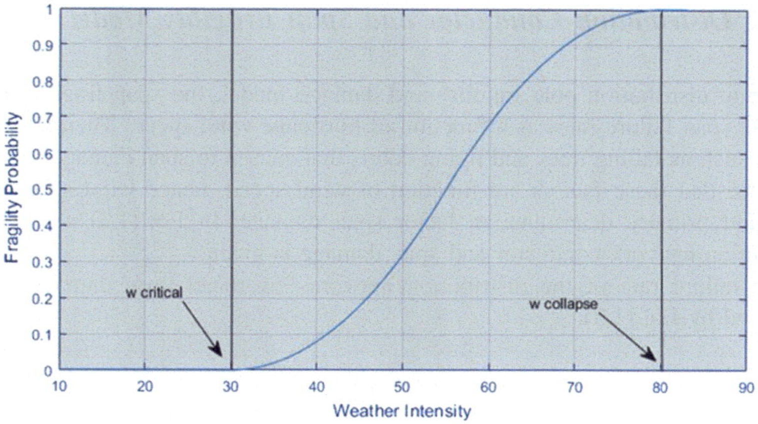

Fig. 1.3 A fragility curve showing the failure probability of a component versus weather intensity

1.3.1 Distribution Pole Fragility Model

To model the distribution overhead lines damage models it is necessary to have an available historical data. Hence the model for the other utilities may be different data set and therefore the damage model for this utilities may be different. The pole failure rate is given by dividing the number of poles issued during storm restoration by the total number of poles exposed to hurricane force. The pole failure rate with respect to hurricane wind speed is given in Eq. (1.1).

$$\lambda_{Pole} = 10^{-4} \mathrm{e}^{0.0421w} \tag{1.1}$$

where w is site-specific wind speed. It is possible to model the pole failure rate with other mathematical function.

1.3.2 Distribution Transformer Fragility Model

The fragility curve for distribution transformer damage model is given in Eq. (1.2).

$$\lambda_{Tr} = 2 \times 10^{-7} \mathrm{e}^{0.0834w} \tag{1.2}$$

In the above equation λ_{Tr} is the failure rate for distribution transformer damage model.

1.3.3 Distribution Conductor and Span Fragility Model

Similar to distribution pole fragility and damage model, the span fragility model estimate span failure rates as a function of hurricane wind speed. Therefore, other factors such as falling trees and flying debris that caused to span damage. It should be notice that these factors are function of wind speed, hence wind speed is the main and primary determinative factor span damage. In Eq. (1.3) a model for distribution network conductor and span damage is given.

The failure rate of the distribution networks overhead lines damage can be explained by Eq. (1.3).

$$\lambda_{Con} = 8 \times 10^{-12} w^{5.173} \tag{1.3}$$

In Eq. (1.3), λ_{Con} is the failure rate of the distribution lines with respect to hurricane wind speed.

1.4 Hurricane Model

In this section, we discuss on hurricane mathematical model, in which different uncertainties related in hurricane caused it necessary to extend a probabilistically model for hurricanes system. Reviewing of hurricane modeling texts indicate that there are some basic methods to modeling hurricane such as statistical models using probability distribution functions, empirical models and sampling approach. In some cases, to reach the cost and benefit model, it is necessary to use a combination of the above modeling methods. Some hurricanes characteristics that are modeled statistically involved hurricane occurrence, landing position, approach angle, translation velocity, central pressure difference, maximum wind speed, radius to maximum wind, gust factor, wind speed decay rate, central pressure filling rate and radial wind field profile.

In this chapter only the hurricane occurrence is modelled probabilistically. Most of the characteristics have a standard probability distribution functions. A random number is used for each modelled hurricane to determine the value for the model. The hurricane model is given in Eq. (1.4). In this equation this natural hazard is modelled using Poisson distribution function.

$$P(h) = \frac{\exp(\lambda) \times \lambda^h}{h!} \tag{1.4}$$

In Eq. (1.4), P is the Poisson probability distribution function that show the annual occurrence of the hurricane. In this equation λ and h are the average number of hurricane and number of hurricane per year respectively.

1.5 Resilient Distribution Network Planning to Reduce the Hurricane Damage

In this section a model for distribution network resilient planning is presented. The presented model generalized conventional distribution network planning problem to a resilient-based planning considering distribution network components damage and fragility models. The proposed method updates the existing reliability and cost-based planning procedure to resilient based planning framework for assessment hurricane induced outages based on equipment damages of the distribution network. The outage model to damages in distribution networks is given in previous section. In general, distribution networks designed looped and operate radially.

1.5.1 High Voltage (HV) Substation Modeling

For a HV substation the load is the sum of all distribution transformer connected to current HV substation through their relevant MV feeder. The load supplied by *k*th S_{HV}^{hv} is given by Eq. (1.5).

$$S_{HV}^{hv} = \sum_{dt=1}^{N_{DistTr}} \frac{P_{DisTr}^{dt}}{\cos\phi_{dt}^{Tr} \cdot AvLd(DT_{dt})} \tag{1.5}$$

Regarding the S_{HV}^{k} capacity the cost of HV substations are given by:

$$Cost_{HV} = \sum_{hv=1}^{HV^N} \left\{ CC_{HV}\left(S_{HV}^{hv}\right) \cdot S\left(S_{HV}^{hv}\right) + CL\left(S_{HV}^{hv}\right) \cdot T_P \cdot 8760 \right\} \cdot \gamma_i \tag{1.6}$$

where

$$CL\left(S_{HV}^{hv}\right) = \left\{ \begin{array}{l} P_{NLL}\left(S_{HV}^{hv}\right) + \\ P_{SCL}\left(S_{HV}^{hv}\right) \cdot HV_{Load}^{2}\left(S_{HV}^{hv}\right) \cdot AvLoss\left(S_{HV}^{hv}\right) \end{array} \right\} \cdot ELCF \tag{1.7}$$

$$HV_{Load}\left(S_{HV}^{hv}\right) = \frac{\sum_{dt=1}^{N_{DistTr}} \left(P_{dt}\right)}{S_{HV}^{hv} \cdot \cos\phi_{HV}^{hv}\left(S_{HV}^{hv}\right)} \tag{1.8}$$

1.5.2 Medium Voltage Distribution Transformer Modeling

The load demand supplied by *i*th distribution transformer is determined as below and then the upper near standard range is chosen as S (the size of distribution transformer).

$$S_{DisTr}^{dt} = \frac{\left(\sum_{z=1}^{N_{LB,dt}} P_z\right)}{\cos \varphi_{DisTr}^{dt} \cdot AvLd\left(Load_z^{dt}\right)} \tag{1.9}$$

$$Cost_{DisTr} = \sum_{dt=1}^{N_{DisTr}} \left\{CC_{DisTr}^{dt} \cdot S_{DisTr}^{dt} + CL_{DisTr}^{dt} \cdot T_P \cdot 8760\right\} \cdot \lambda_i \tag{1.10}$$

where

$$CC_{DisTr}^{dt} = \left\{ \begin{array}{r} P_{NLL}\left(S_{DisTr}^{dt}\right) + \\ P_{SCL}\left(S_{DisTr}^{dt}\right) \cdot TL^2\left(S_{DisTr}^{dt}\right) \cdot ALSF\left(S_{DisTr}^{dt}\right) \end{array} \right\} \cdot ELCF \tag{1.11}$$

$$CL_{DisTr}^{dt} = \frac{\sum_{z=1}^{N_{LB,}} P_z}{\left(S_{DisTr}^{dt} \cdot \cos \varphi_{DisTr}^{dt}\right)} \tag{1.12}$$

1.5.3 Medium Voltage Feeder Modeling

One of the main sub-problem in distribution network planning is known as feeder routing. The route for a feeder can be selected from different point of view, such as minimum length, minimum cost, and best cross section and so on. Considering the resilient planning of the distribution network, in this section the feeder routing problem is modelled to reflect the resilient behavior of the system. There are many techniques to solve the graph representation of the network using graph theory. In this chapter a distribution network is represented using node-edge illustration. The candidate location of distribution transformers is indicated by graph nodes and the candidate feeder connecting the distribution transformer to HV substation is indicated by graph edges. For instance, the graph can be solved using minimum spanning tree (MST) to satisfy the minimum length of the tree and radially structure constraint.

1.5.3.1 Connectivity Check

The criterion used for connectedness is from "A graph G with k nodes is connected if and only if $(A + I)^{k-1}$ has no zero entries, where A is the adjacency matrix." The adjacency matrix for graph G is the $k \times k$ matrix, and its entries a_{ij} are given by Eq. (1.13).

$$a_{ij} = \begin{cases} 1 & \text{if } node_i \text{ and } node_j \text{ are adjacent;} \\ 0 & \text{otherwise.} \end{cases} \tag{1.13}$$

1.5.3.2 Minimum Spanning Tree Algorithm

In MST the goal is to find a tree with minimum cost (minimum total distance between nodes or other user defined costs on distances). There are different methods to solve the MST problem in the literature. In this chapter Prim's algorithm as one of the famous solvers is used to find the MST of graph. In Fig. 1.4 a graphical illustration of prim's algorithm is given.

1.5.3.3 Medium Voltage Feeder Cost

The cost function for the selected feeder is given by Eq. (1.14). The feeder cross section is selected using feeder current that is calculated by power flow analysis.

$$Cost_{MVF} = \sum_{f=1}^{N_{MVF}} \left\{ \begin{array}{l} CC_{MVF}^{f} \cdot d_f + \\ I^2(MVF_f) \cdot R_f \cdot ELCF \cdot T_P \cdot 8760 \end{array} \right\} \tag{1.14}$$

The constraints below must be satisfied

$$I(MVF_i) < I_M(MVF_f) \quad \forall f \in S_{HV}^{hv} \tag{1.15}$$

$$VD_{MVF} < VD_{MV,\max} \tag{1.16}$$

where, the $VD_{MV,\max}$ is chosen as 2% from Iranian standard.

The total fitness function for optimal design of distribution network can be defined as sum of HV, MV substation and MV feeders cost. In Eq. (1.17) the cost function for network planning is given.

$$Cost_F = \sum_{hv=1}^{N_{HV}} Cost_{HVDist}^{hv} + \sum_{dist=1}^{N_{MV}} Cost_{MVDist}^{dist} + \sum_{f=1}^{N_{Feeder}} Cost_{MVFeeder}^{f} \tag{1.17}$$

In the previous sub-section, the total cost function element is explained in detail considering network constraints.

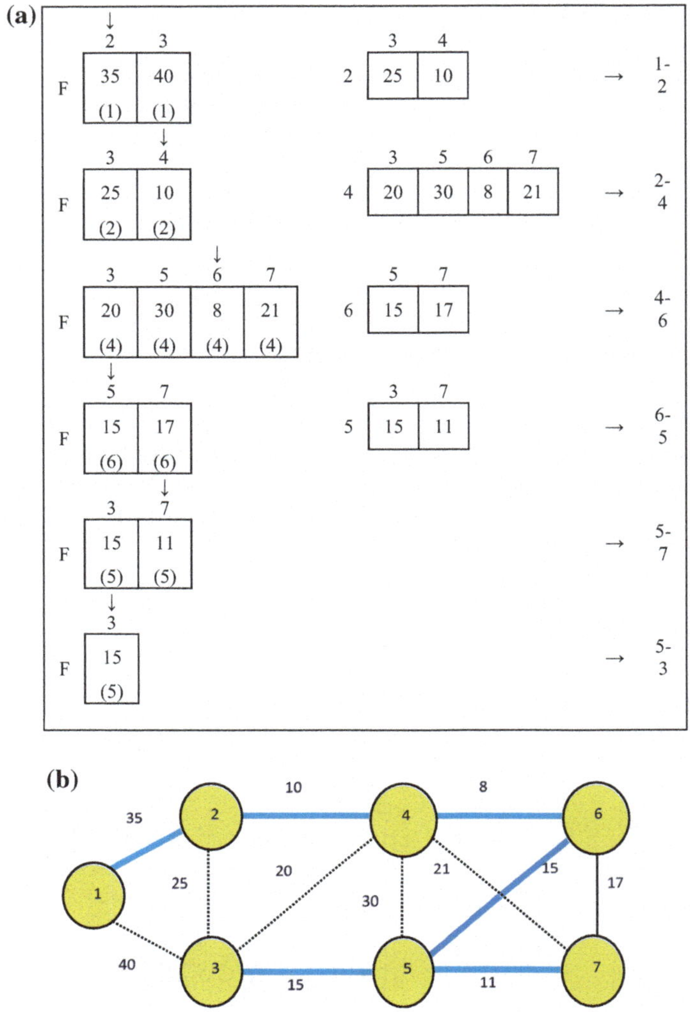

Fig. 1.4 **a** Prim's algorithm illustration, **b** Graph

1.6 Resilient Modelling

In this work the study area is divided into some predefined geographical blocks as hurricane wind speed block as Fig. 1.5. For each block the maximum hurricane wind speed is mapped from climatology database of the study area. In a distribution network infrastructure, the distribution load can be integrated on distribution poles which carry transformers. While the distribution lines routes across the geographical area and defined blocks, hence the distribution poles lie within the predefined blocks. For each block a probability wind speed distribution function based on

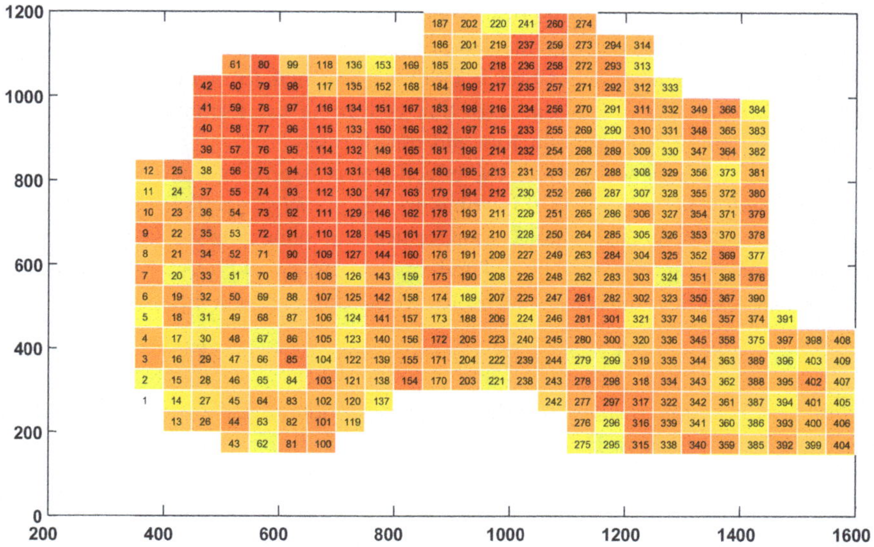

Fig. 1.5 Study area with geographical wind speed data

occurred hurricanes is constructed. On the other hand, for each component in distribution network that affected by wind a fragility curve is used to calculate the damage of network component.

A MST, which originates from the HV substation to the nodes, is built to give main and laterals feeders of the distribution network within its service area. In case of the multiple HV substation planning the study are can be cluster into some distinct HV substation and their associated loads. After clustering for each HV substation zone a graph and its corresponding loads and branches is determined. For each HV substation area the MST algorithms is applied.

In most planning problem the individual distances between the MV substations, and between the MH and HV substations are applied as weights in the MST problem formulating, since they relate to electricity resistance. In case of resilient planning, the distance can be replaced component failure rate of the component and its consequence damage. In this case the goal of the MST is that route the MV feeder with minimum damage due to hurricane speed. To achieve an acceptable plan, it is necessary to map the geographical data and location of the components with their corresponding fragility model. Damage of any component such as poles, conductors and transformers in distribution system due to hurricane can lead to long-term power outage. Consequently, from the availability point of view, distribution network component outage because of reliability and resiliency is the same and lead to energy not supply. The unavailability time because of the resilient base-reasons is very larges than reliability-based reasons.

The number of distribution poles and conductors along a distribution feeder is given by dividing the line length span between two adjacent poles to 30 m.

The resilient-based modelling of the network component and consequently total network resiliency index is provided in this section.

The total feeder section resiliency index is obtained from Eq. (1.18). In this equation there are three different term that can affect the feeder resiliency index namely distribution poles, conductors and transformers.

To model the dependency and the effectiveness degree of each component, a constant coefficient is used. In this equation RI_{Poles}, RI_{Trans} and RI_{Con} are the resiliency index for feeder f, pole, transformer and conductors related to feeder section respectively. On the other hand ω_{Poles}, ω_{Trans} and ω_{Con} are the poles, transformers and conductors related to feeder section f, respectively..

$$RI_{Feeder}(f) = \omega_{Poles}RI_{Poles} + \omega_{Trans}RI_{Trans} + \omega_{Con}RI_{Con} \tag{1.18}$$

For example, the resiliency index for distribution poles are calculated from Eq. (1.19) as modeled in Eqs. (1.1)–(1.3).

$$\lambda_{Pole} = 10^{-4}e^{0.0421w(i)} \tag{1.19}$$

While the number of poles and conductors for a feeder section depend on the feeder section length, consequently for each HV substation service area and for each feeder section connected to its corresponding HV substation, the length of feeder section is given by Eq. (1.20). In Eq. (1.20), $F_{Dist}(s_i, s_j)$ is the direct distance between substation i and j. In this equation, x_{s_i} and y_{s_i} are the X and Y coordination of substation i and x_{s_j} and y_{s_j} are the X and Y coordination of substation j, respectively.

$$\begin{aligned} &\forall S \in HV \\ &\forall F \in Feeder \\ &F_{Dist}(s_i, s_j) = \sqrt{(x_{s_i} - x_{s_j})^2 + (y_{s_i} - y_{s_j})^2} \end{aligned} \tag{1.20}$$

each feeder section the number poles are given by Eq. (1.21) supposing that the distance between two adjacent poles will be 30 m. In this regards, the number of conductors of a feeder section is calculated using Eq. (1.22).

$$Poles_{Num} = \text{round}(\frac{F_{Dist}(s_i, s_j)}{30}) \tag{1.21}$$

$$F_{Con} = Poles_{Num} - 1 \tag{1.22}$$

Rewriting Eq. (1.18) by considering Eqs. (1.20)–(1.22), the resiliency index for a feeder section can be evaluated by Eq. (1.23).

$$RI_{Feeder} = \omega_{Poles}\sum_{p=1}^{N_{pole}} RI_{Poles}(p) + \omega_{Trans}\sum_{t=1}^{N_{Trans}} RI_{Trans}(t) + \omega_{Trans}\sum_{c=1}^{N_{Con}} RI_{Con}(c) \tag{1.23}$$

In a planned network with N_{HV} HV substation and N_{Feeder} for each HV substation, the total resiliency index is given by Eq. (1.24).

$$RI_{Network} = \sum_{h=1}^{N_{HV}} \sum_{f=1}^{N_{Feeder}} RI_{Feeder}(h,f) \tag{1.24}$$

In case of mulit-objective optimization there are two cost-based and resilient based fitness function as Eq. (1.25) that can be selected by user. If $\alpha = 1$, the planning will be based-on cost and if $\alpha = 0$, only resilient based planning is considered.

$$Fitness = \alpha CostF + (1 - \alpha) RI_{Network} \tag{1.25}$$

In this chapter only the distribution network fragility curve is used as resilient-based planning goal.

1.7 Numerical Results

The optimal comparative resilient-based and cost-based planning of the medium voltage (MV) conventional distribution network is the main goal of this chapter. In this work the optimal configuration of MV distribution network is design based-on its corresponding capital cost and then the results are compared with the resilient-based design. For each case both cost and resiliency or component fragility index of the planned network are used to evaluate the network performances.

1.7.1 Test System

The proposed method is applied to a study area with geographical data that is illustrated in Fig. 1.5. In this figure the without loss of generality, study area is divided into some equal 50 × 50 square area or blocks. The number of each block is indicated in this figure. The speed density in the study area is illustrated in current figure (color bar from white to black). To better representation of the wind speed map a three-dimension plot of the wind speed in the study area is depicted in Fig. 1.6. In this figure the X and Y and Z axes related to length, width and wind speed value of the study area. In Table 1.12 the data for wind speed with its corresponding blocks coordinates are given. A counter plot of the study area wind speed amplitude is illustrated in Fig. 1.7.

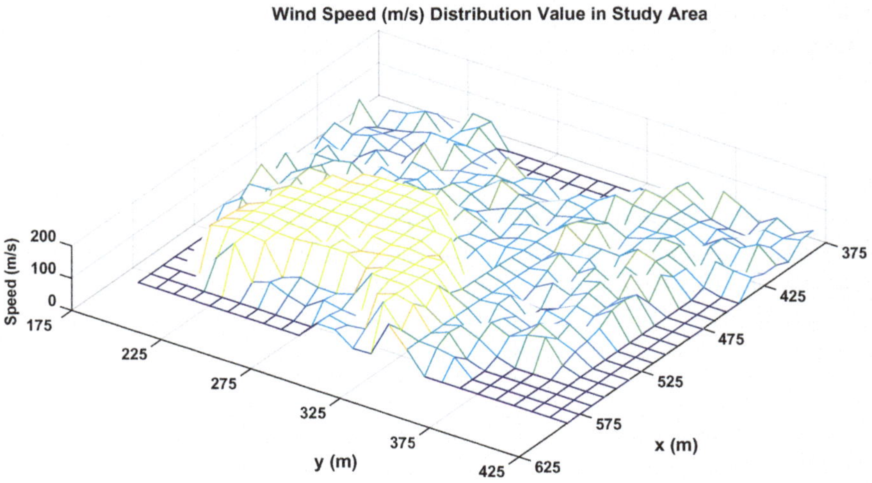

Fig. 1.6 Three-dimension plot of the wind speed in the study area

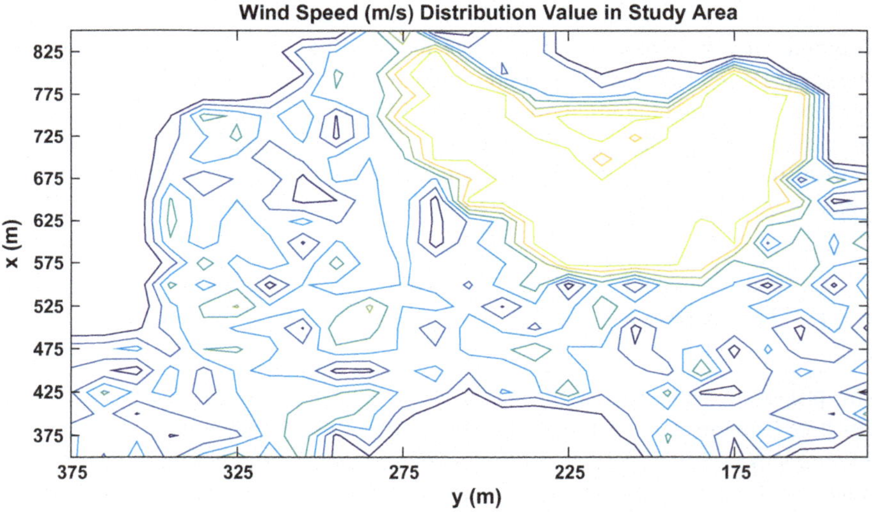

Fig. 1.7 Counter plot of the study area wind speed amplitude

Figure 1.8 illustrates the main candidate feeder's routes that initially checked by the network planner from feasibility view point. The data for feeders are given in Table 1.13.

In this study there are 32 MV substation or load point. The data for MV substation load is coordination is given in Table 1.14.

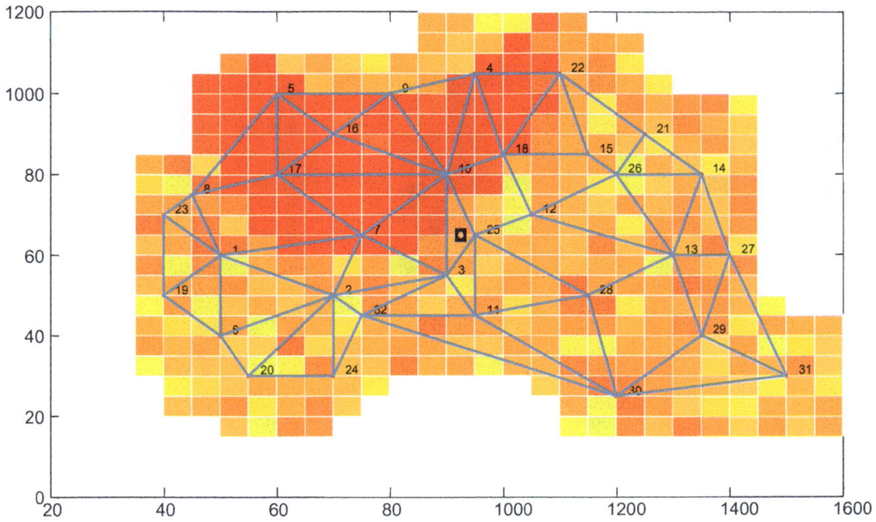

Fig. 1.8 Main candidate feeder's routes

1.7.2 Case 1: Planning with One HV Substation

In this section the optimal network planning algorithm is applied to plan the best feeder's routes considering both cost-based and resilient-based planning. In this study for all cases, both the HV substation (Node 33: blue square in Fig. 1.8) and it corresponding MV substations (Nodes 1–32 in Fig. 1.8) load and sizes fixed to a predefined value.

1.7.2.1 Resilient-Based Planning [8–17, 29–30]

In this case the optimal network configuration is obtained using Minimum Spanning Tree (MST) algorithm. The MST is solved using Prim's algorithm. This algorithm finds a radial network with minimum feeder length. The technical feasibility of the final radial network is evaluated using power flow algorithm. The voltage drop constraints and feeder power limits is calculated to ensure the standard requirement of the planned network.

Figure 1.9 shows the optimal network configuration for resilient-based planning scenario. According to this figure, three main feeders is designed to serve the total network loads. As mentioned before in this study there are 32 MV substations or load points.

In this figure the selected feeders are indicated. Each feeder section connects to MV substation. While the designed network is a radial network, the number of MV feeder sections are equal the number of MV substation minus one.

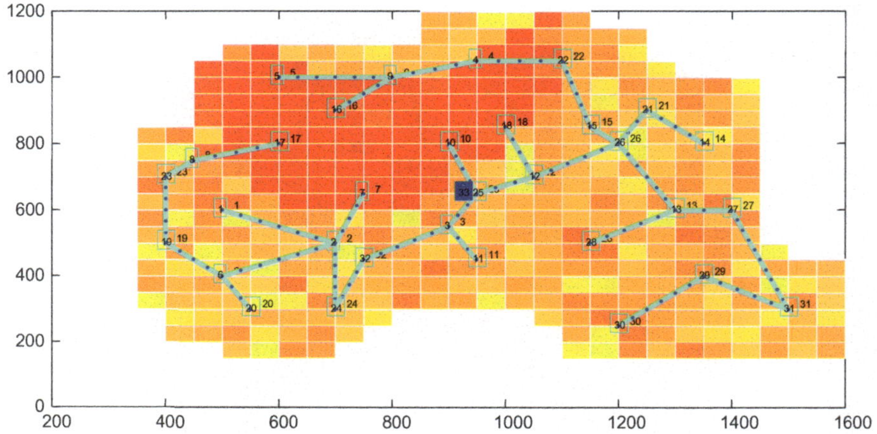

Fig. 1.9 Optimal network configuration for resilient-based planning

Considering the main goal of the chapter, the resiliency index is calculated for this network. Please note that in this case the resiliency index of feeders is used as optimization function for MST. In this regards, for this case the cost is calculated for the resilient-based designed network.

To evaluate the resiliency index for each feeder first of all it is necessary to calculate the fragility value of the network components such as distribution poles, conductors and transformers or fragility index. In this study the distribution poles fragility index is calculated from the fragility curve of the corresponding poles. For a given selected feeder, the number of required poles is calculated considering the role that the distance between to distribution poles are assumed to be 30 m. In Fig. 1.9 the location of poles is indicated on feeder section by blue dot.

At the next step it is necessary that determine the position of the selected poles for each feeder section on geographical map. This is done by searching for the intersection between all square polygons or wind speed bocks in the study area with the X and Y coordination and position of each pole. In fact, the poles that falls in a block is determined. Before final step, a wind speed is labeled for poles within each wind speed blocks. At the final step, the fragility index for each pole is calculated and set. This index is calculated not only for each individual pole but also for entire feeder section containing its corresponding poles as a general index for total feeder section.

For each feeder section the general fragility index is obtained by summing of the distribution poles individual fragility index, inside each feeder section. Feeders with higher fragility index, has less resistant with respect to hurricane. On the other hand, feeders with lower index is more resistant encountering with a hurricane. In facts this is a linear index that reflect the degree of feeder withstand facing with natural disasters.

Table 1.1 Evaluated data for feeder section [16–17] distribution poles

X	Y	Block number	Wind speed	Fragility index
7000	9000	133	115	0.37
6750	8750	114	143	1.13
6500	8500	114	143	1.13
6250	8250	94	147	1.30
6000	8000	94	147	1.30

Table 1.2 The fragility index for each feeder section in resilient-based planning

Feeder section	Feeder cost	Fragility index	Feeder section	Feeder cost	Fragility index
[29,30]	**212.1**	**0.31**	[12,26]	180.3	0.07
[28,13]	180.3	0.22	[12,18]	158.1	1.25
[1,2]	223.6	0.31	[13,27]	100.0	0.13
[2,7]	158.1	2.54	[13,26]	223.6	0.11
[10,25]	158.1	2.63	[14,21]	141.4	0.01
[9,16]	141.4	3.63	[27,31]	316.2	0.22
[4,9]	158.1	2.17	[15,26]	70.7	0.02
[24,32]	158.1	0.07	[15,22]	206.2	0.11
[5,9]	200.0	2.00	**[8,17]**	**158.1**	**4.08**
[3,11]	111.8	0.06	[19,23]	200.0	0.12
[2,6]	223.6	0.08	[12,25]	111.8	0.04
[6,19]	141.4	0.07	[21,26]	111.8	0.01
[6,20]	111.8	0.02	[24,2]	200.0	0.04
[8,23]	70.7	0.30	[25,3]	111.8	0.07
[29,31]	180.3	0.13	[4,22]	150.0	3.86
[32,3]	180.3	0.18	[33,25]	25.0	0.01

Table 1.1 gives the evaluated data for feeder section [8–17]. According to this table there are five distribution poles in along this feeder section. In the table the distribution poles geographical location, block that pole are inside, wind speed related to poles and its fragility are indicated.

Table 1.2 provides the fragility index for each entire feeder section in resilient based planning. For example, in this table the fragility index of feeder [29–30] (connecting load 29 to load 30) is 0.31 and its cost is 212.1. In contrast, the fragility index for feeder section [8–17] is 4.08 that is very high. This feeder section encounter with higher risk during hurricane. In Fig. 1.10 the fragility index for all feeder sections is illustrated. The summary of this case study in given in Table 1.3. In this table the overall resiliency index and cost for resilient based planning is indicated.

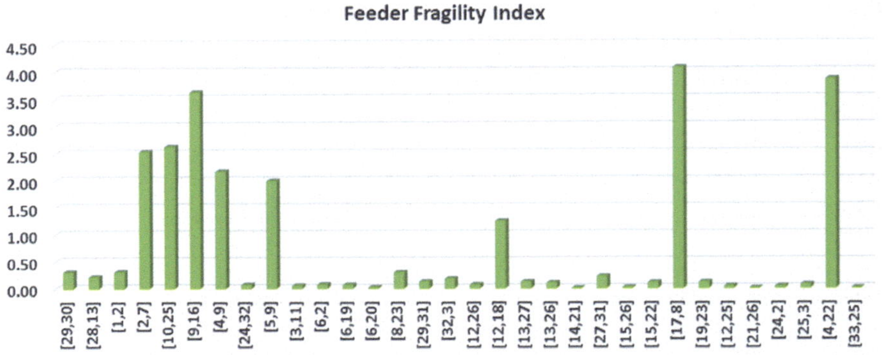

Fig. 1.10 The fragility index for all feeders section

Table 1.3 Overall resiliency index and cost for resilient-based planning

Planning type	Cost	Resiliency index
Resilient network	5074.8	24.84

1.7.2.2 Cost-Based Planning [8–17, 29–30]

In the current case the optimal network configuration is evaluated from cost point of view. Similar to the previous section, the results are obtained using MST algorithm in which MST solve the problem by Prim's algorithm.

Figure 1.11 shows the optimal network configuration for cost-based planning scenario. Similar to resilient-based planning three main outgoing feeder are selected to serve the total network loads.

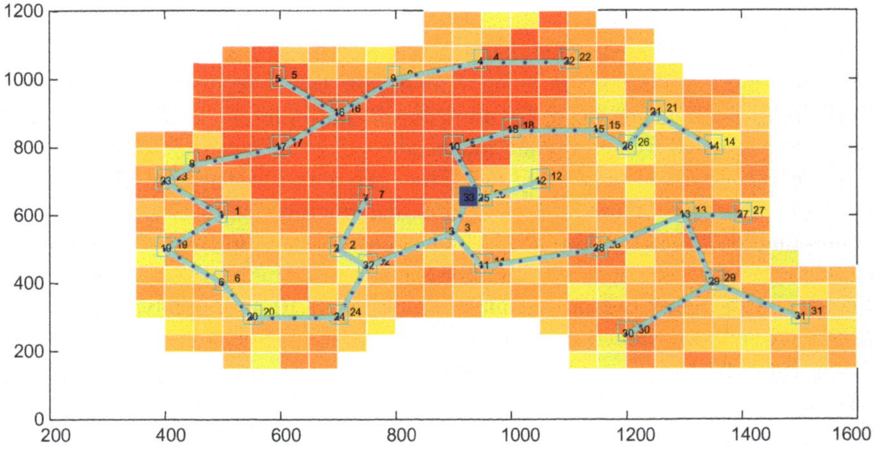

Fig. 1.11 Optimal network configuration for cost-based planning scenario

Table 1.4 Evaluated data for feeder section [16-17] distribution poles

X	Y	Block number	Wind speed	Fragility index
7000	9000	133	115	0.37
6750	8750	114	143	1.13
6500	8500	114	143	1.13
6250	8250	94	147	1.30
6000	8000	94	147	1.30
Sum				5.23

Table 1.5 Fragility index for each feeder section in resilient based planning

Feeder section	Feeder cost	Fragility index	Feeder section	Feeder cost	Fragility index
[1,19]	141.42	0.33	[32,3]	180.28	0.18
[29,30]	212.13	0.31	[13,29]	206.16	0.51
[28,13]	180.28	0.22	[13,27]	100.00	0.13
[2,7]	158.11	2.54	[14,21]	141.42	0.01
[10,25]	158.11	2.63	[1,23]	141.42	0.50
[2,32]	70.71	0.09	[15,26]	70.71	0.02
[9,16]	141.42	3.63	[5,16]	141.42	4.41
[4,9]	158.11	2.17	[16,17]	141.42	5.23
[24,32]	158.11	0.07	[8,17]	158.11	4.08
[3,11]	111.80	0.06	[20,24]	150.00	0.08
[6,19]	141.42	0.07	[12,25]	111.80	0.04
[6,20]	111.80	0.02	[21,26]	111.80	0.01
[8,23]	70.71	0.30	[10,18]	111.80	4.06
[15,18]	150.00	2.50	[4,22]	150.00	3.86
[29,31]	180.28	0.13	[33,25]	25.00	0.01
[11,28]	206.16	0.33	[33,3]	103.08	0.08

The cost is calculated as the primary goal and the resiliency index is calculated as the secondary for planned network. In this case the resilient index of feeders is used as optimization function.

Table 1.4 gives the output data for feeder section [16–17]. According to this table there are five distribution poles in along this feeder section. In the table the distribution poles geographical location, block that pole are inside, wind speed related to poles and its fragility are indicated.

Table 1.5 shows the fragility index for each feeder section in resilient based planning. For example, in this table the fragility index of feeder [29–30] (connecting load 1 to load 19) is 0.31 and its cost is 212.1. In contrast, the fragility index for feeder section [8–17] is 4.08 that is very high. This feeder section encounter with higher risk during hurricane. In Figs. 1.12 and 1.13 the fragility index and cost for all feeder's section are illustrated. The summary of this case study in given in

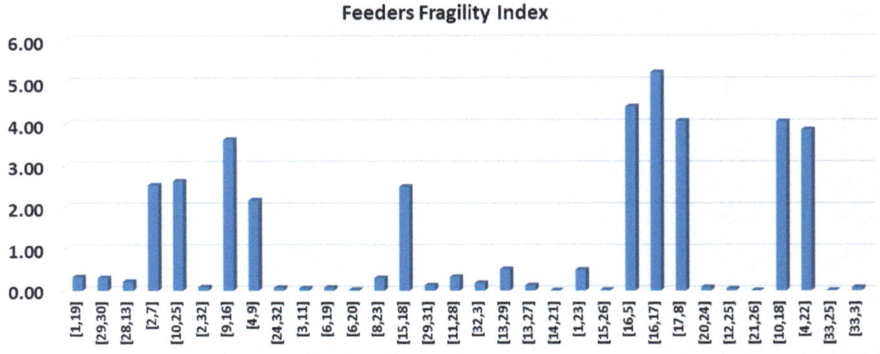

Fig. 1.12 The fragility index for all feeder's section

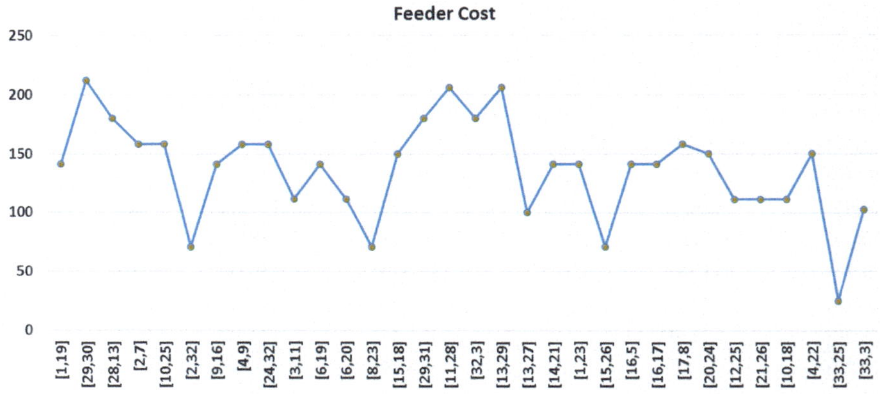

Fig. 1.13 The Cost index for all feeder's section

Table 1.6 Overall resiliency and cost index for cost-based planning

Planning type	Cost	Resiliency index
Minimum cost network	4395.02	38.59

Table 1.6. In this table the overall resiliency index and cost for resilient based planning is indicated. Table 1.7 provides total network cost and resiliency index in Case 1, both for cost-based and resilient-based scenarios considering one HV substation.

In Fig. 1.14, the voltage profile for cost based and resilient based network planning is indicated. By attention to the figure it can be seen that the cost based planning leads to a better voltage profile.

Table 1.7 Total network cost and resiliency index in Case 1, both for cost-based and resilient-based scenarios considering one HV substation

Planning type	Cost	Resiliency index
A-Resilient network	5074.8	24.84
B-Minimum cost Network	4395.02	38.59
A/B	0.866	1.56

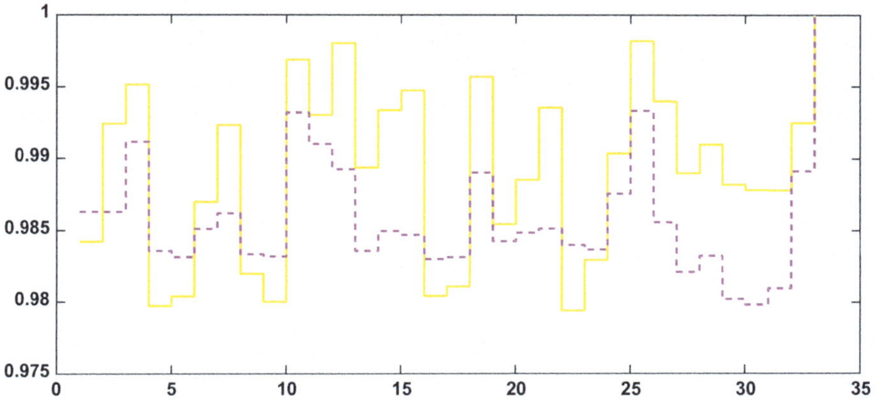

Fig. 1.14 Bus voltage profile for cost-based (solid-line) and resilient based (dash-line) planning

1.7.3 Case 2: Planning with Three HV Substation

Similar to the first case in the second case both resilient-based and cost-based scenario is studied for the same area with three predefined HV substations. Since the procedure is same as the first case, without any extra explanation, only the main results is given for both scenarios.

1.7.3.1 Resilient-Based Network Planning with Three HV Substation [4–8]

The best configuration of the network considering resiliency index as fitness function is plotted in Fig. 1.15. As seen from the figure, some feeders of two HV substations encounter with hazardous area. The feeders near or inside of this area were routed such that they avoid from the area or blocks with high fragility index. In Table 1.8 the overall cost and resiliency index is provided.

In this case there is three HV substation and therefore three radially MV network is optimized considering resiliency risk index. Figures 1.16, 1.17 and 1.18 show the value of fragility index for each selected feeder section. For instance, in Fig. 1.16 this index for feeder section [4–8] is 8. Please note that the higher index shows the

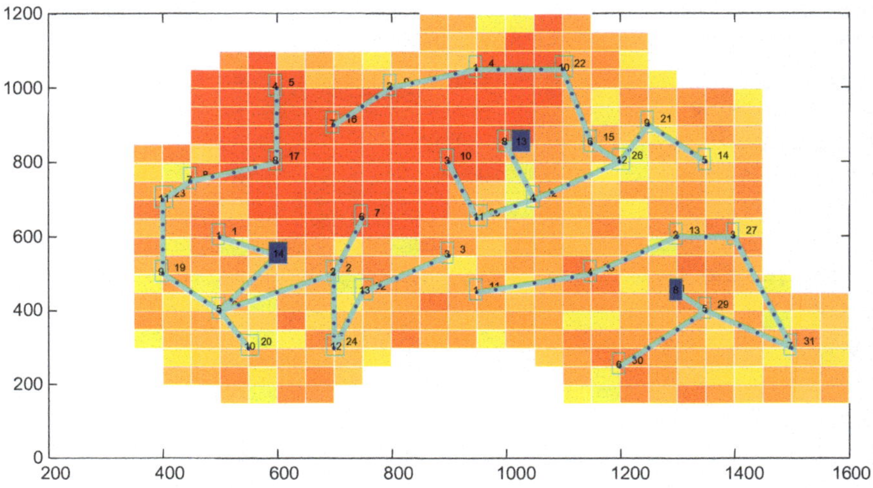

Fig. 1.15 Optimal network configuration for resilient-based planning scenario-three HV

Table 1.8 Overall resiliency index and cost for resilient-based planning

Planning type		Cost	Resiliency
Resilient	HV1	2094.24	15.94
	HV2	1265.78	1.34
	HV3	1612.93	15.01
Sum	Total Network	4972.96	32.29

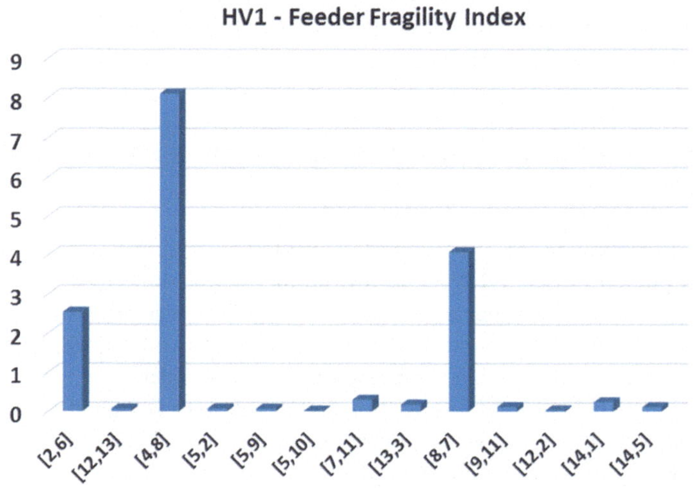

Fig. 1.16 Value of fragility index for each selected feeder for HV1—resilient-based planning

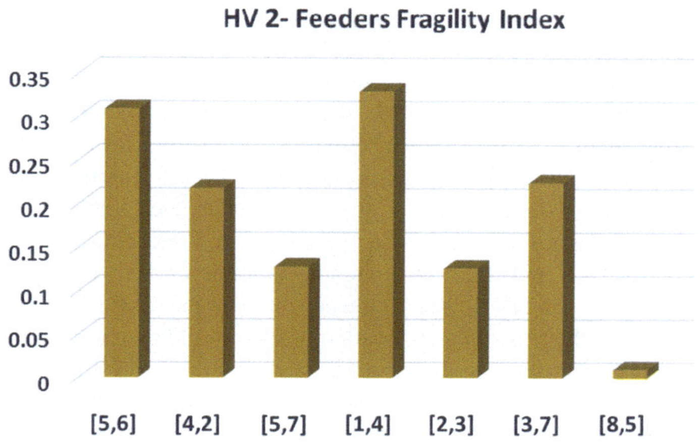

Fig. 1.17 Value of fragility index for each selected feeder for HV2—resilient-based planning

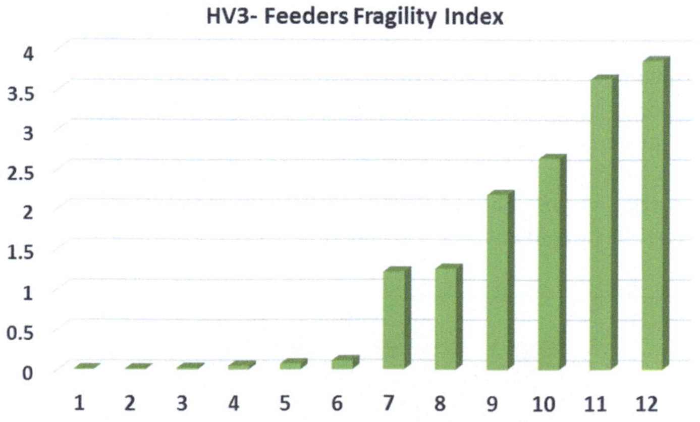

Fig. 1.18 Value of fragility index for each selected feeder for HV3—resilient-based planning

higher risk for network component to damage in disaster condition. This index is the sum of distribution poles fragility for poles along feeder section.

Comparing three figure it can be seen that depending of the geographical location of feeders and their corresponding HV substation, the index may be different. In Fig. 1.16 only some feeders have high fragility index, while in Fig. 1.17 all feeders have low fragility index. Finally, in Fig. 1.18 many of the feeders have high risk index. Table 1.8 summarized the all cost and resilient index for each HV substation and consequently for total network.

1.7.3.2 Cost-Based Network Planning with Three HV Substation

The best configuration of the network regarding the network total cost as fitness function is plotted in Fig. 1.19. In this scenario the effort of the feeder routing algorithm is to reduce the total feeder length as a substitute for cost. In this case more feeder pass from the hazardous area, and hence the fragility index can be increased. The feeders near hazardous area were routed such that they only minimized the feeder length. In Table 1.9 the overall cost and resiliency index are given.

Figures 1.20, 1.21 and 1.22 illustrate the amount of feeder's fragility. The fragilities index for each HV substation feeders are compared. It is obvious that the index depending of the geographical location of feeders and their corresponding HV substation are different. For example, in Fig. 1.21 many of MV feeders have high fragility index, while in Fig. 1.22 almost all feeders have low fragility index. In Table 1.10 the main results from total cost and total resilient index for each HV substation and its corresponding service area is provided.

In this section the summary of obtained results both for resilient-based and cost-based planning with three HV substation is represented in Table 1.10. In Table 1.8 the results are given. Considering Table 1.9 in case of resilient-based planning the network total fragility index is 32.29 that is smaller with respect to cost-based planning that network total fragility index is 43.64. Network with low fragility index is withstand and resilient comparing network with high fragility index. The ration of two value can be a measure of resiliency improvement. According to table with resilient based planning the network resilient is increased to 1.35 comparing the cost based planning, while the percent of cost decreased to 0.87 with respect to resilient based planned network.

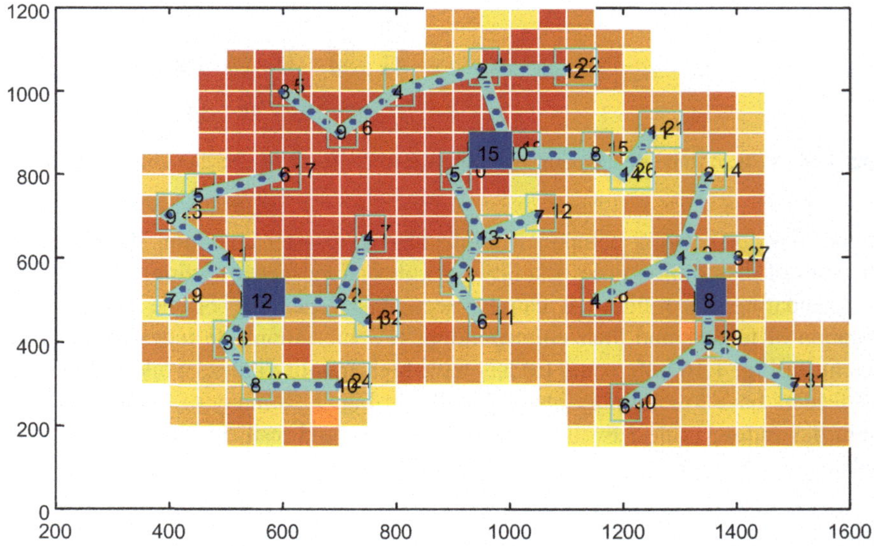

Fig. 1.19 Optimal network configuration for cost-based planning scenario-three HV

Table 1.9 Main results from total cost and total resilient index for each HV substation and its corresponding service

Planning type	Network	Cost	Resiliency
Minimum cost	HV1	1863.88	16.60
	HV2	1369.93	25.67
	HV3	1090.65	1.37
Sum	Total network	4324.45	43.64

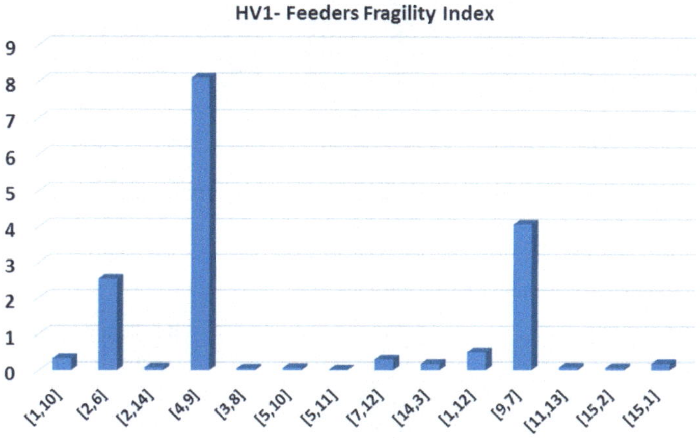

Fig. 1.20 Value of fragility index for each selected feeder for HV1—cost-based planning

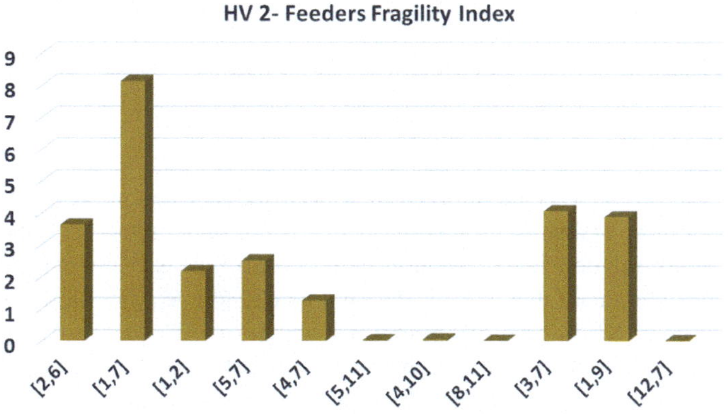

Fig. 1.21 Value of fragility index for each selected feeder for HV2—cost-based planning

Another comparison is done on one HV substation and three HV substation planning. In this regards, the results of Case 1 and Case 2 are compared. The summary of results in Tables 1.7 and 1.10 is reported in Table 1.11. In this table, both resilient-based and cost-based planning considering one HV and three HV substation is represented.

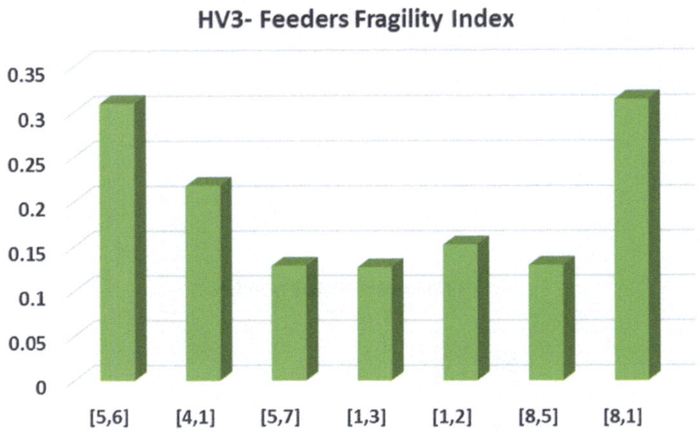

Fig. 1.22 Value of fragility index for each selected feeder for HV3—cost-based planning

Table 1.10 Summary of planning index for three HV

Planning type	Cost	Resiliency index
A-Resilient network	4972.96	32.29
B-Minimum cost Network	4324.45	43.64
B/A	0.87	1.35

Table 1.11 Summary comparison Case 1 and Case 2 (one and three HV substation planning)

One HV	Cost ratio	Resiliency index
A/B	0.866	1.56
Three HV	Cost Ratio	Resiliency Index
A/B	0.872	1.35

Considering Table 1.11 the ratio of total network cost for one HV substation is 0.866 and for three HV substation is 0.872 that is the same, while the total network resiliency index for one HV substation case is 1.56 that is better with respect to three HV substation case that is 1.35. Please note that to HV substations locations is supposed to be fixed. In case of optimal placement of HV substation the reported results may be different. In general, the results show that with optimal planning of distribution network considering resilient index can increase the network total cost, but the rate of cost increase with respect to the rate of resiliency improvement is smaller.

In case of optimal placement of HV substation the reported results may be different. In general, the results show that with optimal planning of distribution network considering resilient index can increase the network total cost, but the rate of cost increase with respect to the rate of resiliency improvement is smaller.

1.8 Conclusion

In this chapter the problem of resilient-based and cost-based planning of conventional MV distribution network is studied and compared from cost and resiliency index point of view. As mentioned before power distribution networks encounter with many unwanted weather natural disasters with high impact and low probability characteristics that may leads to distribution network component. In this chapter a numerical simulation is done to optimal planning of conventional electric distribution network based-on distribution network resiliency enhancement. The optimal topology of network is determined using MST optimization algorithm. A fitness function based on network cost and component fragility during hurricane is proposed. Both network component and topology, and geographical data for hurricane are illustrated graphically. For each network component for example distribution poles that falls within a specific block the fragility index is calculated.

Based-on the comparative results, in case 1 with one HV substation the ratio of total network cost for one HV substation is equal with 0.866 and for three HV substation is 0.872 that is very close to each other. On the other hand, the total network resiliency index for one HV substation case is 1.56 that is better with respect to three HV substation case that is 1.35. In general, the results indicate that with optimal planning of distribution network considering resilient index the network total cost is increased, but the rate of cost increase with respect to the rate of resiliency improvement is less.

Appendix

See Tables 1.12, 1.13 and 1.14.

Table 1.12 AHurricane wind speed map data for study area

Block num	Block X	Block Y	Wind speed	Block num	Block X	Block Y	Wind speed
1	375	275	87	206	975	475	73
2	375	325	12	207	975	525	53
3	375	375	68	208	975	575	37
4	375	425	52	209	975	625	55
5	375	475	15	210	975	675	55
6	375	525	50	211	975	725	42
7	375	575	80	212	975	775	144
8	375	625	45	213	975	825	118

(continued)

Table 1.12 (continued)

Block num	Block X	Block Y	Wind speed	Block num	Block X	Block Y	Wind speed
9	375	675	99	214	975	875	142
10	375	725	65	215	975	925	144
11	375	775	27	216	975	975	146
12	375	825	55	217	975	1025	149
13	425	225	45	218	975	1075	132
14	425	275	19	219	975	1125	48
15	425	325	46	220	975	1175	14
16	425	375	50	221	975	325	15
17	425	425	25	222	975	375	42
18	425	475	69	223	975	425	66
19	425	525	68	224	1025	475	26
20	425	575	15	225	1025	525	68
21	425	625	65	226	1025	575	55
22	425	675	54	227	1025	625	54
23	425	725	68	228	1025	675	10
24	425	775	13	229	1025	725	14
25	425	825	93	230	1025	775	11
26	475	225	54	231	1025	825	58
27	475	275	28	232	1025	875	145
28	475	325	45	233	1025	925	146
29	475	375	65	234	1025	975	147
30	475	425	25	235	1025	1025	148
31	475	475	14	236	1025	1075	150
32	475	525	68	237	1025	1125	150
33	475	575	80	238	1025	325	55
34	475	625	99	239	1025	375	65
35	475	675	95	240	1025	425	43
36	475	725	74	241	1025	1175	16
37	475	775	109	242	1075	275	33
38	475	825	30	243	1075	325	55
39	475	875	120	244	1075	375	57
40	475	925	125	245	1075	425	58
41	475	975	130	246	1075	475	55
42	475	1025	135	247	1075	525	60
43	525	175	43	248	1075	575	61
44	525	225	79	249	1075	625	62
45	525	275	41	250	1075	675	67
46	525	325	45	251	1075	725	70
47	525	375	36	252	1075	775	72

(continued)

Table 1.12 (continued)

Block num	Block X	Block Y	Wind speed	Block num	Block X	Block Y	Wind speed
48	525	425	65	253	1075	825	74
49	525	475	44	254	1075	875	76
50	525	525	80	255	1075	925	90
51	525	575	11	256	1075	975	141
52	525	625	99	257	1075	1025	101
53	525	675	38	258	1075	1075	137
54	525	725	88	259	1075	1125	82
55	525	775	141	260	1075	1175	126
56	525	825	142	261	1125	525	104
57	525	875	143	262	1125	575	47
58	525	925	144	263	1125	625	48
59	525	975	145	264	1125	675	51
60	525	1025	147	265	1125	725	53
61	525	1075	78	266	1125	775	55
62	575	175	11	267	1125	825	56
63	575	225	21	268	1125	875	60
64	575	275	58	269	1125	925	61
65	575	325	13	270	1125	975	62
66	575	375	35	271	1125	1025	63
67	575	425	11	272	1125	1075	64
68	575	475	56	273	1125	1125	65
69	575	525	45	274	1125	1175	66
70	575	575	59	275	1125	175	17
71	575	625	67	276	1125	225	58
72	575	675	140	277	1125	275	65
73	575	725	142	278	1125	325	90
74	575	775	144	279	1125	375	15
75	575	825	146	280	1125	425	70
76	575	875	148	281	1125	475	89
77	575	925	150	282	1175	525	65
78	575	975	148	283	1175	575	56
79	575	1025	146	284	1175	625	89
80	575	1075	142	285	1175	675	62
81	625	175	81	286	1175	725	38
82	625	225	53	287	1175	775	20
83	625	275	55	288	1175	825	66
84	625	325	17	289	1175	875	65
85	625	375	98	290	1175	925	16
86	625	425	65	291	1175	975	19

(continued)

Table 1.12 (continued)

Block num	Block X	Block Y	Wind speed	Block num	Block X	Block Y	Wind speed
87	625	475	41	292	1175	1025	78
88	625	525	48	293	1175	1075	84
89	625	575	81	294	1175	1125	62
90	625	625	135	295	1175	175	11
91	625	675	137	296	1175	225	14
92	625	725	139	297	1175	275	98
93	625	775	145	298	1175	325	78
94	625	825	147	299	1175	375	14
95	625	875	145	300	1175	425	75
96	625	925	143	301	1175	475	98
97	625	975	141	302	1225	525	66
98	625	1025	139	303	1225	575	61
99	625	1075	39	304	1225	625	45
100	675	175	74	305	1225	675	18
101	675	225	83	306	1225	725	68
102	675	275	53	307	1225	775	14
103	675	325	86	308	1225	825	13
104	675	375	29	309	1225	875	45
105	675	425	56	310	1225	925	65
106	675	475	56	311	1225	975	79
107	675	525	75	312	1225	1025	54
108	675	575	74	313	1225	1075	19
109	675	625	142	314	1225	1125	52
110	675	675	144	315	1225	175	86
111	675	725	145	316	1225	225	96
112	675	775	147	317	1225	275	90
113	675	825	145	318	1225	325	71
114	675	875	143	319	1225	375	56
115	675	925	141	320	1225	425	48
116	675	975	139	321	1225	475	19
117	675	1025	45	322	1275	275	68
118	675	1075	57	323	1275	525	63
119	725	225	35	324	1275	575	11
120	725	275	45	325	1275	625	79
121	725	325	46	326	1275	675	66
122	725	375	47	327	1275	725	48
123	725	425	23	328	1275	775	34
124	725	475	12	329	1275	825	55
125	725	525	69	330	1275	875	22

(continued)

Table 1.12 (continued)

Block num	Block X	Block Y	Wind speed	Block num	Block X	Block Y	Wind speed
126	725	575	24	331	1275	925	44
127	725	625	142	332	1275	975	54
128	725	675	145	333	1275	1025	18
129	725	725	148	334	1275	325	73
130	725	775	150	335	1275	375	56
131	725	825	147	336	1275	425	64
132	725	875	143	337	1275	475	57
133	725	925	115	338	1275	175	44
134	725	975	145	339	1275	225	67
135	725	1025	64	340	1325	175	94
136	725	1075	34	341	1325	225	36
137	775	275	20	342	1325	275	54
138	775	325	42	343	1325	325	60
139	775	375	53	344	1325	375	63
140	775	425	57	345	1325	425	83
141	775	475	81	346	1325	475	56
142	775	525	90	347	1325	875	55
143	775	575	69	348	1325	925	86
144	775	625	147	349	1325	975	76
145	775	675	149	350	1325	525	102
146	775	725	147	351	1325	575	75
147	775	775	144	352	1325	625	48
148	775	825	140	353	1325	675	80
149	775	875	110	354	1325	725	74
150	775	925	146	355	1325	775	60
151	775	975	150	356	1325	825	43
152	775	1025	51	357	1375	475	76
153	775	1075	19	358	1375	425	80
154	825	325	97	359	1375	175	68
155	825	375	68	360	1375	225	24
156	825	425	59	361	1375	275	50
157	825	475	66	362	1375	325	28
158	825	525	53	363	1375	375	33
159	825	575	11	364	1375	875	74
160	825	625	145	365	1375	925	56
161	825	675	147	366	1375	975	88
162	825	725	149	367	1375	525	84
163	825	775	147	368	1375	575	52

(continued)

Table 1.12 (continued)

Block num	Block X	Block Y	Wind speed	Block num	Block X	Block Y	Wind speed
164	825	825	144	369	1375	625	93
165	825	875	150	370	1375	675	45
166	825	925	130	371	1375	725	45
167	825	975	150	372	1375	775	45
168	825	1025	50	373	1375	825	20
169	825	1075	56	374	1425	475	53
170	875	325	52	375	1425	425	23
171	875	375	63	376	1425	575	86
172	875	425	99	377	1425	625	27
173	875	475	35	378	1425	675	78
174	875	525	56	379	1425	725	88
175	875	575	90	380	1425	775	82
176	875	625	78	381	1425	825	53
177	875	675	144	382	1425	875	40
178	875	725	146	383	1425	925	34
179	875	775	148	384	1425	975	18
180	875	825	146	385	1425	175	34
181	875	875	142	386	1425	225	19
182	875	925	140	387	1425	275	38
183	875	975	121	388	1425	325	66
184	875	1025	66	389	1425	375	69
185	875	1075	45	390	1425	525	54
186	875	1125	54	391	1475	475	14
187	875	1175	55	392	1475	175	68
188	925	475	57	393	1475	225	57
189	925	525	19	394	1475	275	19
190	925	575	69	395	1475	325	55
191	925	625	58	396	1475	375	11
192	925	675	75	397	1475	425	66
193	925	725	67	398	1525	425	56
194	925	775	148	399	1525	175	38
195	925	825	145	400	1525	225	50
196	925	875	148	401	1525	275	48
197	925	925	148	402	1525	325	84
198	925	975	142	403	1525	375	23
199	925	1025	149	404	1575	175	65
200	925	1075	39	405	1575	275	22
201	925	1125	41	406	1575	225	63

(continued)

Table 1.12 (continued)

Block num	Block X	Block Y	Wind speed	Block num	Block X	Block Y	Wind speed
202	925	1175	54	407	1575	325	38
203	925	325	63	408	1575	425	53
204	925	375	43	409	1575	375	38
205	925	425	73				

Table 1.13 Candidate feeder's route data [24]

Feeder number	Feeder from	Feeder to	Feeder number	Feeder from	Feeder to
1	1	19	37	11	28
2	29	30	38	32	3
3	28	13	39	12	26
4	1	2	40	12	18
5	1	7	41	12	13
6	2	3	42	13	29
7	2	7	43	13	27
8	2	20	44	13	26
9	9	10	45	13	14
10	10	25	46	14	27
11	2	32	47	14	21
12	1	8	48	1	23
13	9	16	49	14	26
14	3	7	50	32	30
15	3	10	51	27	31
16	4	10	52	25	28
17	4	18	53	15	26
18	4	9	54	15	22
19	24	32	55	16	5
20	1	6	56	18	22
21	5	9	57	16	17
22	5	8	58	16	10
23	5	17	59	17	8
24	3	11	60	7	10
25	6	2	61	19	23
26	6	19	62	20	24
27	6	20	63	12	25
28	22	21	64	21	26
29	31	30	65	17	7
30	8	23	66	24	2
31	30	11	67	10	18

(continued)

Table 1.13 (continued)

Feeder number	Feeder from	Feeder to	Feeder number	Feeder from	Feeder to
32	15	18	68	25	3
33	29	31	69	4	22
34	11	32	70	27	29
35	25	11	71	28	30
36	10	17			

Table 1.14 Substation data [24]

Substation number	Substation X	Substation Y	Substation number	Substation X	Substation Y
1	500	600	17	600	800
2	700	500	18	1000	850
3	900	550	19	400	500
4	950	1050	20	550	300
5	600	1000	21	1250	900
6	500	400	22	1100	1050
7	750	650	23	400	700
8	450	750	24	700	300
9	800	1000	25	950	650
10	900	800	26	1200	800
11	950	450	27	1400	600
12	1050	700	28	1150	500
13	1300	600	29	1350	400
14	1350	800	30	1200	250
15	1150	850	31	1500	300
16	700	900	32	750	450

References

1. D. Ward, The effect of weather on supply, grid systems and the reliability of electricity supply. Clim. Change **121**, 103–113 (2013)
2. U.G. Knight, *Power Systems in Emergencies: From Contingency Planning to Crisis Management* (Wiley, Chichester, 2001)
3. M. Panteli, P. Mancarella, X. Hu, I. Cotton, D. Calverley, R. Wood, C. Pickering, S. Wilkinson, R. Dawson, K. Anderson, Impact of climate change on the resilience of the UK power system (2015), pp. 22–24
4. M. Panteli, P. Mancarella, Modeling and evaluating the resilience of critical electrical power infrastructure to extreme weather events. IEEE Syst. J. **11**(3), 1733–1742 (2017)
5. E.V. Badolato, J. Bleiweis, J.D. Craig, L.J. Orace, W. Fleming, Hurricane Hugo: Learned in energy emergency preparedness, 1990
6. R.J. Campbell, Weather-related power outages and electric system resiliency, 2012

7. Executive Office of the President, Economic benefits of increasing electric grid resilience to weather outages, 2013
8. National Grid Electricity Transmission PLC, Climate Change Adaptation Report, 2010
9. Energy Networks Association, Electricity Networks Climate Change Adaptation Report, 2011
10. Northern PowerGrid, Climate Change Adaptation Report, 2013
11. B. Fleming, Climate Change Adaptation Report, 2011
12. H.B. Leonard, A.M. Howitt, Routine or crisis: the search for excellence. Crisis Response **4**, 32–35 (2008)
13. EPRI, The integrated grid: realizing the full value of central and distributed energy resources, 2014
14. P.F. Ribeiro, B.K. Johnson, M.L. Crow, A. Arsoy, Y. Liu, Energy storage systems for advanced power applications. IEEE Proc. **89**, 1744–1756 (2001)
15. J.A.P. Lopes, N. Hatziargyriou, J. Mutale, P. Djapic, N. Jenkins, Integrating distributed generation into electric power systems: a review of drivers, challenges and opportunities. Electr. Power Syst. Res. **77**, 1189–1203 (2007)
16. G. Strbac, Demand side management: benefits and challenges. Energy Policy **36**, 4419–4426 (2008)
17. P. Basak, S. Chowdhury, S. Halder, N. Dey, S.P. Chowdhury, A literature review on integration of distributed energy resources in the perspective of control, protection and stability of microgrid. Renew. Sustain. Energy Rev. **16**, 5545–5556 (2012)
18. S. Mirsaeidi, D.M. Said, M.W. Mustafa, M.H. Habibuddin, K. Ghaffari, An analytical literature review of the available techniques for the protection of micro-grids. Int. J. Electr. Power Energy Syst. **58**, 300–306 (2014)
19. European Distribution System Operators (EDSO), Flexibility: the role of DSOs in tomorrow's electricity market, 2014
20. S.R. Gupta, F.S. Kazi, S.R. Wagh, N.M. Singh, Probabilistic framework for evaluation of smart grid resilience of cascade failure, in *IEEE Innovative Smart Grid Technologies* (2014), pp. 255–260
21. J. Winkler, L. Duenas-Osorio, R. Stein, D. Subramanian, Performance assessment of topologically diverse power systems subjected to Hurricane events. Reliab. Eng. Syst. Saf. **95** (4), 323–336 (2010)
22. FEMA, Hazards U.S. Multi-Hazard (HAZUS-MH) Assessment Tool versus 1.4, Washington, DC, 2008
23. M. Panteli, C. Pickering, S. Wilkinson, R. Dawson, P. Mancarella, Power system resilience to extreme weather: fragility modeling, probabilistic impact assessment, and adaptation measures. IEEE Trans. Power Syst. **32**(5), 3747–3757 (2017)
24. S. Najafi Ravadanegh, R. Gholizadeh Roshanagh, A heuristic algorithm for optimal multistage sizing, siting and timing of MV distribution substations. Electr. Power Syst. Res. **105**, 134–141 (2013)

Chapter 2
Power Systems Connectivity and Resiliency

Horia Andrei, Marian Gaiceanu, Marilena Stanculescu, Iulian Nicusor Arama and Paul Cristian Andrei

Abstract This chapter presents the role of power system connectivity and resiliency under the conditions of vulnerability to natural disasters and deliberate attacks. The importance of introducing the Internet of Things (IoT) concept in developing smart grids will be shown, as well as the methods of increasing the power system resiliency. Case studies from the Romanian power system will be included. At the same time, the authors underline the necessity of introducing standards and developing the protection systems of Big Data in order to design the future smart power system.

Keywords Big data · Connectivity · Internet of things (IoT) · Natural disasters Protection system · Resiliency · Smart grids · Vulnerability

H. Andrei (✉)
Doctoral School of Engineering Sciences, University Valahia of Targoviste, Târgoviște, Romania
e-mail: hr_andrei@yahoo.com

M. Gaiceanu · I. N. Arama
Department of Control Systems and Electrical Engineering, University Dunarea de Jos Galati, Galați, Romania
e-mail: marian.gaiceanu@ugal.ro

I. N. Arama
e-mail: iulian.arama@ugal.ro

M. Stanculescu · P. C. Andrei
Department of Electrical Engineering, University Politehnica Bucharest, Bucharest, Romania
e-mail: marilena.stanculescu@upb.ro

P. C. Andrei
e-mail: paul.andrei@upb.ro

© Springer Nature Switzerland AG 2019

N. Mahdavi Tabatabaei et al. (eds.), *Power Systems Resilience*, Power Systems,
https://doi.org/10.1007/978-3-319-94442-5_2

2.1 Chapter Overview

The challenge of writing this chapter has come knowing very well the vulnerability of the Power System (PoSy) to natural disasters and deliberate attacks. Having analyzed the current state of the power system, the authors are going to highlight the main methods of developing a system which is *robust* to deliberate or involuntary perturbations, *open* in the matter of incorporating renewable energy sources, *integrated* and *efficient* for the whole two-way energy chain. Having in mind to modernize the PoSy, in Sects. 2.2 and 2.3 of the chapter, two properties will be defined: connectivity and resiliency, essential for the development of a safe system.

The connectivity types, the opportunities of developing the connectivity in the chain of the grid connected or autonomous PoSy (production, transport, and distribution) will be studied, aiming to increase their efficiency and performances. The Internet of Things (IoT) concept will be highlighted, as well as its role in the development of smart PoSy. The main factors regarding the PoSy design using IoT are: cost, accessibility and easy use, interoperability and vision. The IoT development must be made in high security conditions, on the basis of Big Data principles, analytic analysis and foresight. At the same time, the necessity of developing some standards/regulations concerning these data has appeared, as well as the development of their protection systems.

The resiliency includes the capacity of strengthening the system against high impact events, with low frequency (natural-tornadoes, earthquakes, fires and severe geomagnetic perturbations etc. or human-physical, coordinated cyber-attacks) and the fast recovery of the system properties. Different methods of increasing the PoSy resiliency for each component (generation, transport, distribution and consumers) will be described. Case studies regarding the Romanian PoSy will be presented. The future Supervisory Control And Data Acquisition (SCADA) system and the future distributed control systems can have an additional diagnosing infrastructure for checking the system's normal operation and the data coherence to detect its manipulation. One will show the necessary actions for increasing the PoSy's cyber security. The last Section has drawn the conclusion. The chapter ends with adequate references.

2.2 Comparative Power Systems: Actual and Potential Route

The sustainable development of a society is strictly dependent of the power sector and represents a strategic target of each country to consolidate a safe and solid future. The state of the art level of knowledge precisely reveals us that the resources and riches of the Earth are running out, the environment pollution level becomes alarmingly high, so it's the time for human intelligence to capitalize on new energy resources, new ways to transmit it as well as a much more efficient energy use.

In the same time worldwide researchers focus on a thorough recycling of all used products in order to keep the environment as clean as it can and to ensure a rational life level without reducing the quality of goods and products which humans developed over time [1, 2].

Of all types of energy sources, electrical energy is the most spread. In the traditional architecture of electroenergetical systems, based on generation, transport, distribution and consumption, important reconfigurations take place, shown in Fig. 2.1. The generation of electrical energy is now achieved also through photovoltaic, wind, biomass and geothermal plants [3, 4]. There are numerous researches in finding new solutions for transforming carbon in electrical energy, advanced researches for building electrical plants based on fusion and some practical solutions for replacing methane gas burning in industry with electro thermal technologies. Since the first solutions used at the middle of the 20th century in the large impoundment hydroelectric plants which ensured the most important balance sources, at any given time, between generation and consumption, new energy storage systems have developed and are in a continuous performance improvement. Battery banks, super capacitors, high power UPS systems or systems which convert electrical energy into water potential energy and vice versa, all of them are successfully used in electroenergetical systems to ensure a reliable and quality supply of the users. The expansion of the storage systems in the electroenergetical system but also for the electrical vehicles supply still needs some important steps to make, especially in what concerns the energy density increase and the number of charge-discharge cycles.

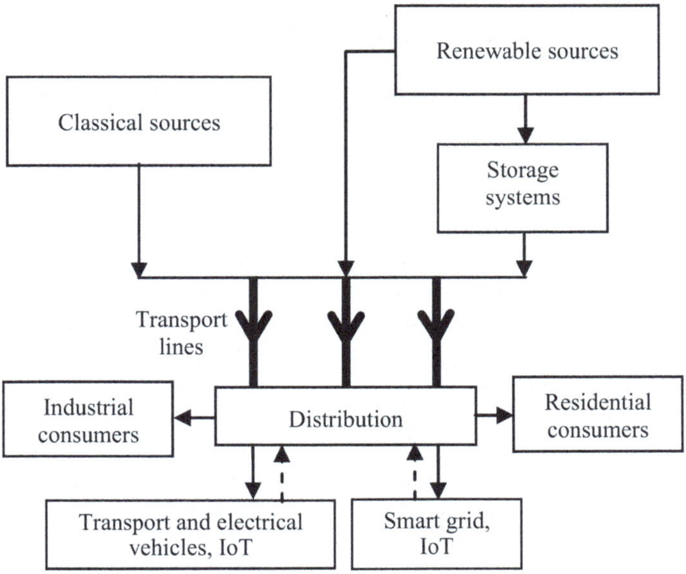

Fig. 2.1 New configuration of the PoSy

PoSys of the future and urban congestions will benefit from numerous smart-grids to ensure the best way of primary energy local resources using, with the decrease of the necessary fossil fuel for producing electrical energy. Also, the other part of the smart-grid concept will lead to a substantial change of the way in which humans will report to the environment [5]. Important investments will be necessary for the reconfiguration of classical PoSys to the structure of smart grids where each user has the possibility to use energy in a rational way, without compromising the comfort level obtained in over 150 years of using electrical energy.

Some other important reconfigurations will aim on the diversification of the consumer types, especially regarding household appliance, electric vehicles and developing of IoT [6]. The connection of all these new equipment to the PoSy will create, on one hand, serious trouble in keeping the electrical energy quality and, on the other hand, new risks of system malfunction, which can be easier hacked through the Internet.

In the field of the electrical energy distribution, automatic control systems have been designed to increase the technical efficiency, with the main objective of increasing the speed in making the necessary decisions for the efficient development of the energy distribution process [7, 8].

The fundamental issues of the energy distribution are the safety of power consumers and energy quality assurance [9]. At the fundamental level, in order to maintain the appropriate operation of the power distribution system, the classical control systems use databases, while modern systems use information bases [10]. This difference allows the transition from old automatic control systems to new ones by introducing intelligence. Data flows are used to perform certain tasks, while information flows are used to ensure intelligent behavior and restart of the PoSy.

At the level of management and organization there is the Supervisory Control and Data Acquisition (SCADA) system, which communicates directly with Remote Terminal Units (RTU) through Operational Control Points (OCP), with the role of monitoring and controlling the distribution network. The SCADA system is subordinated to the Energy Management System (EMS). The power distribution company manages the EMS system. The company has its own automated control system, Enterprise Management Information System, which communicates with the EMS, without a subordination relationship (Fig. 2.2). By introducing automatic control systems, the evolution of energy distribution systems, in which decision making and action were undertaken by the human operator to the current ones, has been achieved. A modern energy distribution system makes switching from automatic control systems to intelligent systems, an evolution reflected in increasing decision-response speeds and action to disturbances in these distribution systems.

The above mentioned issue is also a requirement of the European Union, as defined by the Energy Directives issued. One of the achievements is the Digital Fault Recorders (DFR) system for faults analysis in the power distribution network. It is about diagnosing defects by protecting and locating these defects.

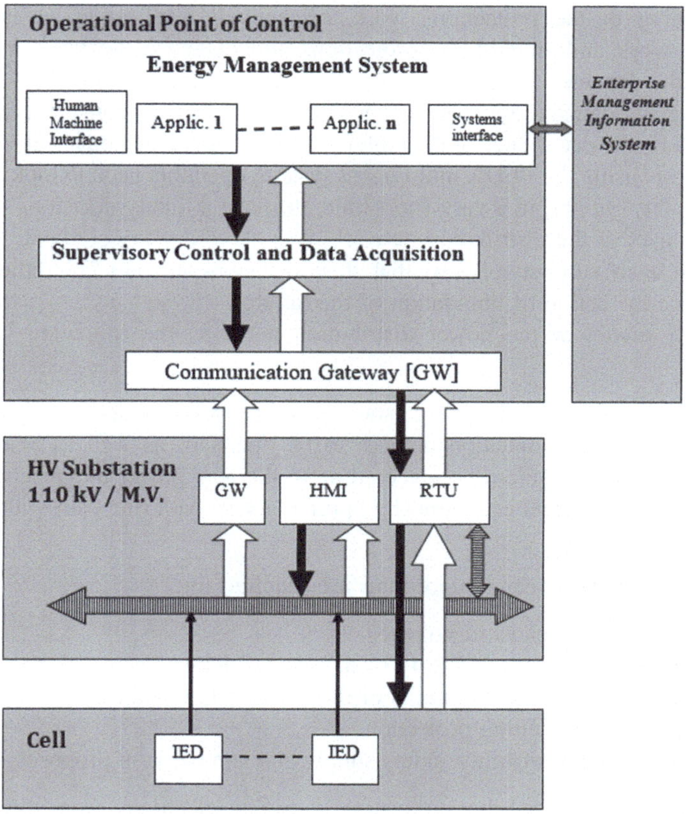

Fig. 2.2 Architecture of the traditional control system [10]

According to [10], the occurrence of the fault is signaled by:

- The operation of a protection, which has the effect of isolating the fault by the specific circuit breakers,
- The work of a protection has the effect of launching a command on the primary commutation device without triggering the device since that switch was triggered by the protection that worked,
- The start of a protection—has the effect of perceiving the fault, but does not move to the stage of work because the circuit-breaker is already triggered.

In order to eliminate the damage, the power distribution company identifies and locates the defect in the distribution network as soon as possible. An example of a smart DFR based fault identification and localization system is also PEDA-Protection Engineering Diagnostic Agents [11]. It uses a rule engine for the Java platform JESS-Java Expert System Shell [12] that interrogates more intelligent agents. These agents identify the incidents and events, receive and record the faults, interpret the recorded faults as a result of the analysis of the operation, the work and the beginning of the protections work, respectively confirm or invalidate the operation, work and start of the protections and diagnose the fault in order to eliminate the damage.

Clearing the damage involves replenishing the electricity consumers that have interrupted by the occurrence of the fault by the operational personnel. Based on the fast diagnosis using the PEDA multi-agent system, operators need to look at how to assure a safety supply, in a very short time, to make a timely decision.

At the level of the distribution network, it is necessary to coordinate the operators from each sub-network, so that their decisions are not contradictory and compromise the action of liquidation of the failures.

At each subnet of the power distribution network, the following issues are analyzed:

(i) The paths of access the voltage sources. This involves identifying power lines and primary commutation devices to be operated, as well as resetting the protections. The latter resets according to the new power movement;
(ii) The technically and economically optimal access path (the path with minimal losses).

As a consequence, search algorithms are required for:

- Identifying access paths to voltage,
- Circulation of powers to identify overloads and losses,
- Voltage drops to check the limits of the admissible voltages,
- Short circuit for resetting protection,
- Static and dynamic stability, safety supply and validation of proposed solutions.

The classic power distribution system is driving electricity from the transport system to consumers. The power distribution passes from the passive grid, which has a radial configuration, to the active grid with a looped configuration. The active network takes power from the distributed generation system (photovoltaic, wind, hydraulic, etc.) and directs it to the consumers. Thus, it is important to manage the distribution of electricity both from the distributed generation system and from the transmission system. The distribution operator has to make decisions about voltage regulation rather than frequency. Frequency regulation is regulated by implementing sequential frequency sequential triggers, but not at lower than nominal values, such as in passive distribution grids, but at values higher than the nominal one, so that frequency control remains under the decision-maker's power.

Voltage regulation is in the decision-making authority of the electricity distribution operator, a problem that is complicated by the emergence of distributed generation active grids and which needs to be addressed with new solutions. The way to solve the issue of bandwidth management is achieved through artificial intelligence.

Passive power distribution grids (distributed generation) such as smart grid and virtual electrical networks can be mentioned [10, 13]. Nowadays, for the purpose of liquidation of failures, applications are used that contain the optimization of an objective function, which includes the balance between the generated and the consumed power.

The electricity distribution network has to face new challenges in liquidation of failures:

1. The realities of the maneuvers which, in view of the liquidation of the failures, aim firstly at the access to the voltage and then the satisfaction of the consumed power, prevailing, therefore, minimizing the costs of the access to the voltage and not maximizing the available power,
2. Restrictions have become more complicated by introducing the distributed generating networks via virtual generators, thus overcoming the restrictions imposed by traditional standards on the power quality, bandwidth compliance and continuity of consumer supply
3. Regardless of the circumstances, the frequency control is in the decision-making authority of the system operator, so that the operative management of the electricity distribution network must proceed in the course of liquidation of the failures so as not to disturb the activity of the system operator.

The challenges of the power distribution network in terms of liquidation of failures must be met by new requirements for the use of artificial intelligence, requirements that are likely to stimulate the evolution of artificial intelligence itself in terms of decision-making.

In the Fig. 2.2 the architecture of the current control systems is shown. The principle of the traditional control system architecture is the pronounced hierarchy, the main data flow being bottom-up, i.e. the data being acquired is communicated from the bottom up, the secondary stream being of the top-down type, i.e. from the top to the bottom, the sense in which it is transmitted remote controls from the Operational Command Point to the high voltage (HV) substation 110 kV/MV primary switching devices. The bandwidth of the HV 110 kV/MV is limited.

The so-called Energy Management System is not modularized and uses the data from the relational database in the Supervisory Control and Data Acquisition (SCADA) structure. Interconnection with the Enterprise Management Information System is done through a special interface.

The SCADA is the only component that acquires the data in real time and contains the configuration information in its relational database. Intelligent Electronic Devices (IED) are in most cases autonomous protections with different data modeling, functions and protocols, each manufacturer having its own specification.

Just because the frequency band is narrow and each manufacturer practices its own models, its functions and firm protocols, it has become a question of standardizing protocols and modeling data. The architecture of an IT system must ensure data acquisition in real-time, complex data processing and self-diagnosis, requirements more or less met with traditional hierarchical control systems.

Instead, autonomous control systems provide services that allow broad and non-hierarchical access to information, real-time data sharing between different cell-level components, and active reallocation of functions to other components in the event of unavailability.

By increasing the local data processing quantity allows the stand-alone control system to significantly reduce information traffic between hierarchical levels, while allowing for increased information traffic between components on the same hierarchical level. All this leads to specific requirements for the implementation of the autonomous control system architecture.

Communication Technology

From the architecture implementation point of view, the distributed computing system [14] is the key strategy of the physical structure of an autonomous control system. The autonomous control system has an “open” logic architecture and runs logical modules that use distributed data stored, the applications being organized according to the logical and operating aspects of the technological process, which is possible due to free access to data and to storage of data in independent locations. The latest communications architecture uses the Open Systems Interconnection (OSI) standard framework [15].

Traditional communication architecture uses only 3 or 4 layers of communication and was designed to ensure secure communication over a narrow frequency band between SCADA from the Operational Point of Command and Remote Terminal Units (RTUs) of HV 110 kV/M.T. Data traffic occurs between anonymous points defined by RTUs. Obviously, switching from the traditional control system operating on 3-layer communications to the new autonomous control system implies the adoption of the OSI communications standard, the architecture of a computer system being conditioned by the architecture of communication. In Fig. 2.3 the topology of modern communication architecture is depicted.

The bus from a HV substation 110 kV/M.T. has been replaced with the Local Area Network (LAN), to which IEDs, GWs and the human-machine interface are directly connected, the Command Operational Point communication and HV substation 110 kV/M.T. in the vicinity of a Wide Area Network (WAN), within the Operational Point of Control, implementing another LAN. WAN is a distributed network of point-to-point interconnections through copper wires, fiber optics, radio or satellite links. How WAN is not always fully available for control systems, which is why traditional backup channels are often used as well.

To meet the bandwidth and security requirements of the real-time control system, the WANs of the electricity distribution companies are separated from the public WAN. This is why the Enterprise Management Information System, which, as any public system needs to be connected to the Internet but for obvious reasons to the WAN of the power distribution company, is more or less certifiable by interposing a server especially between the WAN of the electricity distribution company and the Enterprise Management Information System.

Communication needs to ensure data flow in real time but also efficiently. The layer of the data link is achieved through an ATM—Asynchronous Transfer Mode service. Its operation is asynchronous and multiplexed [16], but can also be used in

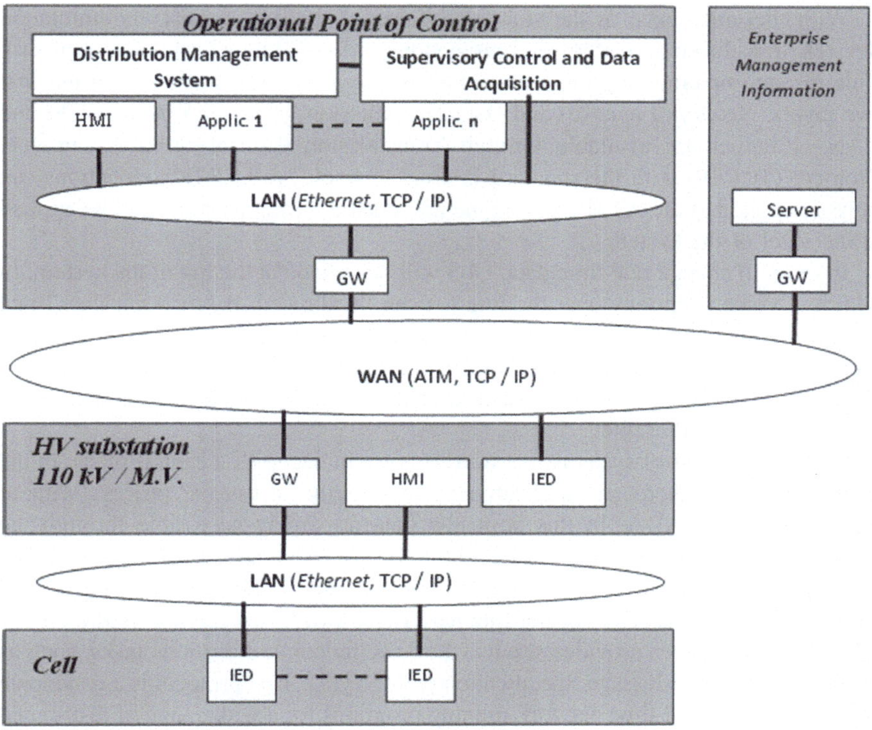

Fig. 2.3 Autonomous control system with modern communication architecture [10]

synchronous communications for fixed length information cells. The cell header contains the information that specifies the connection to retrieve the intrinsic data of the cell. This channel that provides the connection between the source and the access link to the information source is called the *virtual channel* [10].

The information cell is different from the cell in a HV substation 110 kV/M.T., referring to a particular packet of information. The nature of each packet of data, whether real-time or not, is determined by special protocol markers. These markers define the quality of the ATM service (Quality of Service (QoS)) [16]. The ATM communication system speed is 2.4 Gb, and synchronous transmission management (for real-time data) or asynchronous transmission is done through QoS markers. Local communication of a HV substation 110 kV/M.T. connected via routers or switches to WAN can be done through an Ethernet protocol of a LAN.

This protocol can transfer data at a 1 Gb speed. If a broadband bandwidth is provided then real time data traffic can be ensured. The cell data cell of HV substation 110 kV/M.T. provides transmission of short data packets, data from primary commutation devices, transducers, electronic protection devices—IED, AAR protections, RAT.

With the emergence of the smart interfaces of the new primary commutation devices, transducers, counters, and protections, the data bus with LAN is replaced. This means implementing the architecture of the auto-control systems in the first two layers. Network Layer (3) and Data Transmission Layer (4) WAN—ATM and LAN—Ethernet are available through Transmission Control Protocol—Internet Protocol (TCP/IP) with the standard Internet technology. TCP/IP connections can only be applied at the ATM top, and the universal Ethernet protocol can be applied at the level of the system.

In order to change real-time data, QoS will be defined at the top of the system. In addition, voice telephony over the Internet can be allowed. By marking the virtual channels within the QoS marker communications system, they can carry out concrete tasks leading to an autonomous control system.

In Fig. 2.4, an architectural structure of the communication of an autonomous control system is presented.

If in classic control systems the SCADA system occupies a central place, in the new architecture there are applications that directly access the process without having to use SCADA. In this way, the data are freely accessible through the autonomous components. Layers 5 and 6 allow system applications to access data via simple commands: e-mail, File Transfer Protocol (FTP) and the Hyper Text Transport Protocol (HTTP) of the Internet. These two layers are not used by TCP/IP. The seventh layer provides services for specific communications tasks, such as the Manufacturing Message Specification (MMS) [17]. This protocol is a conducive environment for real-time network running programs for a wide range of distributed applications. There are two standard protocols that can be used by MMS. One of these is IEC 61870 Part 6 Telecontrol equipments and systems [18], known as Telecontrol Application Service Element 2 (TASE 2) or Inter-Control Center Protocol (ICCP). MMS uses two standard protocols: IEC 61870 Part 6 Telecontrol equipments and systems [18] and IEC 61850 Communication networks and systems in substations [19], that is, the standard that has now become mandatory for HV substation 110 kV/M.T.

For the exchange of data between Operational Point of Command, the standard IEC 61870 Part 6 protocol or the ICCP or TASE 2 was designed [18]. It is used by MMS. The face of SCADA and IEDs data, TASE 2 also provides the exchange of information messages in the form of unstructured American Standard Code for Information Interchange (ASCII) text or short binary files and structured data items such as power exchange planning, energy balance transfers, and periodic reports on electricity generation of generators. The second standard protocol of MMS, IEC 61850 Communication networks and systems in substations [19], has now become mandatory for MV substation 110 kV/M.T. control. This standard defines interfaces for communications services, so-called Abstract Communication Service Interfaces (ACSI), that provide mapping of communication services on the MMS protocol.

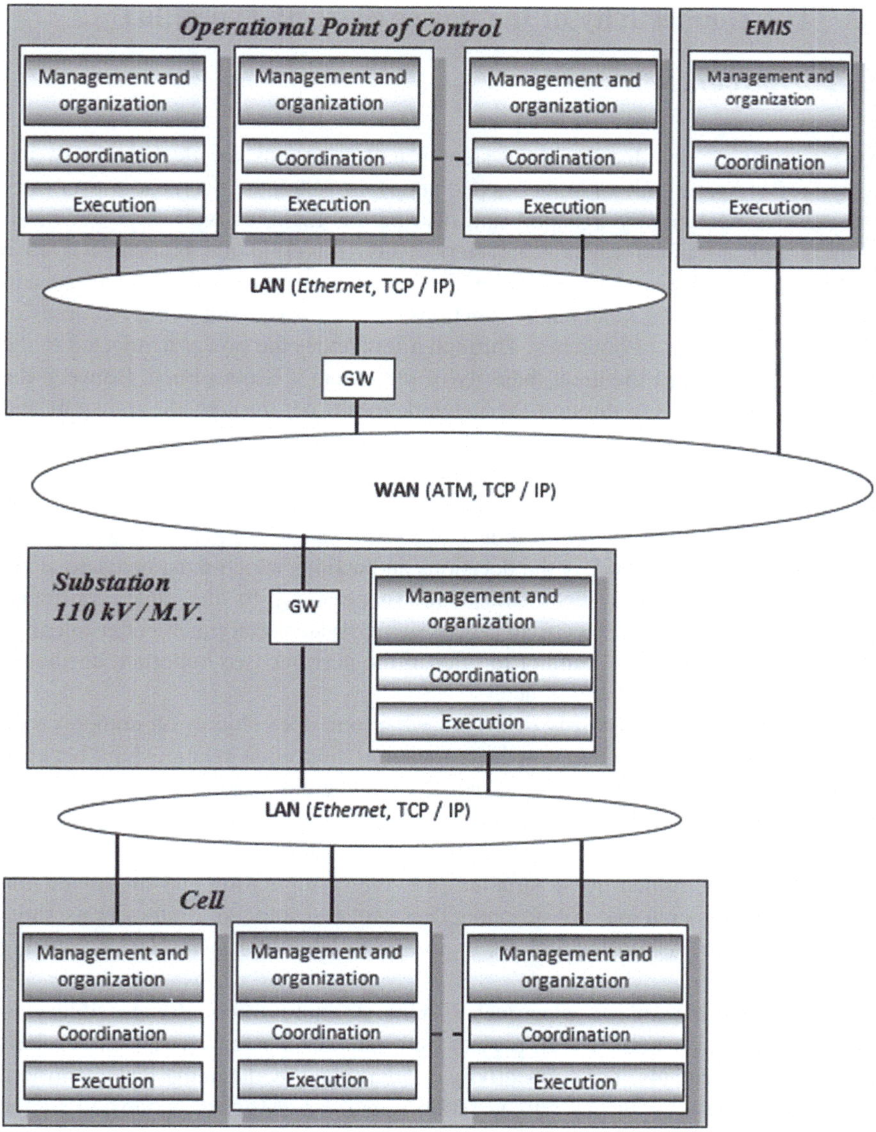

Fig. 2.4 Architectural structure of the communication of an autonomous control system [10]

2.3 The Connectivity of the Power System: Types, IoT

2.3.1 Introduction

An electrical energy distribution system usually contains an electrical network, composed of a multitude of branches and nodes, in which the overhead power lines and the electrical energy transport and distribution cable, as well as power transformers and autotransformers, characteristically, represent the branches. Even a medium electrical power company, which supplies a mix population, urban and rural, of a few million inhabitants, functions as a network, containing hundreds of nodes and thousands of branches. Through a few nodes the power is injected in the network, and through the great majority of the others it is consumed. Between the nodes, the powers flow through the network mesh, i.e. through the transport and distribution lines. Thus, a given set of powers can be obtained from a generator group, in an infinite number of configurations.

From the electrical systems point of view, the *connectivity* term describes all the equipment linked between them, found on a large spread area and integrating data fluxes and functions regarding the decisions and actions which must be taken along the energy chain—from electrical plants (generators) to the final consumer. Through *connectivity*, the electrical power supply system can much better integrate advanced digital functions aiming to offer to the network two important attributes, which are flexibility and endurance.

The structure of a system is described when one does studies on complex systems like the PoSy, because it reflects the most important relationships between individual elements and groups to which they belong. These relationships are almost invariable when perturbations in the respective systems take place and guarantee a normal behavior of the system. The system integrity and its intrinsic behavior are determined by its structure, i.e. the configuration and the interaction between the elements and subsystems. These relationships and interactions show fairly well-defined estimates and it's important to know the method of computing this estimate.

The *connectivity* of an electrical system structure offers the possibility to determine the subsystems which present tight relationships and weak relationships between their elements. The connectivity is determined by the numerical quantity of the distance between the system generators and expresses the extension of the interaction between these in permanent and steady conditions and the one corresponding to transient regimes. From this point of view one can identify three connectivity relationship types: on the same level, on an inferior level and on superior level.

In this chapter one describes issues regarding the connectivity of electrical systems, challenges and opportunities derived from the PoSy connectivity, but also the directions given by this important concept.

2.3.2 Connectivity of Electrical Systems

Through connectivity, the electrical network can integrate advanced digital functions aiming to transform the given network into a more flexible and enduring one. The transition from a one way power flow network to a two way flow involves major changes in the hierarchical history of the network. According to Grid Wise Architecture Council [20], future electrical networks must cope with *transactive energy*, defined as an energy which admits "techniques for managing generation, consumption or energy transfer in an electrical system or network by using economic or market constructions, also taking into account the constraints related to the network reliability". Transactive energy is essentially a two-way flow product. The "transactive" term involves decisions or transactions based on value. Thus, transactive energy can include price implementations for the involved equipments, auction markets, transactive control etc. [21].

Electric Power Research Institute (EPRI) suggests architecture for an integrated network which needs a safe connection between the distributed energy resources and the system operators with the purpose of ensuring an efficient, safe and economic usage of the central and distributed resources [22].

The connectivity has value when the data (information) are employed by the user or by applications to improve operational efficiency, reliability and endurance of the system. Connectivity leads to the occurrence of innovative applications and the awareness that they can be used in all aspects derived from the energy supply and demand.

Investments in aging infrastructure, changes appeared between the supply and demand chain, the increase in renewable energy portfolios, the gain in power efficiency as well as technological innovations, all these aspects imply an increased connectivity (Fig. 2.5).

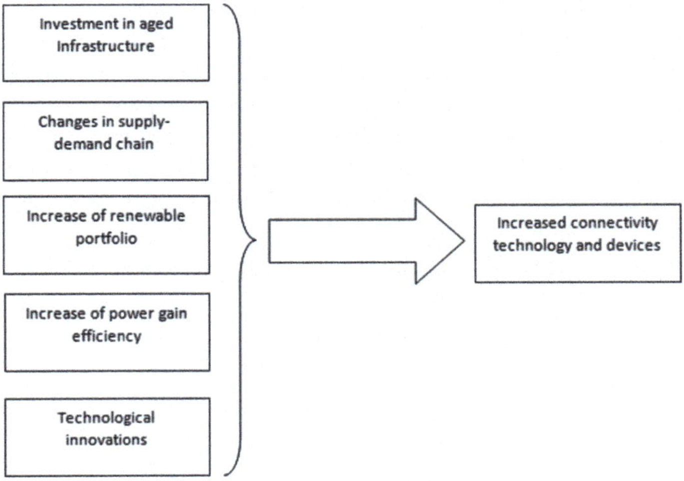

Fig. 2.5 Increased connectivity contributing factors

The concept of connectivity also offers a series of ***challenges***, which we mention:

- the great volume of data involved
- proprietary legacy systems
- the need of an increased system security
- inconsistent lifecycle and timescales
- rapid technological changes
- the effective integration of the technologies in the PoSy requires communication support for intelligent equipment, sensors, advanced measuring equipment and even for those technologies dedicated to the user.

Fortunately, all these challenges represent opportunities at the energetic chain level. For example, an investment in the telecommunication system as support for equipment connectivity can represent a strategically investment because it can be used by more systems and applications, its value rising significantly.

According to Metcalfe's law, the value of a communication network is proportional to the square of the number of users n^2 (n is the number of users) connected to that system. Metcalfe's law describes many of the network effects of modern communication technologies (Internet, social networks, www etc.). One can mathematically compute the number of unique connections in a network with n nodes as being $n(n-1)/2$, which is proportional to asymptotic n^2. Together with technological development, nowadays one considers that the value of a network is $n \times \log(n)$ [23].

2.3.3 Types of Connectivity

From the connectivity point of view, one can consider two major types: vertical connectivity and horizontal connectivity. Vertical connectivity along the measuring chain is applied to sensors, control systems and active equipment. Horizontal connectivity refers to communication between network equipment to allow consumption and energy storage in an intelligent mode (as it appears in the definition of transactive energy). Horizontal connectivity combines information from previous data such as end-to-end information resulting from the chain between the utility company and the last equipment.

Connectivity can also be seen in terms of the Open System Interconnection (OSI) model [24], as described in Fig. 2.6.

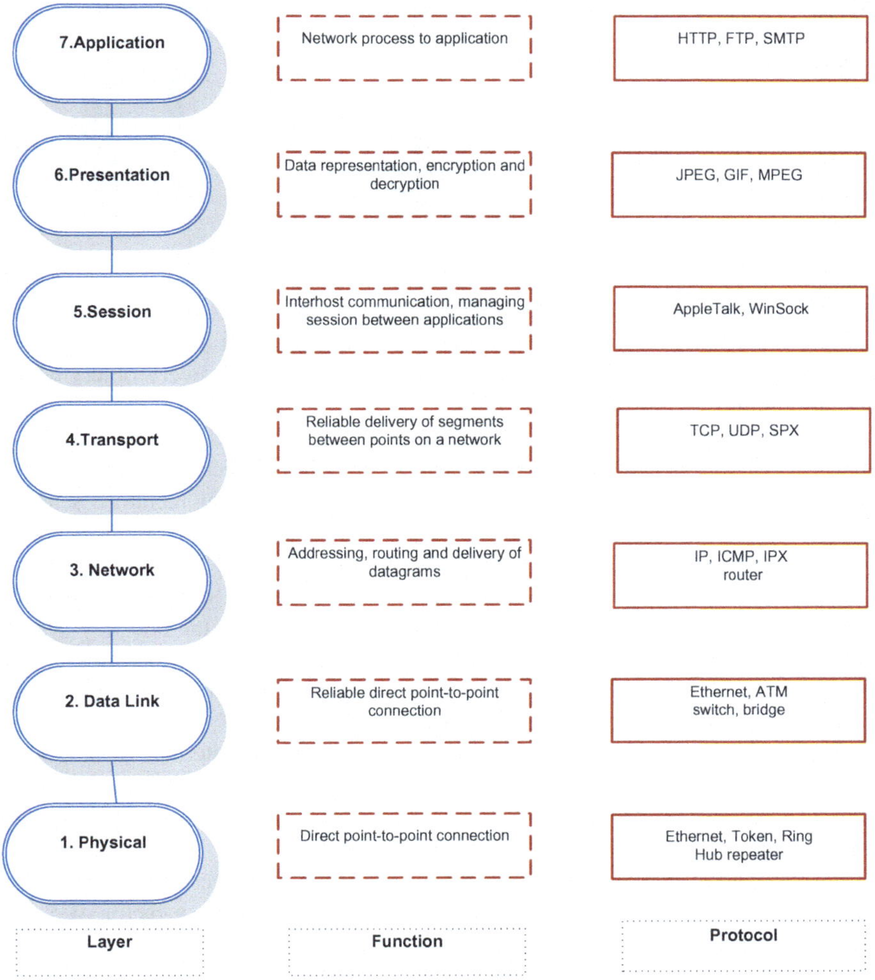

Fig. 2.6 OSI model: layers, function and protocol

2.3.4 Electric Networks and the Internet of Things (IoT) Concept

Internet of Everything (Internet of Things), for short IoT, is the phrase which defines the concept of connecting to the Internet all the objects which incorporate electronic circuits with the purpose of their remote monitoring or control. IoT is a term invented in 1999 by the British Kevin Ashton, and is used more and more in the domain of electronic devices. IoT or the Internet of Objects defines a network of objects which have electronic circuits which allow the communication through the existing infrastructure (the Internet) wireless or wired, with the purpose of remote

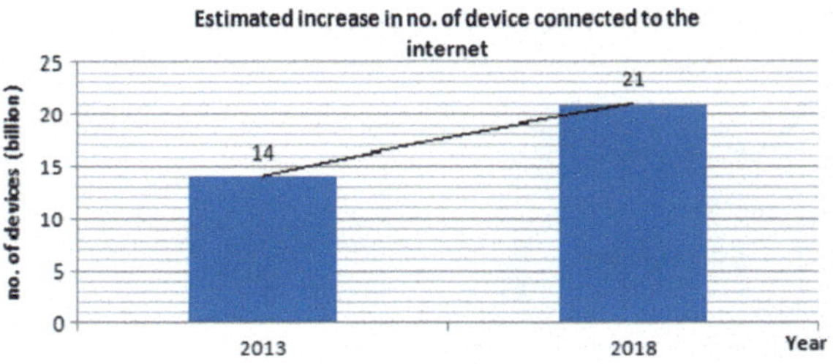

Fig. 2.7 Cisco estimated increase of devices connected to the Internet

monitoring and control [25]. The objects (things) can be meteorological sensors, pollution level measuring sensors, medical equipment, smart vehicles, household appliances, Smartphone, surveillance cameras and in general, every object which has communication capabilities [26].

According to "Cisco Visual Networking Index: Forecast and Methodology, 2016–2021", it is estimated that, by 2018, over 21 billion devices will be connected to the Internet [27] (Fig. 2.7). Among these devices, we can mention: sensors, energetic resources and energy consuming devices, as well as other devices which are components of the electrical energy distribution system.

In this context, the connectivity becomes a key concept, especially when the transition from the classic, traditional system is made, namely the one-way system towards intelligent networks described by active data fluxes and two-way systems. The interest of the consumer for connected products and services is in a continuous growth, due to the impetus of the IoT domain.

According to Cisco, IoT [28] connects objects to the Internet, making data available and offering great possibilities. IoT is a concept which assumes the use of Internet to connect different equipment between them, devices, services and automatic systems, thus forming a network of objects [29].

As mentioned above, Cisco estimates over 21 billion devices connected by the end of 2018. If during the last 5 years the traffic increased by a factor of 5, the estimation for the next five years is supposed to have an increase by a factor of 3 [30, 31].

For an overview of the equipment connectivity, we present in a graphical manner (Figs. 2.8, 2.9, 2.10, 2.11, 2.12, 2.13 and 2.14), the data provided by Cisco [32]. Therefore, Figs. 2.8 and 2.10 depict the global IP traffic 2016–2021 by Type, respectively by Segment. Figures 2.9, 2.11 and 2.13 show the Compound Annual Growth Rate (CAGR) 2016–2021, the CAGR by Segment, respectively the CAGR by Geography. Figures 2.12 and 2.14 present Global IP traffic 2016–202 by Geography, respectively the total IP traffic between 2016–2021.

The following definitions are used, according to CISCO:

- ***Consumer***: includes fixed traffic generated by: universities, internet-cafe, household population.
- ***Business***: includes fixed IP WAN or traffic generated by business and governments

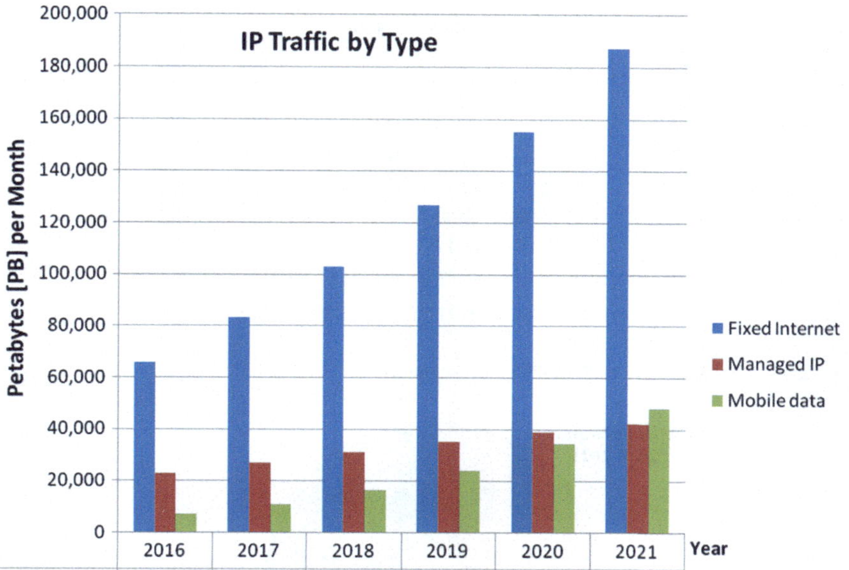

	2016	2017	2018	2019	2020	2021
Fixed Internet	65,942	83,371	102,960	127,008	155,121	187,386
Managed IP	22,911	27,140	31,304	35,226	38,908	42,452
Mobile data	7,201	11,183	16,646	24,220	34,382	48,270

Fig. 2.8 Global IP traffic 2016–2021 by Type

Fig. 2.9 Compound annual growth rate (CAGR)

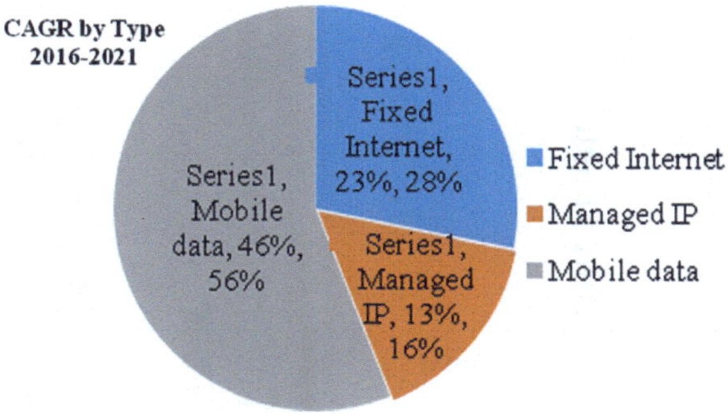

- ***Mobile***: includes mobile traffic (handsets, notebook cards, and mobile broadband gateways)
- ***Internet***: the whole IP traffic which crosses the Internet
- ***Managed IP***: includes corporate IP WAN traffic and IP TV and VoD transport.

The electrical energy consumer's interest for connected products and services is in a continuous growth, as resulting from the Accenture report [33], which estimates an increase of the number of such consumers from 7 to 57% between 2014 and 2019. This increase is justified by the consumer's interest in a lower electricity

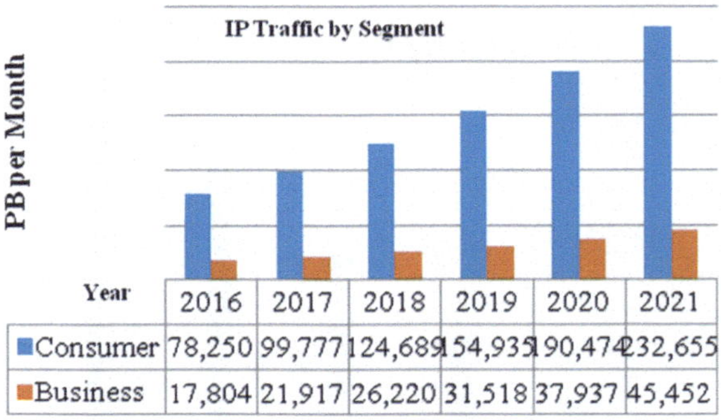

Fig. 2.10 Global IP traffic 2016–2021 by segment

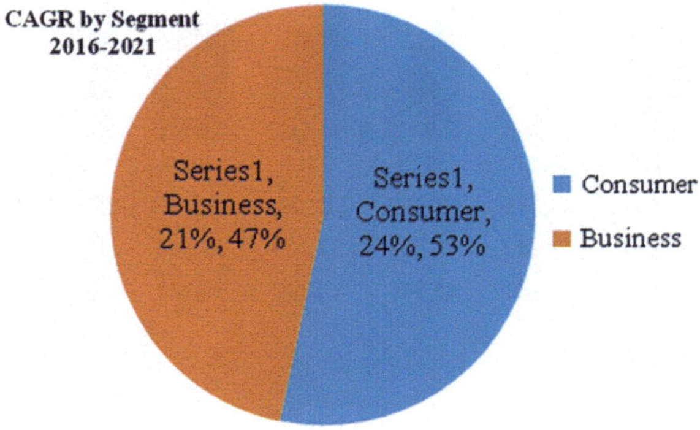

Fig. 2.11 CAGR by segment

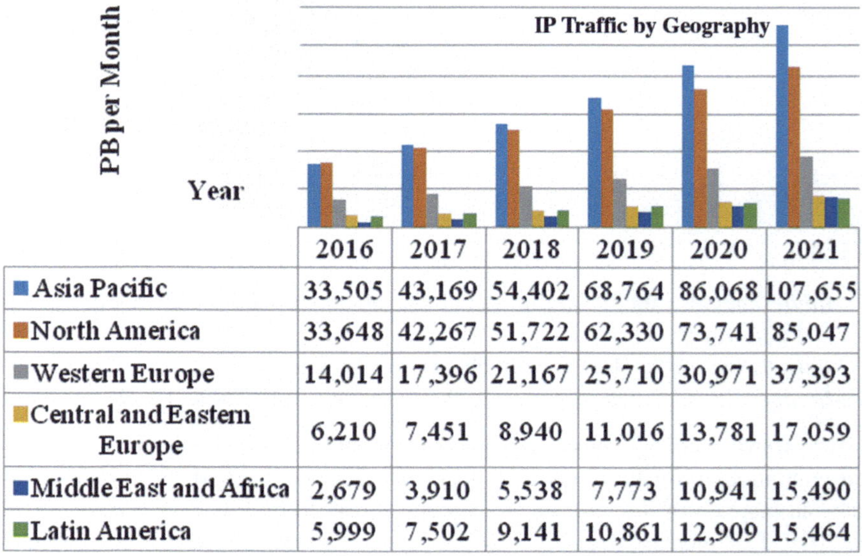

Fig. 2.12 Global IP traffic 2016–202 by geography

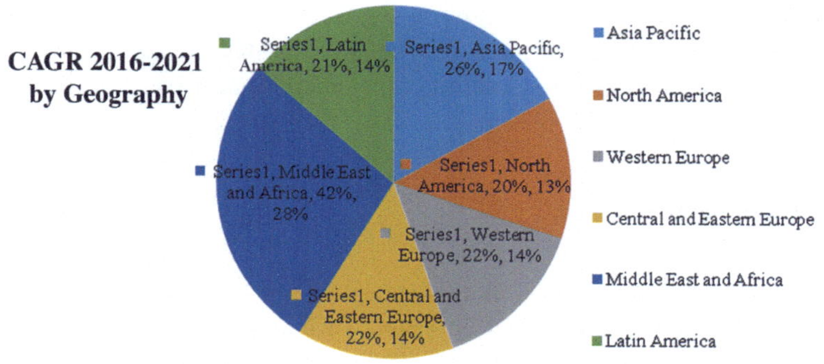

Fig. 2.13 CAGR by geography

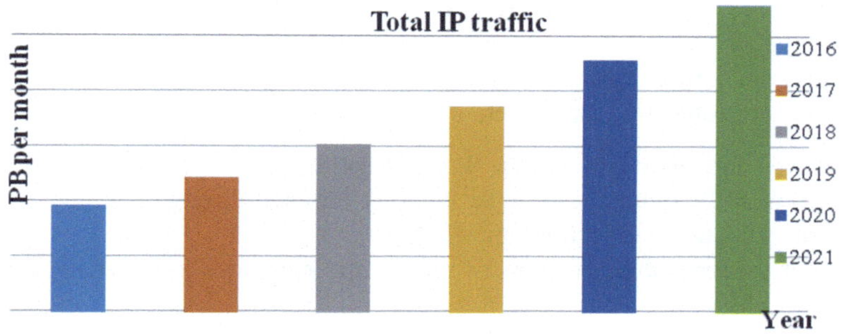

Fig. 2.14 Total IP traffic between 2016–2021

Fig. 2.15 System connectivity determining factors

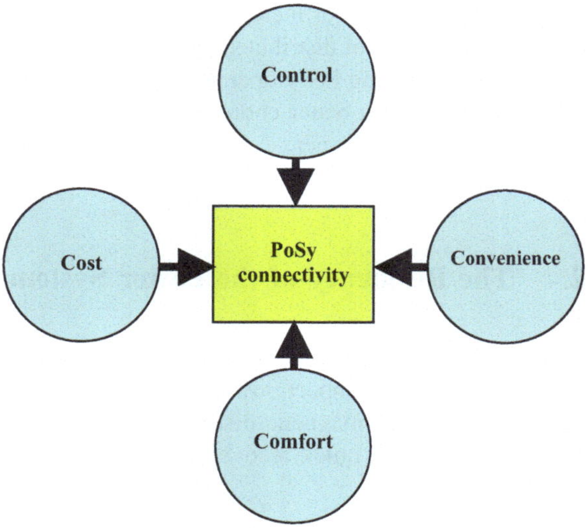

bill, for a higher comfort offered by using such connected systems and the advantage of remote using the named systems. Thus, the clients are led in their choices by the four C: Control, Comfort, Convenience and Cost (Fig. 2.15).

2.3.5 Conclusion

The center of the PoSy connectivity concept is the convergence between IT (Information Technology) and OT (Operation Technology).

Until recently, IT functions were not considered to be part of the functioning of a PoSy. Although different as rhythm and approach, the efforts of aligning IT and OT seem to fall into one or more of the following discussions:

- utilities are considered when one puts the problem of the IT and OT convergence
- utilities are subject to a partial reorganization
- utilities are subject to a complete reorganization.

Whatever the approach is, the PoSy benefits from the results in the ICT (Information and Communication Technology) domain through aligning the organizational structures and technologies [34, 35].

In what concerns the directions of the electric networks connectivity, these are given by:

1. A greater diversity and large scale using of connected end-use intelligent equipment (e.g. plug-in electric vehicles).
2. The enhanced usage of sensors and communication, which will lead to the increase of computing capacity.
3. new opportunities for increasing reliability
4. the rapid growth of distributed energy sources, especially the photovoltaic plants
5. the need of having PoSy operators to monitor the data and results in real time in order to ensure a better coordination between the supply and distribution systems and the consumer.

2.4 The Resiliency of the Power System

In the context of the energy system, resilience includes the ability to strengthen the system against high-impact, low-frequency events (natural-tornadoes, earthquakes, fires and severe geomagnetic disruptions, or human-physical, cyber attacks, coordinates attacks) and quick recovery of system properties [33].

Improving the strength of the PoSy is based on three elements

- Prevention of damage,
- Recovery of the system
- Survival.

Damage prevention involves designing standards, construction, and maintenance, inspection, operating practices, cyber and physical security [34] to reinforce the PoSy to limit damage.

System recovery requires damage assessment, fault management with the goal of restoring the service as quickly as possible.

Survival involves providing basic services, communications, emergency services, new business models to help consumers, communities and institutions continue a certain level of normal operation without having full access to their normal energy sources.

In the technical literature, reactive, proactive and predictive approaches are used to increase resilience [35–37]. Within this framework, system recovery is reactive to a high impact event, but learning from previous system recoveries and incorporating lessons learned can lead to proactive approaches. Damage prevention and survival are proactive.

By anticipating the potential impact of events, measures can be taken to minimize it. The increase in system resistance refers to each component of the system: generation, transmission, distribution and consumers [38].

The widespread connection of distributed energy resources, intelligent appliances and more complex energy markets increases the importance of cyber security and increases privacy concerns [39].

Key points to consider are:

- Industry must adopt best practices in cyber security and develop a culture of risk management; Rules on cyber security are important, but these regulations remain behind evolving threats;
- Information about cyber threats must be transmitted quickly, subject to confidentiality
- Digital security strategy involves the assignment of a dedicated team and equipment to detect and respond to abnormal cybernetic activity, reduce cyber attacker's time and deploy stratified cyber warning;
- The need to understand and enhance the resilience of the system to avoid prolonged interruptions and recover the damage caused by cyber attacks
- Advanced cyber security technologies will be used to respond to cyber incidents in milliseconds.

The future SCADA system and future distributed control systems may have a secondary diagnostic infrastructure to verify the normal system functionality, to verify coherence data for detecting data manipulation [40–46].

The National Institute of Standards and Technologies in the US has set targets for IT security [34, 47]:

- identification: managing the cyber security risk by organizational systems;
- protection: the process of implementing the appropriate means of defense;
- detection: identify the occurrence of a cyber security event and allow responses in a timely manner;
- answer: the set of activities required to be taken to detect a cyber-event;
- reaching: restoring critical infrastructure capabilities or services that have been affected by an IT event.

The European energy scene has changed considerably, with current problems including dependence on imports, the vulnerability of European economies to rising oil and gas prices, the possibility of disruption in gas supplies, insufficient investment in energy infrastructure and difficulties in completing market liberalization, the magnitude of terrorist attacks and meteorological phenomena becoming more and more difficult to control.

In both Europe and the rest of the world, the need to ensure the continuity of energy flows has increased as a basis for preserving/promoting security and national or regional economic interests.

The current instability of energy markets has increasingly focused international economic and political community's attention on existing resources in the Caspian Sea and Central Asia as a viable alternative to the main current sources.

The current environment is characterized by the high level of dependence of the Central and Eastern European countries on the Russian energy resources and the efforts of these state entities to identify viable alternatives and the denunciation of bilateral agreements at regional or European level [48].

It could be noted the actions of the Russian Federation to maintain the status of the main supplier of hydrocarbons for the European states [49].

In this context, the development of oil and gas transport projects for a number of diverse and diversified markets is favored, which could be inclusive of the beginning of the Russian Federation's integration in energy operations developed in the Central and Eastern European region on exclusively commercial criteria (Fig. 2.16).

Sustainable and secure energy supply can be ensured mainly through competitive energy markets and by developing energy policy solutions at European level to the extent that consensus is reached on these issues.

National and international security is heavily dependent on critical infrastructures, which are increasingly vulnerable to the increasingly sophisticated means of attacking those [50–52].

Two axioms are accepted in the analysis of this field:

- practically, the full protection of a critical infrastructure is not possible;
- there is no a unique way to solve this problem.

While natural disasters grow in magnitude and frequency, and the terrorist phenomenon is in ascending way, critical infrastructures need increased protection against threats and dangers. This has led to a particular concern for governments at global level to ensure a state of security for the population and state authority. In

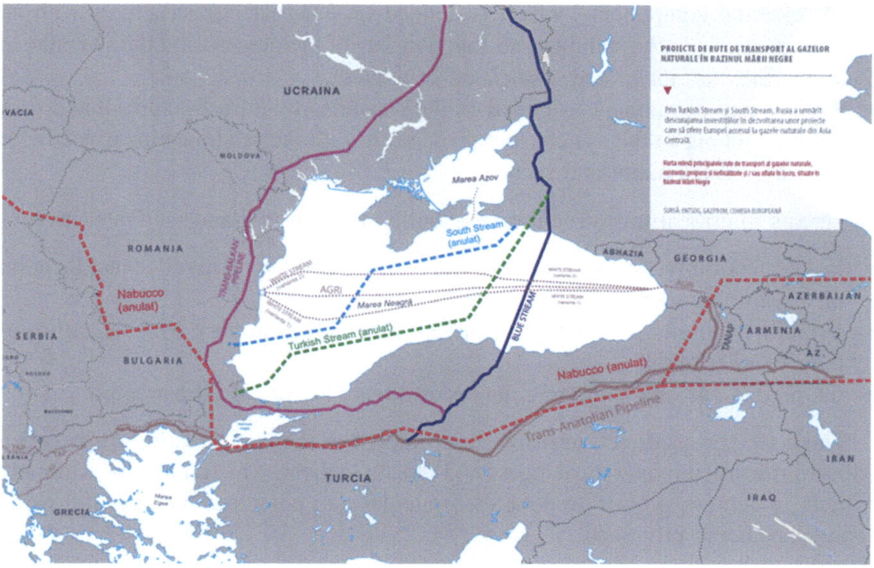

Fig. 2.16 Projects and routes of energy independence in the Black Sea Basin [47]

this respect, the first step was to assess vulnerabilities and the impact on society in the event of infrastructure and service failures.

In general, the European countries have established as critical objectives: telecommunications, water sources, distribution networks, energy sources, food production and distribution, transport systems, health institutions, financial and banking services, defense institutions (military), and public order institutions (gendarmerie, police).

In this respect, a critical infrastructure is a good material or complex objective that is vital to the overall functioning and, as a rule, interconnected with other infrastructures of the economy and society. The protection of a critical infrastructure is made up of all the measures set out to prevent and reduce the risks of blocking or destroying it, which in turn would affect other economic processes that would cause victims or have a major impact on good governance and the morale of the population.

National and international security is highly dependent on critical infrastructure in society. But they are increasingly vulnerable to the increasingly sophisticated means of attacking them.

Infrastructures are or become critical, primarily because of their vulnerability to those threats that directly affect them or are directed against their systems, actions and processes, on the following coordinates:

- internal component—defined on the increase (direct or imposed) of infrastructure vulnerabilities with an important role in the functioning and security of the system;

- the external component—which is defined on external infrastructures with an important role in the stability and functionality of the process and the systems, in which they are integrated, associated or related;
- the interface component—defined on the set of nearby infrastructures that do not directly belong to the system, but provides it with the relationships it needs for stability, functionality and security.

Threats to critical infrastructure are conditioned, favored and facilitated by:

- lack of flexibility—given by the fixed character and the relatively exact location of the infra-structures, including the critical ones;
- flexibility, fluidity of threats and threats, as well as the very wide spectrum of their manifestation;
- the predictable and surprising nature of threats to critical infrastructure.

There are always direct, intrusive, well-known or random relationships between critical infrastructure and the dangers and threats to them, which makes it extremely difficult to achieve protection policies, strategies and practices, with the following approaches being promoted:

- Considering only the security of connections, the powers of physical protection being dissipated between various state or private bodies;
- Ensure uninterrupted operation of the critical components (components) of critical infrastructures;
- Establish a mandatory minimum system of protection of certain vital components, including state bodies.

In addition, energy systems face a number of threats and the emergence of unconventional sources of electricity production (wind, photovoltaic, biomass, etc.).

The transport and distribution of the electricity is currently facing great challenges (Fig. 2.17) [51]. The reason for this is the passage from the passive distribution network to the active network as a result of the generation of distributed generation. The situation is all the more complicated by the fact that wind power parks and micro-hydropower plants are the largest share of the installed power in the distributed generation, characterized by the eminently random nature of weather-generating generation. The challenge of the electricity distributor is to manage the active power distribution network, including the discharged distributed electricity. Existing management needs to respond to both medium and long-term challenges at the level of network planning as well as in the short term in the operational dispatching activity.

Similar concerns exist in many countries and are generally promoted by renowned electricity companies [51]. In the context of the emergence of distributed generation in the form of wind and solar parks, the activity of Scotish Power Ltd. [52] can be mentioned, the studies of this company illustrating, on the one hand, the need for the prior implementation of an effective way of liquidation of the failures as a sine qua non condition for switching from the network passive to the active

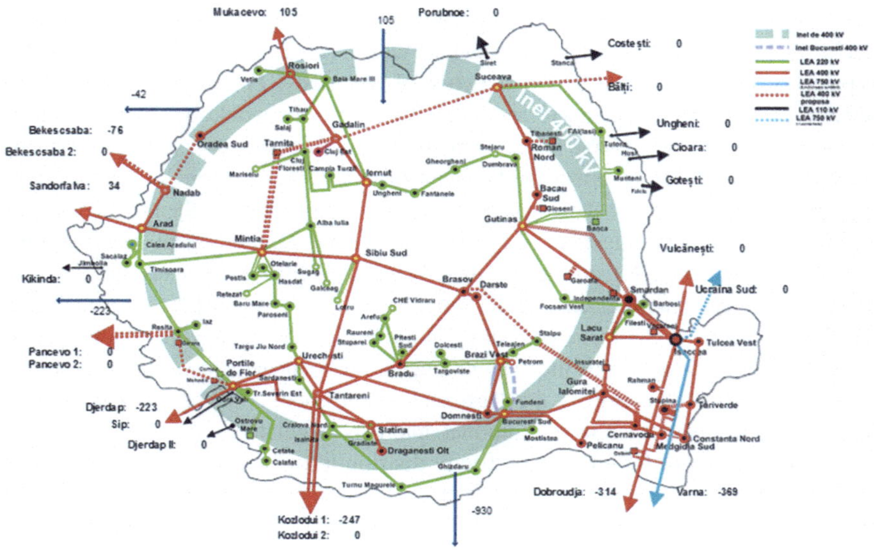

Fig. 2.17 The national energy system of Romania [51]

network and on the other hand the management of the distributed generation with the help of the 3rd generation SCADA systems, which are expressly upgraded to control frequency and bandwidth. Remarkable are the achievements of General Electric in the management of faults liquidation, in this regard the SCADA & DMS system called Power Fusion [53] can be mentioned. This system assists the dispatcher during the liquidation of the failures by validating the proposed maneuvers for the liquidation of the failures, the way of human-machine intercommunication being as effective as it is "friendly" [54].

These are efficient but traditional working tools. In order to evolve, the implementation of "artificial intelligence" is needed. In this regard, the Hiroshima Institute of Technology and the University of Hiroshima on the use of "artificial intelligence" in the Hokkaido University of Sapporo's disaster recovery on the use of "artificial intelligence" for maintaining voltage into admissible limits are important [55–57]. The problem to solve in the electricity distribution is the management of the active power distribution network [58].

Summarizing only the issue of operational management, it can be shown that on the one hand the sine qua non condition of the effective destruction of damage has to be ensured with the help of "artificial intelligence" and on the other hand the efficient management of the active and reactive powers generated and consumed in the electricity distribution network so that, by balancing the above mentioned power efficiently, the power quality is ensured, on the one hand by guaranteeing the frequency constancy and on the other by voltage maintaining in the bandwidth.

It turned to "artificial intelligence" in order to increase the efficiency of damage cessation and the management of active and reactive power balances. The valences of "artificial intelligence" consist in prompt finding of solutions by a "intelligent agent", that is, the "object" that simulates human behavior in certain situations, in this case the dispatch of the active distribution network of electricity. No matter how well the precepts of "artificial intelligence" would apply, the results would be disappointing in the case of an "agent" whose behavior is more or less "aberrant", a phrase used by psychologists to define an abnormal behavior of a person. Therefore, a more accurate physical modeling of this behavior is required, so that by applying then the techniques specific to "artificial intelligence", it is best to animate the "smart agent" that plays the role of the dispatcher. Regarding the management of distributed generation, we are talking about balancing the active and reactive power balances, shaping the "intelligent agent" to describe an automatic behavior. Unlike the loss of damage, in which case the focus is on the description of the "agent" object, the management of distributed generation emphasizes the use of the "artificial intelligence" valences to increase the efficiency of behavior such as a traditional automatic regulator.

The liquidation of failures in electricity distribution networks is now faced with new challenges, challenges that have already been perceived by actors. At the theoretical level, however, they are still "descriptive", not yet systematized and theorized. This is the "stimulating demand" of this approach to artificial intelligence, providing the necessary foundation by promoting ways of simplifying and intensifying wishes, empowering intentions, and "fanaticizing" the faith, thus finding a particular approach to defining the objective function, the method optimization and code writing so as to reduce the need for resources for the multi-agent system and to ensure the promptness for liquidation of failures [58].

The question of analyzing, designing, developing the software component and implementing it remains open in this context.

Resolving the issue of the active distribution of electricity as a result of the generation of the distributed generation and threats discussed above requires the solution of the self-orientation of the graphs, i.e. the definition and hierarchy of the source nodes as well as the node discrimination in general by the nodes communicating to all their real status.

It is noted the necessity of introducing the smart decision-making function of the decision to shift the operative management from the competence of an operative step to the competence of another, trying to realize the distributed intelligence through distributed intelligence functions (Fig. 2.18):

- Self-orientation of the graph by modeling the electricity distribution network—the solution of the mechanisms such as send-by-receive,
- The transfer of the operational management competence, the decision being made by comparing the costs of the liquidation of the failures at each voltage level or the sub-network at the same voltage range, competing with the decision-making authority regarding the liquidation of the failure-another class of intelligent agents than the nodes of the distribution network of electricity,
- Clearance procedures by seeking access to the source, which is intelligently accomplished by responding quickly to requests, minimizing the need for resources using the "restrict single" research method aggregating all the restrictions.

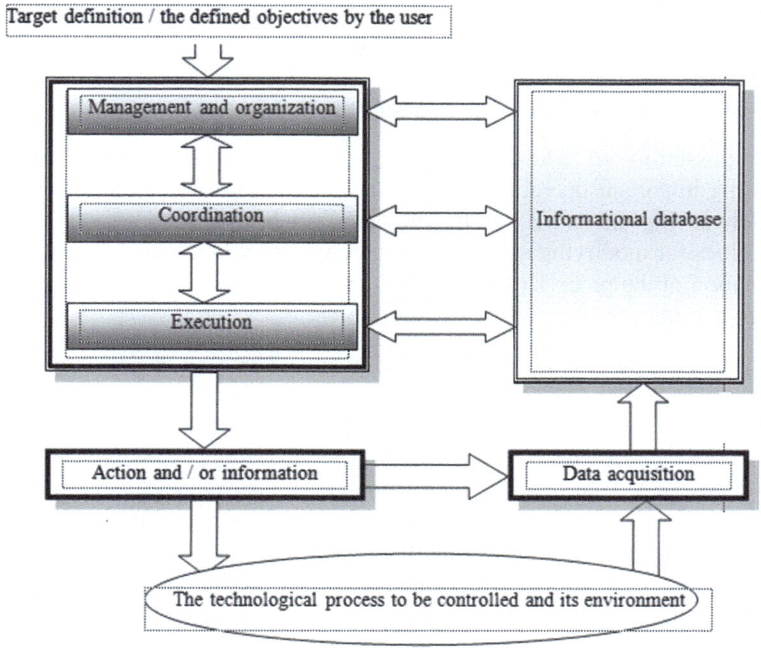

Fig. 2.18 The functional scheme of an autonomous system for the transport and distribution of electricity [10]

Essential are, on the one hand, the "stimulating demand" manifested by the challenges of public distribution of electricity on artificial intelligence and, on the other hand, the valences of artificial intelligence to meet the imperatives of public distribution of electricity in terms of timely decision making, with minimal consumption resources, regarding the liquidation of the failures—an essential aspect of the electricity distribution requirements, requirements imposed by the Energy Directive issued by the European Union [59].

2.5 Case Studies—SCADA System in Romania, Smart Grid Power System

In this Section the integrate SCADA system with remote reclosers controllers is described. The Distribution Management System (DMS) [60] is a computer-based system that provides support for the operative management of electrical distribution networks using SCADA (the process by which real-time information from geographically located locations to a control center for real-time processing, analysis and remote control purposes), user graphical interface and various application programs [61, 62].

The SCADA/DMS system has two fundamental features:

- The possibility of adding, replacing, redistributing the equipment without causing important operating disturbances (open system architecture) and completely saving the entire system software (portability);
- Facilities for modifying and integrating new system functions, as a result of the evolution of the process (development) and the changing of the requirements in time (evolution).

SCADA integration allows service and maintenance staff at the station to track maneuvers or on-site inspections, increasing the efficiency of SCADA/DMS and Periodic Inspection and Testing of Electrical Installations of tele-signaling, tele-control and telemetry.

Basic SCADA features are: data acquisition and exchange, sequential event logging, data processing, instant data recording, historical information system, remote control, marking, user interface, alarm processing and management, on-screen display, SCADA/DMS system status supervision as well as basic DMS functions: network topology processing (permanently builds and updates the electrical network model using real-time information from electrical stations, short-circuit analysis, implementation of a program for performing current calculations short-circuit), tracing the quality of the consumer feed, simulator for the training of the operators [10].

The use of single wire synoptic schemes in the application for reclosers' remote control.

The interaction between the SCADA system and the operational staff is achieved through graphical interfaces that feature various remote-controlled equipment or different MV network areas [63].

Initially, the main interface was made that includes links to the following synopsis:

- synoptic of the list of remote-controlled separators and synoptic of the list of remote-controlled reclosers;
- synoptic for each substation integrated into the SCADA system;
- synoptic according to the PA/PT list integrated into the SCADA system.

The synoptic corresponding to the list of remote control reclosers (Fig. 2.19) contains an enumeration of the name, position of the switch, the state of the different types of protections for each recloser. Within this synoptic it is possible to visualize the remote reclosers list organized on MV networks (Fig. 2.20).

Starting from the MV network organization model presented in the case of the synoptic of the remote recloser list (Fig. 2.20), single bar synoptic schemes were made as necessary for:

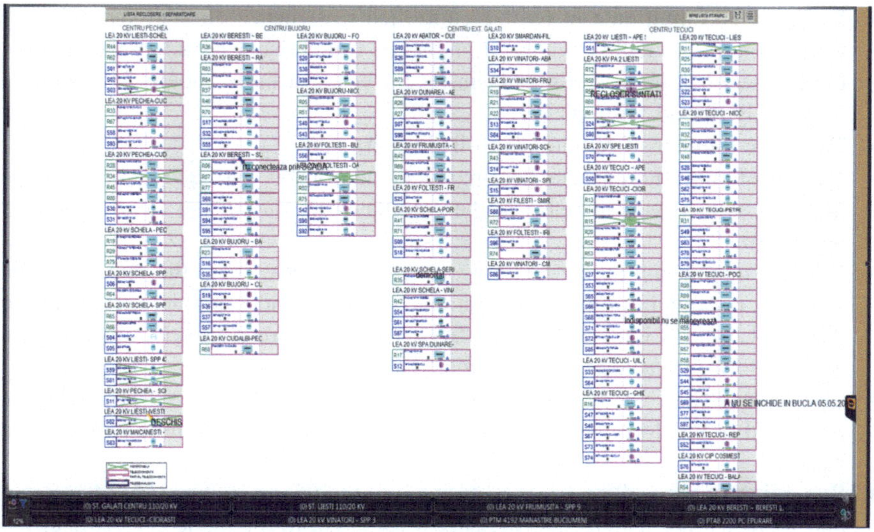

Fig. 2.19 The synoptic for the list of remote-controlled reclosers [64]

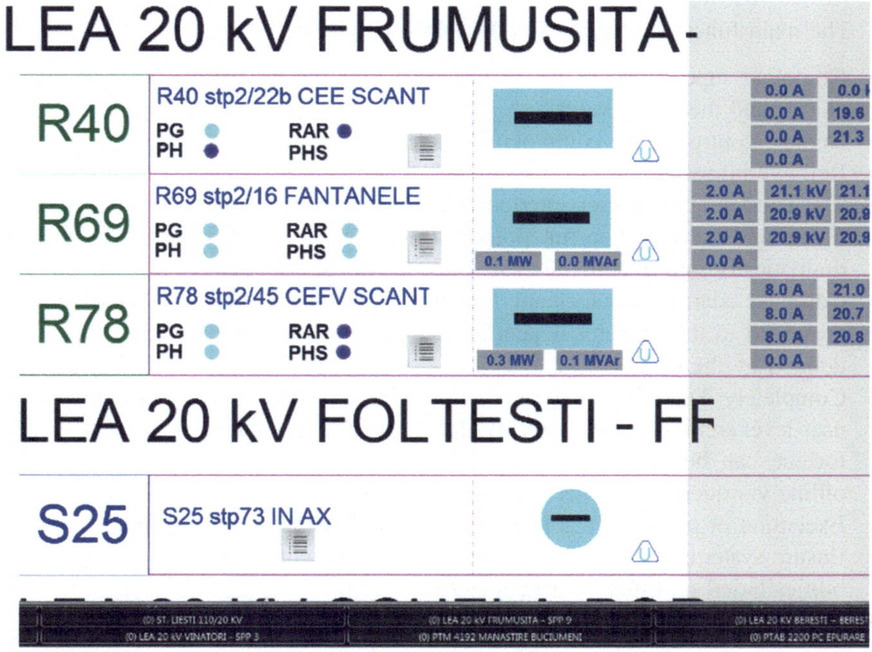

Fig. 2.20 Synoptic for the remote recloser list on network area [64]

- Diversity and multitude of remote-controlled re-switches; though synoptic
- Corresponding to reclosers is well organized (on network areas), as the number of reclosers has increased greatly in recent years, it has difficulty in operating (especially during failures);
- The large number of remote-controlled separators;
- The substations number to be upgraded, automated and integrated into the SCADA system will increase in the coming years due to the development of wind farms;
- Efficient monitoring of the medium voltage aerial network, as the same synoptic schemes present information on remote reclosers and separators, transformer electrical stations integrated in the SCADA system, respectively, signaling (in graphic form) transmitted by the smart fault indicators;
- Operative dispatcher personnel are more familiar with the synoptic schemes presented in this form, and for this reason operation is much more efficient and easier;
- Synoptic diagrams allow a fast assessment of the failure status by the dispatcher and by analyzing the signals (messages) received, to make the correct decision regarding the detection/isolation of the defective area;
- Network reconnection of the users and directing operational staff of the operation center to the defective area for faulty disposal/repair and return to the normal schedule.

The main functions of the SCADA-DMS system are:

- Complete supervision of the distribution network, as position of equipment, alarms and measured values;
- Remote control of the equipment installed in the transformation stations;
- Retrieve information from existing RTUs;
- The latest graphical user interface based on windows with support for zoom, layout and decongestion functions, with contextual help functions, designed to dramatically improve the efficiency of system use by operators;
- Complete alarms management function, which limits the number of alarms presented to the operator depending on the defined operating areas and filters defined by the user;
- Completely keep lists of events and system actions in system-defined or user-level archiving tables. Archived data can be accessed in graphical or tabular format, can be included in user-defined reports or accessed either online or offline via queries that meet the SQL standard;
- Execution of parameterizable and user-defined automation functions;
- Ensure system access control based on user passwords and limit control capabilities based on different access levels;
- GPS synchronization of the entire system.

In the SCADA system there are different events and they are processed according to their type and depending on the configuration. They can be printed on paper or can be recorded in digital format and can initiate associated processing functions. The main types of events are:

- Changes in values (digital or analog);
- User actions. All actions of the user, such as knowing the alarms that lead to data changes in the system, are considered events;
- System internal activities. The main internal activities of the system are also considered events.

The event list has multiple filtering criteria, including date, priority, station, cell, event type, and so on. The SCADA system is provided with a performance alarm system designed to optimize operator response time when abnormal situations are encountered. The ability to filter alarms allows the removal of alarms that are unrelated to the main causes that triggered the alarm state, and also allows the user to select relevant alarms at each moment. Only the alarms corresponding to the area of responsibility defined for the active user are displayed on the workstations, thus drastically reducing the number of alarms displayed.

The alarm processing system is a component of the SCADA base system and is asynchronous to the SCADA data acquisition system. It also can not have adverse effects on the control system. Normally, all events in the distribution network receive the time stamp on the acquisition equipment. This time stamp is also used as the time stamp for the alarm. If this is not true, such as alarms generated by server operation, the current system time will be used as the best approximation of the event time stamp. Thus, in the event of avalanche events, there may be a delay in creating alarms in the alarms list for a certain status change, but this will not have a negative impact on the basic telemetry and command process.

Similarly, any event recording activity is asynchronous to the data acquisition process and alarm processor. These relatively slow processes use buffers during intense activity periods or when the printer can not work.

Types of alarms:

- Digital signal events;
- Analogue signal above or below defined alarm limits;
- System or communication errors;
- Deferred alarms.

The commands from the system to the remote control can be of the analogue type (voltage or current) or of the digital type (relay outputs). A command is executed with user-defined command-line interface interfaces so that the user can view the state of the equipment before and after the command is executed. Security measures are implemented both on hardware and software to avoid the execution of incorrect commands. Special operating privileges are required for operators to execute orders. In addition to simple commands, command sequences can also be defined.

The following command mechanisms are supported:

- External control mechanisms for two or more states with or without feedback (selection-check back);
- External state mechanisms with or without feedback (selection-check back);

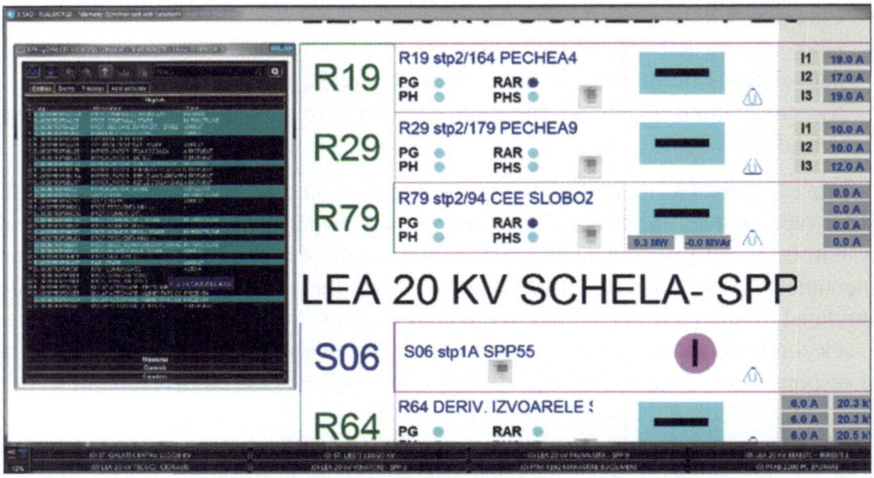

Fig. 2.21 List of digital signals and their current status, corresponding to the R79 recloser [64]

- Command for analog or digital signals;
- Prepare parameters for automation functions;
- Inhibition;
- Check the final order status;
- Supervising the execution time of the order;
- Internal commands to entities in the database.

The above described system was implemented at ELECTRICA SA—Galati, Electric Distribution Branch and it is functional (Fig. 2.21) [64].

2.6 Conclusion

Connectivity and resiliency as essential properties of a safe power system are presented. The opportunities of connectivity developing will be studied aiming to increase the efficiency. Also the interdependence between IoT and smart power systems is highlighted in order to show the necessity of define some standards/ regulations concerning the development of new data protection systems. Different natural or human power systems against high-impact are described. In this context the increasing of power systems resiliency is extremely necessary by using modern data acquisition system as SCADA.

References

1. R.C. Dugan, M.F. McGranaghan, S. Santoso, H.W. Beaty, in *Electrical PoSys Quality* (McGraw Hill Professional, 2012)
2. J.H.C. Pretorius, J.D. Van Wyk, P.H. Swart, An evaluation of some alternative methods of power resolution in a large industrial plants, in The 8th International Conference on Harmonics and Quality of Power (ICHQP), VIII, vol. 1 (Athens, Greece, 1998), pp. 331–336 October 7–8
3. O. Erdinc, M. Uzunoglu, Optimum design of hybrid renewable energy systems: overview of different approaches. Renew. Sustain. Energy Rev. **16**(3), 1412–1425 (2012)
4. IEA, Solar Energy Perspectives 202 (International Energy Agency, Paris, France)
5. D. Biggar, M. Hesamzadeh, in *Introduction to Electric PoSys* (Wiley, IEEE Press, 2014)
6. X. Feng, L.T. Yang, L. Wang, A. Vinel, Internet of Things. Int. J. Commun Syst. **25**(9), 1101–1115 (2012)
7. Directive 2006/32/EC of the European Parliament and of the Council of 5 April 2006 on Energy End-Use Efficiency and Energy Services and Repealing Council Directive 93/76/EEC
8. ICT for a Low Carbon Economy Smart Electricity Distribution Networks, Office for Official Publications of the European Communities, Luxembourg. European Communities, July 2009. ISBN 978-92-79-13347-3, https://doi.org/10.2759/19105
9. L. Ren, Y. Qin, Y. Li, P. Zhang, T. Gong, Enabling resilient distributed power sharing in networked microgrids through software defined networking. Appl. Energy. Available Online 16 June 2017
10. I.N. Arama, Applications of Multiagent Systems in the Distribution of Electricity, Ph.D. Thesis, Dunarea de Jos University of Galati, Romania, 2001 (in Romanian)
11. E.M. Davidson, S.D.J. McArthur, J.R. McDonald, T. Cumming, I. Watt, Applying multi-agent system technology in practice: automated management and analysis of SCADA and digital fault recorder data. IEEE Trans. PoSys **21**(2), 559–567 (2006)
12. E.J. Friedman-Hill, F. Jess, *The Java Expert System Shell, SAND98-8206, Distributed Computing Systems* (Sandia National Laboratories, Livermore, CA, 1997)
13. P.H. Larsen, K.H. LaCommare, J.H. Eto, J.L. Sweeney, in *Recent Trends in PoSy Reliability and Implications for Evaluating Future Investments in Resiliency, Energy*, vol. 117 (2016), part 1, pp. 29–46
14. T. Ding, Y. Lin, Z. Bie, C. Chen, A resilient microgrid formation strategy for load restoration considering master-slave distributed generators and topology reconfiguration. Appl. Energy **199**, 205–216 (2017)
15. EPRI, Electric PoSy Connectivity: Challenges and Opportunities, February 2016, https://bit.ly/2q2IO0I
16. How ATM Works, March 28, 2003. https://bit.ly/2q4BdOd. Accessed 2017
17. A. Elgargouri, Implementation of IEC 61850 in Solar Applications, Master's Thesis, Faculty of Technology Telecommunication Engineering, University of VAASA, Finland, 2012
18. D. Becker, EPRI, ICCP Protocol—Threats to Data Security and Potential Solutions, 2001
19. https://bit.ly/2HbDhfq
20. https://bit.ly/2GQXJEX
21. E.C. Boscoianu, D. Popa, Internet of Things, Buletinul AGIR (2), 2016. Accessed on 9 June 2017
22. https://bit.ly/2GwUjb8
23. Electric PoSy Connectivity Challenges and Opportunities, EPRI, Data Analytics Initiative for Transmission and Distribution: Year One Update, 2015, https://bit.ly/2EjHpaq
24. EPRI, The Integrated Grid: Realizing the Full Value of Central and Distributed Resources, 3002002733, 2014, https://bit.ly/2q58DNA
25. https://bit.ly/2lQ8Pjd
26. https://bit.ly/2EjLAmu

27. Cisco Visual Networking Index: Forecast and Methodology, 2016–2021, https://bit.ly/2wmdZJb
28. Utility Chief Information officer (CIO) Outlook, EPRI, Palo Alto, CA: 2013, product number 3002000085, https://bit.ly/2JjXtwD
29. M. Wakefield, EPRI, Information and Communication Technology (ICT)—A Key Enabler for the Future PoSy, Electric Energy T&D Magazine, 2014, https://bit.ly/2GyBswi
30. Accenture, Interest in Connected-Home and Alternative Energy Solutions to Increase Six-Fold, Accenture Research Show, 2014, https://accntu.re/2q6Qu1c
31. GridWise, Interim Report: Transactive Valuation Methodology, 2015, https://bit.ly/2JjKOK3
32. B. Mitchell, Visual Index of Computer Networking Topics, The OSI Model of Computer Networks, 2016, https://bit.ly/2Hb8mjJ
33. EPRI, Electric PoSy Resiliency, Challenges and Opportunities, https://bit.ly/2GC49US
34. National Institute of Standards and Technology, Information Security Handbook: A Guide for Managers, Gaithersburg, MD 20899-8930, 2006
35. M.S. Dinu, C. Bahnareanu, Actualities and Perspectives in the European Security and Defense Policy, Publishing House of the "Carol I" National Defense University Bucharest, Romania, 2006 (in Romanian)
36. C. Mostoflei, Defense and Security Strategies at the Eastern and NATO Border of NATO and EU, vol. I, Publishing House of the "Carol I" National Defense University Bucharest, Romania, 2006 (in Romanian)
37. N. Paun, A.C. Paun, G. Ciceo, *United Europe. Our Europe* (Europa Unita, Cluj University Press, Cluj-Napoca, Romanian, 2003)
38. A. Grigore, G. Vaduva, Critical Infrastructure, Dangers, Threats to Them. Protection Systems, Publishing House of the "Carol I" National Defense University Bucharest, Romania, 2006 (in Romanian)
39. I. Bidu, C. Troncota, *Security Coordinates* (Publishing House of the National Information Academy, Bucharest, Romania, 2005). (in Romanian)
40. G. Toma, T. Liteanu, C. Degeratu, *Evolution of Security Architectures under the Impact of Globalization* (Publishing House of the National Information Academy, Bucharest, Romania, 2007). (in Romanian)
41. https://bit.ly/2q7WSVN
42. https://bit.ly/2q757Bs
43. https://bit.ly/2HafzQP
44. https://bit.ly/2uOTG7e
45. https://bit.ly/2EhulCd
46. https://bit.ly/2GTOXpt
47. L. Rosoiu, Black Sea and Romania—points of reference for EU energy security, https://bit.ly/2Isk48L. Accessed 2017
48. D. Kreutz, O. Malichevskyy, E. Feitosa, H. Cunha, D.J. de Macedo, A cyber-resilient architecture for critical security services. J. Netw. Comput. Appl. **63**, 173–189 (2016)
49. K. Shuaib, Z. Trabelsi, M. Abed-Hafez, A. Gaouda, M. Alahmad, Resiliency of smart power meters to common security attacks. Proc. Comput. Sci. **52**, 145–152 (2015)
50. H. Sun, C. Peng, T. Yang, H. Zhang, W. He, Resilient Control of Networked Control Systems with Stochastic Denial of Service Attacks, Neurocomputing, Available Online 17 June 2017
51. https://bit.ly/2q7X6fB. Accessed 2017
52. Scottish Power Ltd Study on Active Power Distribution Network Management, 2006
53. PowerFusion User Manual of General Electric, 2008
54. EFACEC's ScateX User Manual, 2002
55. T. Nagata, H. Sasaki, A Multi-Agent Approach to PoSy Restoration, 2003
56. H. Kita, J. Hasegawa, Operation of Quality Control Center Based on Multi-Agent Technology, 2003
57. M. Chirita, C.V. Costescu, F. Gabor, Method for Calculating Power Circulations in Electrical Networks, 2003

58. I.N. Arama, C.V. Costescu, Use of Artificial Intelligence in Electrical Networks—Liquidation of Failures (in Romanian)
59. Directive 2006/32/EC of the European Parliament and of the Council of 5 April 2006 on Energy End-Use Efficiency and Energy Services and Repealing Council Directive 93/76/EEC
60. How the Distribution Management System (DMS) is Becoming a Core Function of the Smart Grid, Reducing Risks and Costs by Optimizing Distribution Network Operations, Point of View (POV) Executive White Paper Series, 2012, https://sie.ag/2uHECYQ
61. EPRI, Common Functions for Smart Inverters, Version 3, 3002002233, 2014, https://bit.ly/2GTQGLt
62. V.D. Krsman, A.T. Saric, Verification and estimation of phase connectivity and power injections in distribution network. Electric PoSys Res. **143**, 281–291 (2017)
63. Y.V. Pavan Kumar, R. Bhimasingu, Electrical machines based DC/AC energy conversion schemes for the improvement of power quality and resiliency in renewable energy microgrids. Int. J. Electr. Power Energy Syst. **90**, 10–26 (2017)
64. https://bit.ly/2H6Nrhr

Chapter 3
Power System Flexibility and Resiliency

Ersan Kabalci

Abstract The flexibility of the power system can be described with its ability on providing dynamic and adaptable structure against the various circumstances. It requires balancing the power supply and demand in terms of intervals such as minute or hourly. The flexibility of power system, which is a critical driver, is the fast and assorted deployment of distributed sources such as hydro, wind, solar etc. The early challenges of the flexibility researches are focused on rapid deployment of distributed generation while the followings are related to pricing, standards, policies, and microgrid (MG) integration to power system that includes customer adoption. The environmental policies, subsidies and similar factors may constrain the power system management in terms of generation. Therefore, the power system characteristic changes to distributed generation searches instead of conventional generation. It may also shift consumers to energy generators by using their micro sources as solar plants, hybrid electric vehicles and smart appliances. The alteration to more flexible power system involves novel technologies and methods to sustain the security of the network. On the other hand, the resiliency of a power system requires the ability to increase the security of power system against extreme conditions. The main interest on resiliency is caused by significant weather conditions such as hurricanes, earthquakes and floods. Such weather events are defined as high impact and low frequency events. Moreover, increased communication and monitoring infrastructures have raised the cyber security concerns in the aspects of resiliency. Therefore, the resiliency of power systems requires to be handled in terms of physical and cyber damages. The resiliency is researched in three main topics as damage prevention, system recovery, and survivability of power system. These topics are studied in generation, transmission, distribution and consumer sections with its all drivers.

E. Kabalci (✉)
Department of Electrical and Electronics Engineering, Faculty of Engineering and Architecture, Nevsehir Haci Bektas Veli University, Nevsehir, Turkey
e-mail: kabalci@nevsehir.edu.tr

© Springer Nature Switzerland AG 2019
N. Mahdavi Tabatabaei et al. (eds.), *Power Systems Resilience*, Power Systems,
https://doi.org/10.1007/978-3-319-94442-5_3

Keywords Flexibility · Demand side management · Distributed generation Information and communication technologies · Microgrid · Cyber-physical system Resiliency

3.1 Introduction

When the recent trends and tendencies examined in the context of power systems, it is expressed that current power networks require to be converted to more flexible, resilient and connected network due to several issues such as increments on conventional sources, alternative source integration, load type increment, and distributed generation issues [1–5]. Although the power system is a vital infrastructure, curtailments or blackouts have been experienced at any time up to now. The blackouts can affect just residential and regional customers or industrial plants that costs to high financial losses. Flexibility, as a widely accepted and widespread definition, is the ability of any power system for adapting to changing conditions while sustaining to provide energy for consumers in a secure, reliable, and affordable way [1, 6].

The expected effect of flexibility on a power system is to extend its sources to manage load changes. The net load means load changes that are required to respond but not deployed by the generation service due to inadequacy. Therefore, the power system including variable generations with long starting times and low ramp rates will not facilitate to increase flexibility of power system. In case of this situation is handled in terms of system integration of an islanded plant to utility grid, the successful integration should be planned at the installation stage of power plant. The distributed generation (DG) or variable generation (VG) concepts define a generation infrastructure where the sources are wind power, solar power, tidal power or similar that is dependent to environmental conditions. The integration of such sources the utility grid targets flexibility and generation capability of power system with load. Therefore, flexibility of any power system is depended to generation planning, source characteristic analysis, and operation certainty.

Some extraordinary power system deficiencies have been reported including high level penetrations of wind turbines and the Electric Reliability Council of Texas (ERCOT) event [6]. In the event which is occurred in 2008, the load ramp extended by ramping up to 3800 MW day-ahead load forecast of ERCOT that has been forecasted as to be around 3550 MW. The evening load ramp began 25 min earlier than expected that has caused to imbalance between generation and load. Therefore, system frequency first declined to 59.94 Hz and then to 59.91 Hz and 59.85 Hz in a few minutes. ERCOT started an emergency electric curtailment plan to prevent complete blackout and recovered the system in three hours with curtailments [7]. Three major factors of this event were large ramp-down of wind generation, the unpredicted loss of conventional generation, and unexpected load ramp-up in the evening. This was a certain lesson exhibiting the importance of long term planning and flexibility. The high penetration of renewable energy sources

(RESs) forces operators to reconfigure flexibility and reliability of entire power system. In addition to this, rapidly responding generators with high ramp up/down features is required to improve the flexibility of system. Flexibility becomes more critical not only for generation but also for transmission, distribution, and consumption levels. Therefore, it is important to increase system flexibility to prevent unexpected curtailment and blackouts.

Another crucial aspect to install a robust power system is resiliency that is defined as the ability of rapid recovery against rarely occurred but high impact events. Most apparent events are extreme natural events such as earthquakes, hurricanes, tsunamis, floods and so on. The most important cause of outages in United States is extreme weather damages. In 2011, Hurricane Irene resulted to a blackout in Washington DC affecting more than 6.5 million people. The most recent ones were Hurricane Harvey in Texas and Hurricane Irma in Florida resulted in more than 6 million people to lose power at each state. The cost of outages that are caused by weather is around $25-70 billion to United States. In addition to US, the weather-related outages are seen all over the World. For instance; curtailments experienced in China due to an ice storm in 2008; a storm passing from North Europe from Ireland to Russia caused several outages in several countries including Denmark, Norway, and Sweden in 2005; strong storms and lightning strikes triggered an outage caused more than 1.5 million people to lose power in Australia in 2016 [8].

On the other hand, cyber-physical system (CPS) improvements also cause some resiliency problems. The power system has become much more vulnerable to cyber-attacks with the increased use of information and communication technologies (ICT) in automation, monitoring, measurement, and control operations. The most prominent cyber-attack types include denial-of-source attacks, manipulating device attacks providing wrong data to system, surveillance attacks that inherit the system information on vulnerabilities and operational features, eavesdropping attacks to violate confidentiality, unauthorized intrusions, malicious code and snippets such as virus, worm or Trojan [3]. The cyber-attack can be potentially hibernated, widespread, and processed at the planned date by attacker. Despite physical attacks, impacts and damages of any cyber-attack on operational and ICT system cannot be easily detected at a glance. It may take long time to detect and recover damages of cyber-attack since it requires diagnosing and remedying a large infrastructure. Another prominent challenging task on resiliency is related to security assessment that is the ability of coping with CPS disturbances. The short circuits triggering blackouts in a power system are followed by response of protection equipments. Disconnecting the critical devices and components of the system during blackouts enhances the damaging effect of contingency, and thus the stability is lost. Therefore, security and stability assessment should be repeated at short intervals in order to improve the system security [3, 9, 10]. The resiliency requirements have been presented in the following chapters in detail.

The connectivity means the communication and remote monitoring devices of a power system. These devices provide measurement data to determine policies and management approaches regarding to behaviors of generators and consumers.

Connectivity improves smart infrastructure in the context of grid that are essential to enhance flexibility and resiliency of power system. The increased connectivity requirements and improved applications resulted smart grid (SG) term, which defines the next generation grid with the capabilities of intelligent management, bi-directional or briefly two-way communication, two-way power transmission, connection of wired or wireless sensor networks (WSNs), interconnection of mobile devices at industrial and residential plants. The connectivity is located at any level of SG infrastructure including generation, transmission, distribution, and consumption. Connectivity of grid is being promoted by numerous approaches as enhanced use of mobile devices, widespread deployment of intelligent systems, increased use of ICT, data transfer on cloud networks, and WSNs, enhanced growth of DG with renewable energy sources (RESs). The integration of ICT and power system increases the CPS security, homogenous architecture, energy monitoring, and Internet of Things (IoT) support to enhance connectivity. The IoT connects devices, machines, systems, and operators as being a widespread and most recent concept. It facilitates the connectivity of all components and beneficiaries. IoT facilitates numerous SG applications interfacing connections such as machine-to-machine (M2M), vehicle-to-grid (V2G), grid-to-vehicle (G2V), WSN infrastructure, smart meters (SMs), gateways, and databases [5, 11].

The three important attributes of any power system are illustrated in Fig. 3.1 with the factors that are handled by flexibility, resiliency, and connectivity issues. Flexibility is required to manage increasing fuel prices, various consumer habits, power market share and incentives, DG tendencies, and regulations, policies, and arrangements. The resiliency copes with rarely occurring but highly dangerous factors that include natural or weather events, earthquakes, floods, hurricanes and

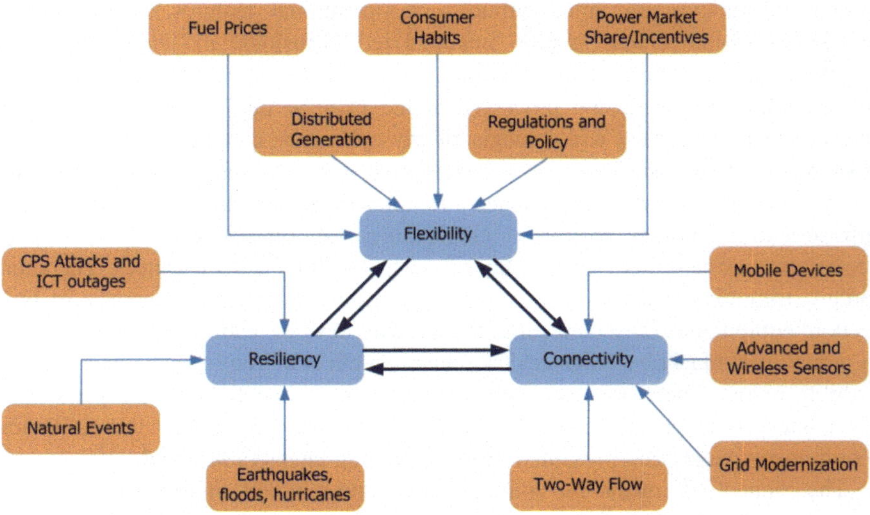

Fig. 3.1 Attributes of power system and handled factors

CPS attacks. The connectivity facilitates mobile devices, advanced WSNs, grid modernization and two-way information and power transmission that enhance flexibility, resiliency, and stability of power system [1, 12].

This chapter deals with the feasibility and resiliency of power networks. The basic principles of power system are presented in the context of flexibility and resiliency. The third and fourth sections address flexibility as a vital driver for DG deployment that is enhanced with high integration of RESs such as wind, solar and tidal power. Moreover, this integration caused a number of unpredictability on fuel prices including fossil based and natural gas. These factors and requirements to cope with them are presented in third section while fourth section introduces improving applications for flexibility of power system. The DG integration, increased use of electric vehicles (EVs), SG applications and microgrid (MG) arrangements transformed consumers to active users generating and consuming energy. The resiliency requirements and driving factors are described in fifth section and improving and developing applications are presented in sixth section that is followed by conclusions.

3.2 Improvements on Power System Flexibility and Resiliency

The architecture and components of a power system can be found in any textbook with generation, transmission, distribution and consumption levels. The conventional power generation system is composed of cogeneration systems such as combined heat and power (CHP) plants, hydroelectric plants, gas turbines, nuclear plants, and recent RES plants. The generated power is provided across transmission and distribution system including lines, substations, equipments such as transformers, circuit breakers, switches, and a number of monitoring and control infrastructures. The consumption level or consumer side includes industrial and residential loads at first glance. However, the improved and widespread grid requires ICT system at any level of future power system due to enhanced generation and consumption scenarios. Moreover, control and management of transmission and distribution systems are also included in these scenarios.

Although regulations, technical improvements, and market share aspects surround the next generation power system, it is driven by a number of enhancements. The regulatory approaches are more robust among others, but the technical improvements are also rather important. The prominent technical and regulatory enhancements for power systems can be listed as follows; cost reductions due to use of RESs, developments in ICT and data management technologies, growing concerns on energy security, reliability, and resiliency, gradually changing consumer profile, environmental concerns, changing revenue and investment predictions [12–15].

The cost reduction of RESs enhances rapid installation and operation of DG plants that increases the capacity of power systems. The cooperation of conventional sources and RESs requires accordance with bulk generation. Furthermore, the decreased cost of RESs facilitates residential users and small consumers to install their MG plants that increase the resiliency of entire system. In a Barclays report in 2014, it is noted that cost of a U.S. residential photovoltaic (PV) system is decreased from 12 to 6 $/Wdc between early 2000s and 2012 [12]. Since the design and installation of generation, transmission and distribution infrastructures in a power system, it is crucial to perform a detailed planning activity by predicting and considering the flexibility and expansion of RES systems at the beginning. While the power system is being advanced by incorporating RESs, system operators should recognize and improve the flexibility in generation, transmission, and demand side management (DSM) processes. If the required precautions are not taken at installation stage, the curtailments are experienced during operation [13].

The developments in ICT and data management technologies present ubiquitous data for monitoring applications. A large DG system may be seen as a complex power network. However, an appropriate data acquisition infrastructure could facilitate to meet the resiliency requirements. The recent applications promoted cooperation of ICT and power systems with SG and IoT integration. These gradual improvements enhance remote monitoring, metering, and control systems to be wiser owing to resiliency and connectivity advances. On the other hand, CPS comprised by integration of ICT and power system requires robust cyber security, privacy protection, data integrity, and authentication precautions. The security, reliability and resiliency are interdependent growing concepts in terms of power systems. The fuel prices and decreasing reserves raise energy security concerns. The financial capacity of RESs depends to geographical structure of countries. However, it is apparent that the use of RESs empowers power system security [12].

The decrements of installation costs and widespread incentives increase installation and integration to the utility grid of DG sources at consumer sites as well as energy suppliers. It is reported that the installed global photovoltaic (PV) plant capacity has been increased up to 128 GW in 2013 that is around 4 GW in 2003. The installed plants and improved technologies in terms of generator and electric systems are also advancing the wind power. The improvements on power system flexibility are driven by transforming consumers to active users that generate energy with their own MG including several sources such as PV panels, small wind turbines, EVs, and electric storage systems (ESSs). Moreover, energy management or home management systems included in the context of SG applications facilitate to control energy consumption, which improves flexibility of consumers and power system. The SG and IoT enhancements increase number of such active users. The net-zero or low energy building concepts are some of the most recent concepts defining active users integrating RESs, EVs, energy management systems, and smart appliances to utility grid. These improvements enable energy suppliers to obtain better DSM and demand response (DR) control. The aforementioned improvements are performed in two areas that one is local flexibility, which regards to active users at distribution and consumption levels. The second is system flexibility that corresponds with connectivity, ICT systems, remote control, and additional management services for DGs at generation, transmission, and distribution levels [12, 13, 15].

Planning is the first and one of the most important aspects to increase power system flexibility and resiliency. The planning phase of a power system requires some technical data such as sufficient flexibility performance, capacity and resource adequacy, and reliability-investment match to improve reliability [16]. The flexibility metrics are required to define indices and to assess the flexibility of power system. These metrics are briefly introduced in this section but are presented and discussed in detail in the following section. Main reasons to improve assessment metrics are related to deregulations seen in electricity market, increased energy trade between countries that are based on different potentials, and environmental concerns [6, 17]. The environmental concerns drive some important targets to reduce carbon emission that is significantly increased by energy generation systems. Therefore, several suggestions have been reported all over the world, and especially by the developed countries and unions including European Union (EU) and US. These reports forces energy suppliers and system operators to improve energy efficiency, promote the EV usage and electric transportation systems, generation with near-zero carbon emission, and integrating RESs to electricity generation.

The planning studies are extended to 2050 agenda to improve power system reliability that is driven by flexibility and resiliency. Thus, the energy plants in operation and planned plants should have the ability to interact future power sources and load types to ensure the suggestions of 2050 roadmap. Environmental regulations concern conventional energy plants as well as renewable plants and use of RESs. The water intake during aridity, sufficient water level for environment and wild life should be taken into consideration for a hydroelectric plant while carbon emission is crucial for CHP and thermoelectric power plants. Wind and solar power plants are more effective on achieving system flexibility comparing other RESs. The effect of a wind energy system on steeper ramps, deeper turn-downs, and shorter peaks in system operation is illustrated in Fig. 3.2 [14]. The ramps indicate increment or decrement requirements of generation to respond the demand changes. Once demand is increasing and wind power generation is decreasing, the ramp elevates to steeper-ramps. The turn-down sections refer to low level operation of generators in the power system. Despite the steeper ramps, if high generation capacity obtained from wind plants during low demand times, the plants are limited to remain available for increased demand times. The shorter peaks occur at the times that generation exceeds demand at higher levels. The enhancements of power systems incorporating with RESs let the system operators and regulators to recognize flexibility and to take precautions to ensure flexibility at each level as generation, transmission, demand side management (DSM), and system operations.

The generation flexibility enables generator plants to efficiently ramp up and down on demand and provide low output levels when deep turn-down is required. The transmission system is required to provide sufficient capacity for energy sources and to share power between power systems. Moreover, transmission network should enable smart ICT infrastructures for optimization. Accordingly,

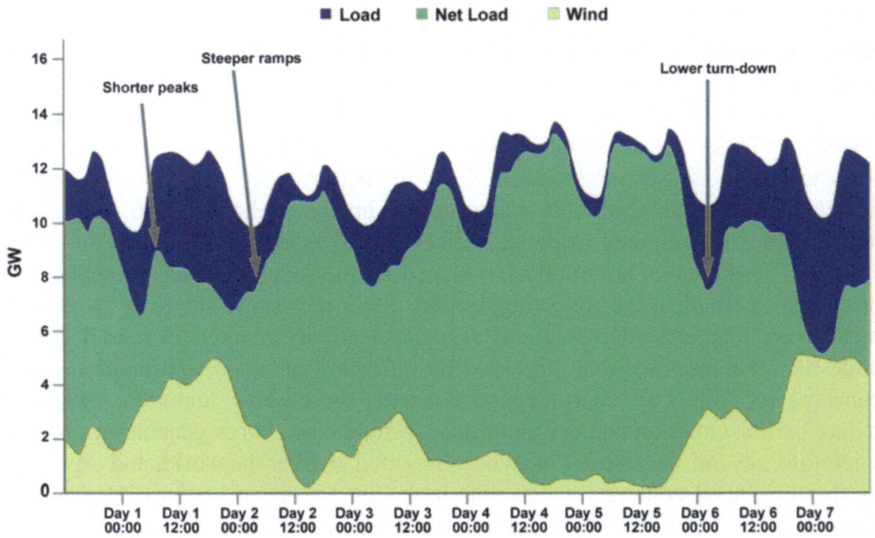

Fig. 3.2 RES effect on DSM of a power system

flexible DSM systems requires ICT and SG infrastructure to enhance DR, ESS, and DG incorporation.

The insufficient flexibility control in a power system forces operator to decrease the generated power of RESs such as PV or wind system. It is reported that curtailments less than 3% can be accepted as cost effective source of flexibility, higher curtailment shares may cause to degradation of installed power and agreement conditions for providers [13]. The flexibility concept requires defining capacity of wind and solar generation systems integrated to a power system. This research enables system operators to detect how much flexible and effective power system they have and how they manage RES integration to this system. The flexibility of a system is determined by the own features of system where DG sources and ESSs integrated power system will be more flexible according to power systems that are based on one and dominant source such as hydro or CHP plants.

It was discussed earlier that resiliency is mostly disrupted by rarely occurring but high impact events. The weather-related blackouts have affected power system significantly in the past decade. Severe weather-related deficiencies are reported with their economic impacts in [18]. The initiatives, electricity service providers, and governments become more aware on resiliency concept due to power and capital losses. The precautions taken against weather-related blackouts include MG, DG, energy storage, community-wide energy storage, hybrid systems of wind, PV, fuel cell, and diesel sources. The improvements of such systems prevent blackouts during severe natural disasters and enable the power system to sustain on providing electricity to consumers [19–22]. The PV systems deployed in an ESS or MG with different generator plants contributes to the resiliency of entire system during blackouts. The ESSs based on batteries are one of the most commonly utilized storage system in small and DG systems. Regardless of storage-type based on batteries or fuel cells and hybrid systems, a number of governments incentive storage systems in the context of resiliency. PV systems can also be cooperated with gas or diesel generators to generate electricity during extreme weather conditions to prevent blackouts and to sustain the power system. One of the applications of PV is islanded operation mode that PV system provides energy to utility grid during regular situations and is get disconnected from grid to feed local loads during blackout of power system [18]. The defensive islanding system introduced in [20] manages power system by classifying operation modes into stable and self-adequate systems to prevent effect of critical loads on other loads. This effect is known as cascading effect that causes large blackouts and may occur more frequent at extreme weather conditions.

The resiliency of any power system is enhanced in two kind of analysis approach that is short period and long period aspects. In the short period analysis, the significant parameters of power system are investigated in the intervals before, during, and after the event that provide resistance, redundancy, and restoration features of power system, respectively. The inherited data is required to prepare an operational approach and make proper decision regarding to available source and load situations. On the other hand, the long period investigation requires adaptation capability of the system to new events where previous events and obtained data are meaningful while improving strategy on power system. Therefore, resiliency improvement of a power system requires several measurement and assessment procedures. The strengthening of a system is associated with some novel precautions such as improving underground distribution system, enhancing towers and transmission lines with stronger materials, and relocating transmission line to pass from routes that are more reliable [21].

MG is a system composed of generation, transmission, ICT, and ESSs and could be designed to operate in grid-tied or island mode. It can be planned regarding to user-centric requirements and can include several different generation sources. MG is seen as affordable and practical solution on improvement on power system resiliency. An MG provides enhancement on reliability, cost efficiency, environmental contributions such as decreasing the carbon emission, decrement on vulnerability and so on. Due to rapid response of MG to natural disasters, there have been governmental regulations in US on incentives promoting MG usage. Almost all of critical plants such as military bases, hospitals, water and gas facilities have installed their own MG systems operating either grid-tied or island mode. Incorporating DG sources to MG systems improves resiliency of local and community wide power systems. Thus, MG enables power exchange between distribution and transmission systems that facilitates to supply electricity to community usage during blackouts caused by extreme weather events [18, 21]. Nevertheless, MG provides limited capacity due to several international standards such as IEEE 1547-2003. This limitation can be tracked by splitting loads to small groups that are less than 10 MVA and connecting to a number of MGs. Thus, smart DG system can be implemented to manage multi-MG infrastructures where each MG can be controlled independently. Several improvements on MG management systems include control algorithms controlling generation and storage capacity, computational methods for management, resource management, decentralized control, and outage management system [21].

The resiliency researches are listed into four aspects as emergency planning, physical behavior analysis, blackout prediction, and resource allocation. The emergency planning includes increasing the power system resiliency against upcoming weather events. Several estimation and optimization algorithms have been proposed in the literature. The physical situation of power systems and responding features during hurricanes, storms and similar events are analyzed in the context of behavior analysis. The studies investigate solutions for minimizing damages contingent on events. The blackout prediction is an important method to analyze data of past hurricanes and ice storms that caused severe damages on power system resiliency. The improved tools are used to simulate possible natural disasters and to predict their probable impact on power system. The resource allocation researches seek solution to recovery and restoration scenarios for transmission and distribution lines. The strategic operation plans are improved by using computational methods such as mixed-integer model, decision-making models, and decision support tools in this context [23].

3.3 Requirements and Factors for Power System Flexibility

It is expected that a general agreement on power system flexibility to be obtained between industry and researchers. The consensus exhibit that current reserve requires to provide enough flexibility while handling the uncertainty caused by high integration of RESs into power system. The conventional power sources limit the flexibility researches on power system and studies are focused on capacity and load assumptions. However, the flexibility should be analyzed in more dynamic aspects that are based on source measurements, net load detection, and generation-transmission cycles and so on. Hence, it is required to go beyond the traditional approaches to reveal the higher flexibility characteristics.

The driving factor of flexibility are listed as DG, environmental regulations, uncertainty of fuel prices, and load demands in [1]. There a number of solutions are proposed in the context of applicable cycling of conventional sources, DG, and RES, ESSs, transmission and distribution, power system planning and customer behaviors. The solutions can be achieved by applying accepted flexibility measurement methodologies. This can facilitate to improve policies, to detect required improvements, and increased confidence to power system [1, 24].

Any power system provides a characteristic flexibility level regarding to its designed structure. The uncertain structure of RES and DGs may worsen flexibility due to curtailments. If the detailed planning is not handled for power system, it will not be possible to achieve sufficient flexibility. The unbalanced demand and supply causes to frequency variations while power balance cannot be assured. The measurement of system flexibility is get started by determining source types and capacity the sources along power system. Therefore, a short-term analysis can be performed to visualize major generator types in the power system. However, the flexibility measurement is a time-dependent process and requires long-term analysis and monitoring to detect generation structures, load characteristics, and time-variant changes of RES and other sources. Large integration of RES and DG sources to power system increases variability of system that forces to provide much more flexible power system. The flexibility resources are defined as own sources obtained by generation flexibility or external sources such as demand and storage aspects. Therefore, flexibility measurement should consider several parameters as physical and inherent structure, forecasting parameters, financial issues, and integrated energy sources. Some flexibility measurement tools have been proposed in the literature with names of Grid Integration of Variable Renewables (GIVARIII) project and Flexibility Assessment Tool (FAST2) which are widely used among others [4–6, 13].

The estimation analysis is one of the long-term methods that evaluate flexibility of power system and allows assessments to determine the system flexibility. The flexibility estimation enables to decision making on economic and technical issues such as investments, extension planning, capacity management, and demand management. The flowchart of flexibility estimation and planning cycle is illustrated in Fig. 3.3 where orange boxes depict flexibility planning options while blue boxes show additional options that can be considered to increase the system flexibility [13, 16]. The most recent flexibility metrics are based on multi-level

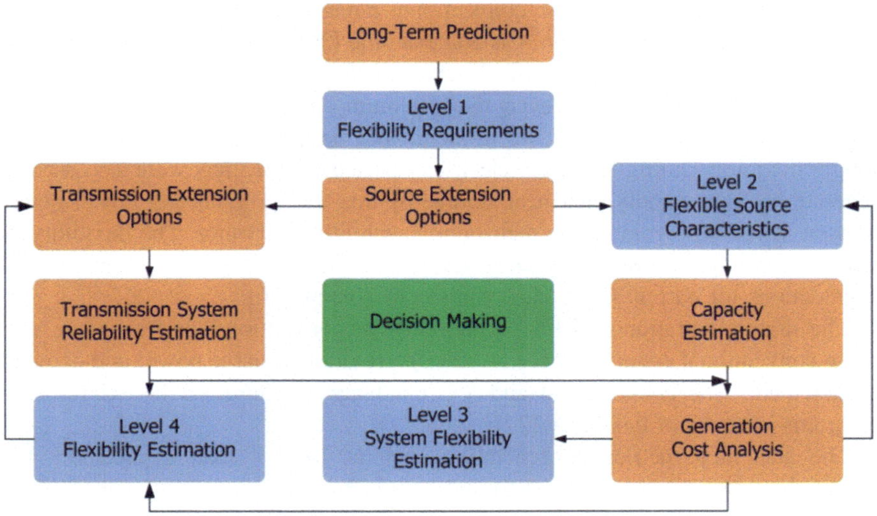

Fig. 3.3 Flexibility estimation and planning flowchart

approach that is also shown in Fig. 3.3 with four levels. The first level deals with VG analysis and flexibility requirements while the second level is focused on resource flexibility calculations.

These both levels are assumed as metering metrics while third and fourth levels are known as detailed metrics that include system flexibility and transmission and fuel compelled flexibility, respectively. The presented four metrics are utilized to determine periods of insufficient flexibility, state of healthy power system, unexpected ramping, and insufficient ramping sources. When the complexity of power system is taken into consideration, these methods can be more appropriate to analyze system where the flexibility is important. The flexibility improvements can be achieved with higher integration of RESs in the systems where low levels of RESs are used in power system. Moreover, environmental concerns and limited use of water can encourage the use of complex and high-level flexibility metrics that there have been some studies review such methods [17, 25].

The flexibility measurements are researched in terms of generation flexibility, transmission flexibility, demand side flexibility and storage systems flexibility that each are introduced in the following subsections.

3.3.1 Generation and Capacity Planning

The conventional energy sources are being utilized to provide the required flexibility of power system. The flexibility characteristics of a power plant can be listed as its ramp rate and minimum operating times for start, stop, ramp-up and ramp-down. The hydroelectric plants and gas-based generators are assumed to be more flexible comparing to CHP and thermal plants due to rapid starting opportunities. On the other hand, the conventional plants can be improved to provide demanded flexibility. In Ontario, Canada example [26], coal plants were improved to provide faster ramp responses and achieved reliable operation at its minimum capacities down to 10%. However, the minimum operating capacity of a regular coal plant ranges around 40–70%. In Denmark, the CHP plants are retrofitted to supply flexibility during low price electricity periods. Thus, they were also used to generate heat and supply their heat storage by this way. The nuclear power plants provide increased regulation by their ramp capabilities in France. The flexibility of any conventional power plant may be improved, but it would cause several degradations for current CHP and hydroelectric plants [26].

The high penetration of RESs to the existing conventional grid structure will cause conventional energy sources to lack on providing reliable power and demand management. The conventional and VG challenge is addressed in EU research programs in terms of flexibility [27].

The adequacy metrics of generation such as loss of load expectation, the expected energy not served (EENS) and well-being analysis has been used up to now [6]. These methods provide some acceptance on capacity management to detect insufficient capacity or adequate capacity for generation. These methods provide some acceptance on capacity management to detect insufficient capacity or adequate capacity levels for generation. The methods include computational methods, Monte Carlo simulations, unit commitment (UC) formulation, linear programming (LP) formulation [6, [28–30]. Thus, long-term planning methods incorporate with conventional and RESs integration to manage generation flexibility. The implemented metrics accomplish generation adequacy to make decision on planned capacity is sufficient or not to meet short-term requirements. The wind and solar plants provide fast response capacity to power system. The ramp-up and ramp-down periods can be managed regarding to their generation at lower capacities or excessive capacities. Once the flexibility requirements and source availability detected, then the estimation between them can be performed. The high-level metrics are helpful to carry out this estimation.

3.3.2 Flexibility Requirements on Transmission and Distribution Grids

Despite of the conventional transmission grid structure, the SG enables bi-directional flow of electricity and information. The SG infrastructure includes transmission and distribution systems in addition to generation. The integration of ICT to power system requires standards and interoperability. The contributions of ICT are listed as remote metering in the context of advanced metering infrastructure (AMI) and control in distribution system, and automation, substation control, and management in transmission system. The widespread use of EVs is a novel factor for distribution system flexibility where it can provide a storage system during discharge and load during charge cycles. Therefore, V2G and G2V concepts are taken into account for flexibility researches.

Two main issues that are voltage level and congestion address the transmission system. Automatic tap transformers and capacitors are used to provide voltage level regulation. The congestion issue is coped with improving transmission cables to higher power rates since almost 70% of capacity can be securely used. The transmission system operators (TSOs) and distribution system operators (DSOs) improve their compensation schemes on cost-based and time-based methods that are associated with their financial plans. The DSO and TSO prefer grid fortification rather than improving DSM strategies. However, microgrid improvements and increased use of EVs force DSOs for increased investments on active DSM systems. The distributed and improved flexibility investments are identified as proactive DSO operations [31, 32]. Regardless of energy source type that is used in a power system, the conventional operation requires more proper management and activation to provide the required flexibility.

3.3.3 Storage Systems for Flexibility

The natural structure of storage systems provides a flexible infrastructure with its fast response, modular and portable structure. The power and energy supply obtained by any storage system can be extended from hours to several months owing to ESS type. The conventional batteries and flow batteries provide energy up to a few days under loaded operation while the power of such systems can be increased to several kWs. On the other hand, fuel cells and hydrogen gas tanks are other energy storage systems that have discharge duration lasting to a month. Pumped hydro storage and compressed air energy storage (CAES) systems provide extreme discharge time and power rates among other ESSs that also increase their contribution to power system flexibility. There are particular researches are being conducted to improve efficiency and power characteristic of batteries, CAES, and fuel cell technologies [26].

The ESS can be used to manage demand and generation ramps to smooth, to provide a buffer between generation and consumption, to sustain power systems during blackouts, and to respond the peak demands. Furthermore, storage systems enable power system to include increased number of DG and RESs together by increasing system flexibility. The emerged storage systems have earned enormous interest in recent years where a 1.3 GW capacity is planned to be installed in California until 2020. Coordinated and cooperative installations in industrial and community areas mostly tackle the technical and financial barriers against storage. Most of ESSs are unsubstantiated in terms of reliability, security, and efficiency that are being researched extensively. Power system planners and invertors provide special supports to improve reliability and efficiency of ESSs. The Energy Storage Integration Council (ESIC) research program that is formed by Electric Power Research Institute (EPRI) aims to provide ESS with higher efficiency by mitigating grid integration drawbacks, insufficient industrial performance, inadequate test operations, and increasing system integration success [1].

The storage is particularly handled by EU for developing an electricity system with lower carbon emissions. It is also addressed to eliminate low system stability due to curtailments of RESs in generation and transmission systems. The researchers are also focused on thermal storage owing to its high potential on efficient integration at DSM and reliability topics.

The innovative storage technologies are also being researched and applied in several trial projects. Gas based back-up systems and gas storage systems are also addressed in this context. In addition to storage technologies, the ICT integration of ESSs is also being studied to monitor state of health (SoH) and state of charge (SoC) of storage system [27].

3.4 Requirements and Factors for Power System Resiliency

The literature surveys show that the reliability studies on extreme weather conditions and disasters are very limited. Arab et al. addresses literature studies into four categories in [23] that are emergency planning, physical behavior, outage prediction, and resource allocation. Improving the power system resiliency against natural events is a vital duty in the context of power system enhancement. As discussed earlier, the hurricanes, floods, and tsunamis cause extreme power outages and operational costs. Moreover, the slow recovery processes leave people and industrial plants powerless for long hours. The enhancements as overhead cable modification, underground cabling, renewing the distribution poles and so on are researched in the context of transmission and distribution grid improvements. The distribution lines moved to underground improve system resiliency by eliminating wind-induced faults, lightning, and short circuits. On the other hand, rising the substations, integration of MGs, and reviewing locations of plants and modifying processes can minimize effects of floods [19].

The emergency planning is required to sustain resiliency of a power system where several models for transmission and distribution lines are implemented for regular and emergency operation conditions [23, 33]. Therefore, the risk estimation methods should be used for planning infrastructure resiliency against extreme weather conditions. In [34], different power systems are analyzed to implement a resilient infrastructure for a given physical and electrical environments. The three power system technologies are analyzed in terms of conventional power system, conventional bulk generation with local ESSs, and MG system with local DG sources. The computational analysis methods are spread to four phases as preliminary, during the critical event, immediate aftereffects, and long-term aftereffects to generate the results. The damage assessment and restoration approaches are selected from one of the most widely accepted methods that are improving a decision algorithm to define which restoration method will be executed after damage, determining the coverage of damage and restoring the system or a hybrid approach that executes damage determination and restoring the system. The studies and researches show that the first approach provides determination that is much more successful and restoration results among others.

The partial or complete restoration of a power system is a very complex and hard process. There are many factors should be taken into accounts such as operating state of system, restoration duration and success of the restoration process. In addition to decision-making, it also requires analysis and feedback of operation. It is defined as a multi-objective, multi-stage, and multi-variable optimization process with non-linearity and uncertainty in [35]. The restoration process which require to force the power system to return its normal operation conditions in a secure and reliable way by minimizing the losses, restoration time, and unexpected effects on community. Thus, the computational methods are required to be used for rapid analysis, assessment, and action to restore the power system in a short while. The genetic algorithms, fuzzy logic and artificial intelligence methods, and heuristic algorithms can be used in system restoration. The aforementioned computational methods are widely researched in the related literature where decision support systems increase the response of operational system [35].

The physical behavior assessment includes the resilience of power system for distribution infrastructure and its reaction with the physical environment. Although the infrastructure does not provide a particular effect on blackout time, interaction of physical environment and infrastructure provide an effect on blackout duration. The damage minimization researches show that the arrangement of transmission and distribution system in cities and community environments affects the resiliency.

The blackout estimation is a crucial tool to ensure efficient response to natural disasters where several estimations are performed on restoration time after hurricanes or earthquakes. The estimations are done considering large datasets of previous disasters and their effects on power system. The estimation parameters also include number of substations, transformers, wind speed and similar physical data that are used to generate a decision data by using computational methods. The system restoration is also related with resource allocation. It is a crucial phase to improve recovery planning and system restoration after any natural disaster strikes the power system. Therefore, the restoration strategy is based on regarding actual structure of power system, determining the system connectivity and power flow graphs, allocating resources, and scheduling restoration periods. In this point of view, three restoration models have been proposed as tactical model, short-term restoration model, and long-term strategic model [23, 36–39].

The tactical model is based on determining restoration stations and transmission locations in a power system, and then defining the difference of distribution nodes in normal operation and extreme weather conditions. Afterwards, the assumptions are performed to perform an analytical approach by considering each restore unit is identical and each node presents a transmission location. The short-term restoration model is performed by using the implemented tactical model as a sub-model in power system network. At this step, the restoration units are determined by locating and dispatching a fixed number of nodes. When a severe weather condition is forecasted, the tactical model is operated and the restoration units are dispatched. The short-term strategic model assumptions include the outputs of tactical model and particular restore data owing to severe weather condition estimations. The short-term model provides restore data for one day or fewer periods. The long-term model assumptions are based on short-term strategic model parameters since it provides data for long-term strategy. Moreover, the total cost for additional restore units are also assumed in the context of long-term strategic model since it cannot exceed the defined budget [36].

One of the most essential topics is planning in power system restoration. Although it is not capable to solve all restoration problems, it provides a wide infrastructure for improving restoration strategies and determining possible solutions to restoration problems. In addition to restoration solutions, the restoration planning provides stability improvement in weak power systems under normal operating conditions. Several advanced planning tools are implemented to improve the efficiency and response time of restoration planning.

The complete restoration of a power system is performed in a multiple step operation. The first step is get started by using a blackstart generator that does not require an external electricity source. After cranking up a number of blackstart generators, the entire power system is gradually restored to its normal operating conditions. The following step of blackstart is energizing the transmission lines, transformers, substations, and balancing loads. Once the transmission lines started supplying the non-blackstart generators that require electricity to generate energy, generators start supplying the remainder of power system in the following steps. Thus, the restoration process continues and generators, transmission lines, substations, and loads are systematically restored [37, 38].

Emerging technologies provide extensive impact on restoration process owing to integration of RESs and MGs to the utility and improvement of SG. Some novel challenges such as MG, voltage source converter-based systems, and wide area measurement systems (WAMS) increase the restoration response of power system [35]. The integration and improvement of a MG system improves self-healing capacity of power system and enables the distribution systems to restore faster than regular operations. The islanded operation mode of any MG provides isolation from power system during a blackout. The control operations can be arranged for a MG system performing blackstart and following islanding operations that are inherited and implemented regarding to computational methods. There are stochastic methods are analyzed to solve DG and MG operation in restoration conditions [35]. The large integration of wind plant to power system has been accelerated by the implemented control methods and developments in wind systems. However, enough researches on restoration capacity of wind plants cannot be found in the literature. The wind turbines can be used as blackstart generators during restoration due to their starting duration is shorter than non-blackstart generators. Moreover, the hybrid blackstart can be performed by integrating CHP, gas turbines, and static compensators to wind plants. The static compensators that are comprised by voltage source converters (VSCs) provide better efficiency comparing to line commutated converters. Therefore, VSCs can be a robust blackstart source with high capacity and reliability.

The improvement of SG allows increasing resiliency of power system with support of measurement systems. WAMS is used to achieve reliable and synchronized measurement data to power system operators. In addition, the phasor synchronization, observation and monitoring of islanded MGs can be performed by using remote measurement systems to improve restoration success. The WAMS provide phasor measurements of widespread-islanded plants and provides a list for energizing priority during restoration phases. The phasor measurement unit (PMU) presents control and monitoring measurements to perform power system restoration.

The CPS attacks are based on system vulnerabilities against intrusions. There have been several attacks causing blackouts in different regions. First blackout caused by cyber-attacks is reported in Ukraine in 2015 where almost half of whole residential users are left powerless by hackers [35]. Since the power restoration depends on ICT system, CPS security is crucial in power system resiliency. Therefore, the cyber security, communication architecture, energy consumption monitoring and communication infrastructures such as SG and IoT are related to power system connectivity that facilitates to meet flexibility and resiliency in the communication base. Integration of many devices into power system operations require a well-planned and secure infrastructure. The IoT is a novel communication medium that is similar to SG but differs in terms of M2M communication. Due to its internet access and cloud services, IoT enables any system operator and planner to acquire required data over databases in a rapid and secure way. The cyber security is one of the most crucial issues in IoT communication as well as in SG. The power system applications that are planned to operate in IoT infrastructure will inherit security solutions regarding to user devices and communication network suppliers [5].

3.5 Conclusion

This chapter presents fundamental and increasing interests on power system flexibility and resiliency in terms of requirements. The technological improvements, policies and social requirements increase the flexibility concerns of a power system. The flexibility of a power system defines its adaptation ability to dynamic and varying conditions such as generation sources, DR of consumers, and DSM. The essential solutions provided by a flexible power system include rapid integration of DG sources, preventing the cost fluctuations in terms of generation and consumption, eliminating the uncertainty, and adopting to novel technologies. The increment on variable generation, environmental regulations, power market requirements, load type variations and social needs shape the flexibility requirements of a power system. Besides the flexibility, a power system should meet resiliency and connectivity requirements that are associated with similar driving factors of flexibility.

The resiliency of a power system is defined as its ability to cope with extreme weather conditions or natural disasters, and rapidly restoring after blackouts. The resiliency requires hardening and strengthening power system against hurricanes, earthquakes, floods, tsunamis and so on. The CPS outages should also be dealt with resiliency of power system as well. Although a large number of researches have been conducted to increase flexibility and resiliency of power system, the widespread blackout risks are not handled many of them. The researches on power system restoration present an increasing interest in the context of planning, strategy improvement, and ICT applications. Connectivity is another aspect of flexibility and resiliency where CPS and ICT are integrated by this concept. The advanced metering, monitoring, and control operations that are required to increase system flexibility and resiliency are provided by connection abilities of the power system.

The data acquisition and processing functionality of a power system allows improving decision strategies against generation issues or outage management cycles.

The solutions to increase power system flexibility and resiliency have been presented in the chapter. The requirements to increase power system flexibility have been presented in the context of generation and capacity planning, transmission grid requirements, and storage systems. Besides, power system resiliency can be enhanced by three approaches based on the damage prevention, system recovery, and survivability that are discussed in the related sections.

References

1. Electric power system flexibility challenges and opportunities. Electric Power Research Institute EPRI, Palo Alto, California, USA, Technical Report, Feb 2016
2. C. Abbey et al., Powering through the storm: microgrids operation for more efficient disaster recovery. IEEE Power Energy Mag. **12**(3), 67–76 (2014)
3. Electric power system resiliency challenges and opportunities. Electric Power Research Institute EPRI, Palo Alto, California, USA, Feb 2016
4. K. Schneider, F. Tuffner, M. Elizondo, C.C. Liu, Y. Xu, D. Ton, Evaluating the feasibility to use microgrids as a resiliency resource. IEEE Trans. Smart Grid **8**(2), 687–696 (2016)
5. Electric power system connectivity challenges and opportunities. Electric Power Research Institute EPRI, Palo Alto, California, USA
6. E. Lannoye, D. Flynn, M. O'Malley, Evaluation of power system flexibility. IEEE Trans. Power Syst. **27**(2), 922–931 (2012)
7. E. Ela, B. Kirby, ERCOT Event on February 26, 2008: Lessons Learned, National Renewable Energy Laboratory, Colorado, US, Technical Report NREL/TP-500-43373, July 2008
8. C. Chen, J. Wang, D. Ton, Modernizing distribution system restoration to achieve grid resiliency against extreme weather events: an integrated solution. Proc. IEEE **105**(7), 1267–1288 (2017)
9. T.L. Vu, K. Turitsyn, A framework for robust assessment of power grid stability and resiliency. IEEE Trans. Autom. Control **62**(3), 1165–1177 (2017)
10. S. Chanda, A.K. Srivastava, Defining and enabling resiliency of electric distribution systems with multiple microgrids. IEEE Trans. Smart Grid **7**(6), 2859–2868 (2016)
11. T. Godfrey, Redefining 'internet of things' (IoT) for the utility industry. EPRI Commun. Connect. Technol. Newsl. **3002004089**, 4–8 (2014)
12. O. Zinaman et al., Power systems of the future: a 21st century power partnership thought leadership report. National Renewable Energy Laboratory (NREL), Golden, CO (2015)
13. J. Cochran et al., Flexibility in 21st century power systems. National Renewable Energy Laboratory (NREL), Golden, CO (2014)
14. M.R. Milligan, Capacity value of wind plants and overview of us experience. National Renewable Energy Laboratory (2011)
15. International Energy Agency (ed.), in *The Power of Transformation: Wind, Sun and the Economics of Flexible Power Systems* (International Energy Agency, Paris, France, 2014)
16. Electric Power Research Institute, in *Power System Flexibility Metrics: Framework, Software Tool and Case Study for Considering Power System Flexibility in Planning*, Palo Alto, CA, 3002000331
17. A. Capasso, A. Cervone, M.C. Falvo, R. Lamedica, G.M. Giannuzzi, R. Zaottini, Bulk indices for transmission grids flexibility assessment in electricity market: a real application. Int. J. Electr. Power Energy Syst. **56**, 332–339 (2014)

18. National Renewable Energy Laboratory, in *Distributed Solar PV for Electricity System Resiliency Policy and Regulatory Considerations*, National Renewable Energy Laboratory, Colorado, US, Nov 2014
19. S. Ma, B. Chen, Z. Wang, Resilience enhancement strategy for distribution systems under extreme weather events. IEEE Trans. Smart Grid **32**(2), 1440–1450 (2017)
20. M. Panteli, D.N. Trakas, P. Mancarella, N.D. Hatziargyriou, Boosting the power grid resilience to extreme weather events using defensive islanding. IEEE Trans. Smart Grid **7**(6), 2913–2922 (2016)
21. H. Farzin, M. Fotuhi Firuzabad, M. Moeini Aghtaie, Enhancing power system resilience through hierarchical outage management in multi-microgrids. IEEE Trans. Smart Grid **7**(6), 2869–2879 (2016)
22. M. Panteli, C. Pickering, S. Wilkinson, R. Dawson, P. Mancarella, Power system resilience to extreme weather: fragility modeling, probabilistic impact assessment, and adaptation measures. IEEE Trans. Power Syst. **32**(5), 3747–3757 (2017)
23. A. Arab, A. Khodaei, Z. Han, S.K. Khator, Proactive recovery of electric power assets for resiliency enhancement. IEEE Access **3**, 99–109 (2015)
24. H. Nosair, F. Bouffard, Reconstructing operating reserve: flexibility for sustainable power systems. IEEE Trans. Sustain. Energy **6**(4), 1624–1637 (2015)
25. B.S. Palmintier, Incorporating operational flexibility into electric generation planning: impacts and methods for system design and policy analysis, Ph.D. thesis, Massachusetts Institute of Technology (2013)
26. H. Holttinen et al., The flexibility workout: managing variable resources and assessing the need for power system modification. IEEE Power Energy Mag. **11**(6), 53–62 (2013)
27. Energy Research Knowledge Centre, *Research Challenges to Increase the Flexibility of Power Systems* (European Union, Belgium, 2014)
28. I. Pierre, Flexible generation: backing up renewables. Union of the Electricity Industry—EURELECTRIC, Brussels, Belgium, Oct. 2011
29. A.R. Bruce, J. Gibbins, G.P. Harrison, H. Chalmers, Operational flexibility of future generation portfolios using high spatial-and temporal-resolution wind data. IEEE Trans. Sustain. Energy **7**(2), 697–707 (2016)
30. B.S. Palmintier, M.D. Webster, Impact of operational flexibility on electricity generation planning with renewable and carbon targets. IEEE Trans. Sustain. Energy **7**(2), 672–684 (2016)
31. K. Knezovic, M. Marinelli, P. Codani, Y. Perez, Distribution grid services and flexibility provision by electric vehicles: a review of options, in *50th International Universities Power Engineering Conference (UPEC)*, pp. 1–6 (2015)
32. M. Khederzadeh, Defining and realizing flexibility in distribution grid, in *CIRED Workshop*, Helsinki, Finland, pp. 1–4 (2016)
33. T.C. Matisziw, A.T. Murray, T.H. Grubesic, Strategic network restoration. Netw. Spat. Econ. **10**(3), 345–361 (2010)
34. A. Kwasinski, Technology planning for electric power supply in critical events considering a bulk grid, backup power plants, and micro-grids. IEEE Syst. J. **4**(2), 167–178 (2010)
35. Y. Liu, R. Fan, V. Terzija, Power system restoration: a literature review from 2006 to 2016. J. Mod. Power Syst. Clean Energy **4**(3), 332–341 (2016)
36. M.J. Yao, K.J. Min, Repair-unit location models for power failures. IEEE Trans. Eng. Manag. **45**(1), 57–65 (1998)
37. J.N. Jiang et al., Power system restoration planning and some key issues, in *Power and Energy Society General Meeting*, IEEE, New York, pp. 1–8 (2012)
38. D. Hazarika, A.K. Sinha, Power system restoration: planning and simulation. Int. J. Electr. Power Energy Syst. **25**(3), 209–218 (2003)
39. K. Balasubramaniam, P. Saraf, R. Hadidi, E.B. Makram, Energy management system for enhanced resiliency of microgrids during islanded operation. Electr. Power Syst. Res. **137**, 133–141 (2016)

Chapter 4
Resilience Metrics Development for Power Systems

Hossein Shayeghi and Abdollah Younesi

Abstract The purpose of this chapter is to explain the metrics were used for quantifying the resiliency of power system. Also, will determine how which metrics are calculated for which system under what conditions. Distribution and transmission infrastructure that is expanded over a wide geographic area, is always affected by weather-related disasters which occur continuously. Therefore, a safe and reliable operation is essential to have a resilient power system, which survives in hard conditions. The metrics investigated in this chapter are quantitative, which are defined based on the topology, hardware, and the efficiency of the system, reliability indices, and also the type and severity of the threat. The accurate assessment of each of these metrics can help to properly understand the concept of resilience in power systems. Also, we can obtain an appropriate assessment of the power network resilience by selecting the proper set of these metrics according to the type of threat and our goal.

Keywords Power system restoration · Quantitative resilience · Reliability indices
Resiliency metrics

Nomenclatures

B	Brittleness
C_B	Betweenness centrality
C_{dn}	Cost of lost demand d at bus n
C_{ei}	Load curtailment in event e_i
C_n	Clustering factor
$d(n_i, n_j)$	Equivalent distance between nodes n_i and n_j
$D(t)$	Percentage of the infrastructures damage
D_G	Diameter of the considered complex grid (graph G)

H. Shayeghi (✉) · A. Younesi
Department of Electrical Engineering, University of Mohaghegh Ardabili, Ardabil, Iran
e-mail: hshayeghi@gmail.com

A. Younesi
e-mail: younesi.abdollah@gmail.com

© Springer Nature Switzerland AG 2019
N. Mahdavi Tabatabaei et al. (eds.), *Power Systems Resilience*, Power Systems,
https://doi.org/10.1007/978-3-319-94442-5_4

e_i	ith extreme event
e_n	The number of joint couples between all neighbours of node n
f	Brittleness distribution
f_c	Critical section of a complex network
K	Total number of lines that are on the outage
k_n	Total number of neighbours of node n
L	Laplacian matrix
l_G	Length of the graph G
M	The number of graph nodes
n	An event which caused violation in voltage level
N	The number of loads in a particular area of distribution system under consideration
n_0	The number of costumers which experienced an outage
N_q	All the similar PNs for the qth FN
P_d	Conditional probability
P_{ei}	The probability of power grid experiencing event e_i
q	Total number of FNs
r_l	Resiliency of a single load
s	The number of graph sections
$S(t)$	Percentage of supplied power
S_e	Set of all disasters in which caused the system loads to exceed the generation capacity
T	Time period
t_{down}	Down time
$t_{down,i}$	A portion of period T, that the load i cannot receive power
T_s	Capacity of local energy storage systems
t_{up}	Up time
$t_{up,i}$	A portion of period T, that the load i can receive power ($t_{up,i} = T - t_{down,i}$)
V	Graph connector weights
$v_r(t)$	Recovery speed
w_i	Weight of the ith factor that affects the grid recovery process
θ	Outage index
$\theta_{\max}$	The time that all costumers experienced an outage
ϕ	Resistance
σ	Measure of the severity of the extreme event
λ	Failure rate
μ	Repair rate
Λ_2	Algebraic connectivity of the power distribution network
Υ	Amount of intensity of a natural disaster
η_i	Value of the ith factor that affects the grid recovery process
P_d^T	Total active power of the power system in normal operating condition
$P_{dn,i}^{t\|\varepsilon}$	Active power at load point n after the restoration plan i regarding disturbance ε at time t
$P_{dn}^{t_d\|\varepsilon}$	Active power demand of bus n at the end of disturbance ε

$\Psi^{\lambda}_{i,n,d,t}$ Flexibility of the demand d at the load point n for the ith plan of restoration at time t

$\Psi^{\mu}_{i,n,d,t}$ Outage cost restoration of demand d at bus n for ith reconfiguration plan at time t

$\Psi^{\sigma}_{i,n,d,t}$ Restoration capacity of load d at bus n for ith reconfiguration plan at time t

4.1 Introduction

It is necessary to track resiliency metrics to be able to determine that, in the operation of the power system under low-probability high-impact events, which goals have been achieved and which one not been achieved. Resiliency metrics are used at different levels for different intentions. Some of the purposes are relevant to the national or regional macro policies and some others to a local or tools aspect. As an example, what is the effect of the resiliency on the economic damages caused by natural disasters at the national or regional level? For a power plant operator, it can be essential to know how many and what types of spare parts are available. Considering each purpose of the system needs a unique set of metrics. Because one set of metrics does not support all the goals of the system. Then this chapter first reviews the existing metrics for measuring resiliency of electrical systems, then a strategy will be developed that can be used to determine the appropriate set of resiliency metrics according to the goals pursued from the system. In the event that national or regional macro intentions are considered, greater focus will be on strategic aspects of the metrics set. In this case, matters like budget, availability of equipment, number of generators and operators, speed and accuracy of response teams, schedules, existing technologies such as smart grids, etc. will be considered. But if local aims are taken into account, the operational aspects of the electrical system are used to define the resiliency. In this case, matters like timely detection of the outages, fast recovery after disasters, convenient repairs, system efficiency, reliability indices, system hardening, improving social welfare, etc. are considered.

The necessity of quantifying resilience metrics is an important challenge, which mostly depends on how to define the resiliency.

Resiliency may sometime be considered as the time for recovery of power system after a disaster. In a complete definition, in addition to the time required for recovery, the capability of the system to withstand malicious events, system adaptability, and desirable extensibility can be considered as principal resiliency characteristics. Resiliency can be calculated mathematically as the area under system's performance curve.

Resilience metrics must have some basic features. These features are essential for the development of a comprehensive metric. In other words, a general metric must [1]:

- *Be useful.* A comprehensive metric must be helpful for decision making incorporate system planning, real-time actions, and policy determinations.
- *Provide a comparable structure.* Calculation of this metric for different systems should provide comparable information.
- *It must be usable in operational and planning contexts.* Operational contexts such as pre-configuration the system before a disaster and planning contexts such as implanting of electric conductors.
- *Be comprehensive and extensible.* The appropriate index should be extensible over time and must be calculated with the advancement of technology and equipment in complex computational methods.
- *Be quantitative.* The appropriate metric should be quantitatively quantifiable.
- *Consider uncertainties.* It is very important that the resilience metric should reflect the system's uncertainties.
- *Consider the recovery/restoration time of the system after a disaster.* An appropriate metric of resiliency should somehow take into account the duration of outages.

4.2 Resiliency Metrics, Different Definitions

The resilience metric may have terms of a threat or a set of threats. In fact, this criterion answers the question of "resilient to what?". Usually, resiliency is considered against natural disasters such as earthquakes, storms, floods, etc., but these studies can be generalized to sudden human-caused events such as accident and war. It can be observed that the natural disasters tend to follow cycles. The time interval between the onset of an event up to the occurrence of another can be classified into four phases [2]:

- *Phase* 1 (*During the event*): The length of this phase (Δt_1) can be a few minutes to a few days. In this situation, the main purpose is to reduce the damages and loss of services.
- *Phase* 2 (*Immediate aftermath*): This stage takes a few days to several weeks. The main goal of this period is to start recovery and repair actions. This phase lasts Δt_2 and ends when these activities are almost completed.
- *Phase* 3 (*Intermediate aftermath*): This phase usually lasts from a few weeks to several months and sometimes interference with phase three. In this phase, the main objective is to investigate the disaster's effect on a specific part of the power system by calculating the system efficiency indices and assessing the extent of the damage.
- *Phase* 4 (*Long-term aftermath*): This stage may take a few months to several years. In this phase, the main goal is to prepare for the occurrence of the next disaster using the results obtained in phase three. These preparations include corrective actions, modification of the operational strategies, and the strengthening of infrastructure. this phase ends with the onset of the next event.

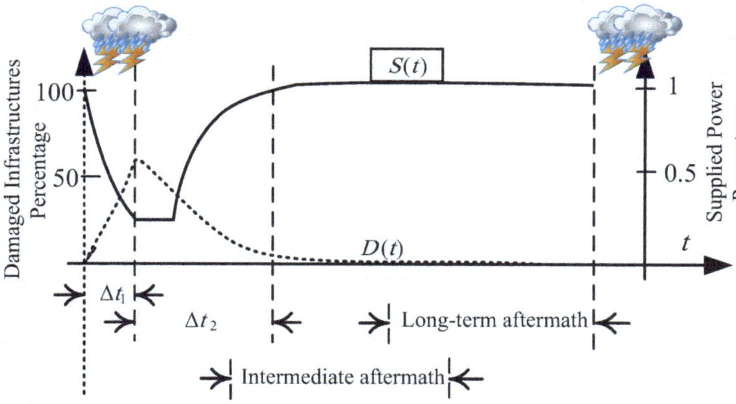

Fig. 4.1 Representation of different phases of the extreme event [2]

Figure 4.1 shows the different phases of the time horizon immediate following a disaster occurrence until the next disaster.

Although the resilience measures are used to evaluate the consequence of a disaster, it must also be used to assess the ability of power grid in cover its objectives. This means that the performance of the system affects the resiliency measures directly. For example, the area under $S(t)$ in Fig. 4.1, which is a measure of the loads supplied by the power grid during and after the disaster, is a performance-based metric for resiliency [2]. Equation (4.1) describes this metric mathematically.

$$R_1 = \int_t S(t)dt \tag{4.1}$$

Another measure based on the quality of the power network service described by (4.2), which is the number of events that, as a result of their occurrence, the network voltage falls outside of the standard range.

$$R_2 = \sum n \tag{4.2}$$

where n is an event which caused the voltage level of the power grid to violate the standard ranges.

According to U.S. Presidential Policy Directive 21 [3] (PPD-21), the resiliency is defined as: "the ability to prepare for and adapt to changing conditions and withstand and recover rapidly from disruptions." In this definition, resilience includes "the ability to withstand and recover from deliberate attacks, accidents, or naturally occurring threats or incidents." Therefore, based on the four main characteristics stated in the definition of the resilience, i.e. withstanding capability, recovery speed, planning capacity, and adaption capability [4], a quantitative metric for the resiliency of a single load in a period of time ($T = t_{up} + t_{down}$) can be mathematically modeled by (4.3).

$$r_l = \frac{t_{up}}{(t_{up} + t_{down})} \tag{4.3}$$

In (4.3), downtime (t_{down}), which is related to the hardware aspects of the power system and human-related processes, shows the system's recovery speed. The ability of the power grid to withstand the disaster is related directly to its hardware and equipment characteristics, which t_{up} shows this index. It should be noted that several references have proposed similar relationships to measure resiliency in other systems, such as communication sites [5], supply networks [6], and urban infrastructure systems [7]. According to [2], it is possible to define the resiliency of the power generation resources for N loads as:

$$R_L = \frac{\sum_{i=1}^{N} t_{up,i}}{\sum_{i=1}^{N} (t_{up,i} + t_{down,i})} \tag{4.4}$$

Equations (4.3) and (4.4) are similar to the equation of availability in reliability theory, but an infinite number of repair and failure sequences are used for calculating of the availability measurement where the measures of the resiliency of (4.3) and (4.4) can be based on a single sequence in duration T.

Suppose that n_0 number of the total customers (N) in a given region under study at the time interval T experienced an outage. In this case, the outage index is calculated by (4.5) [2].

$$\theta = \frac{n_0}{N} \tag{4.5}$$

This is the equivalent to the SAIFI in IEEE Standard 1366 that is widely used to assess the outages of the power systems.

In Ref. [2], the recovery speed (v_r) for the N number of customers is defined as (4.6).

$$v_r(t) = \frac{d\theta}{dt_r}, \quad t_r = t - t_{|\theta=\theta_{\max}} \tag{4.6}$$

For one customer $N = 1$ and thus n_0 is 1 or 0. Assume the customer has experienced an outage ($n_0 = 1$). In this case, since all the customers experienced the outage, dt_r can be taken equal to T_{down}, as a result:

$$v_{r,i}(t) = \frac{1}{t_{down}} \tag{4.7}$$

In a similar manner disruption speed for a group of customers and a single one can be calculated as (4.8) and (4.9) [2], respectively.

$$v_d(t) = \frac{d\theta}{dt}, \quad \text{for } t < t_{|\theta=\theta_{\max}} \tag{4.8}$$

$$v_{d,i}(t) = \frac{1}{t_{up}} \tag{4.9}$$

It is clear that the (4.7) and (4.9) are analogous to the concepts of repair rate (μ) and failure rate (λ) in reliability theory, respectively.

The power network's ability to withstand a destructive event can be measured by a metric called resistance, which is defined for a single customer by using (4.10) [2].

$$\varphi_I = \frac{t_{1,u}}{\Delta t_1}\sigma \tag{4.10}$$

where $t_{1,u}$ is Δt_1 in Fig. 4.1, which the customer still receives power before the outage. σ can be specified as a function which represents the severity of the destruction of the extreme event and can be defined for different types of hazards such storm, flood, earthquakes, etc. [7]. It has to be noted that $\sigma > 0$. In order to better understand the time intervals, Fig. 4.2 shows the details of the time periods of the first two phases of a disaster (phases 1 and 2 shown in Fig. 4.1).

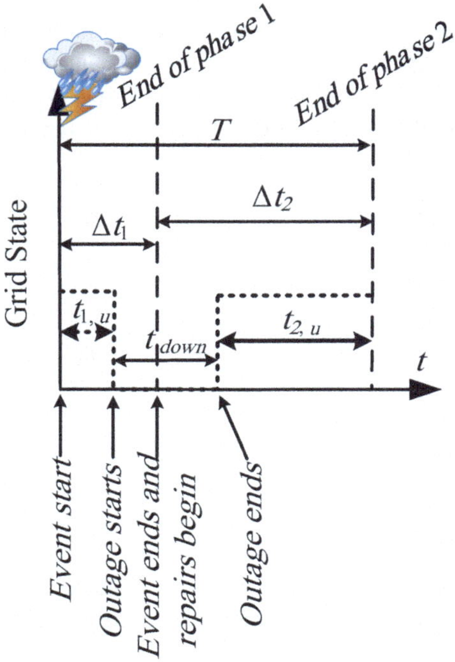

Fig. 4.2 Details of the time periods of the first two phases of a disaster

Also, the resistance of φ for N loads is defined by (4.11) [2].

$$\varphi = \frac{\sum_{i=1}^{N} t_{1,u,i}}{\theta_{\max} N \Delta t_1} \sigma \tag{4.11}$$

In this state, in addition to the importance of the time which loads can receive power under the disaster conditions ($t_{1,u}$ in Fig. 4.2), the maximum amount of lost power of the customers experienced the outage is also very essential.

Brittleness is the amount of damage which the power system receives from a disruptive event and is calculated for N loads using (4.12).

$$B = \frac{\theta_{\max}}{D} \times 100 \tag{4.12}$$

where D is highly related to the characteristics of the infrastructures.

The dependency of one infrastructure to the other ones is defined as [8] "a linkage or connection between two infrastructures, through which the state of one infrastructure influences or is correlated to the state of the other". In accordance with Refs. [2, 8], it is possible to quantitatively measure the dependence of the loads to the power grid by resilience-oriented adjusting the amount of energy storage resources. The level of dependency of a load from the power grid may be calculated based on r_l of (4.3) as [2, 8]:

$$R_L = 1 - (1 - r_l)e^{-\mu T_s} \tag{4.13}$$

According to (4.7), μ is equal to the inverse of t_{down}.

Reference [2] represents the intrinsic relation between dependence with the concept of resilience and how energy storages may or may not lead to a loss of power for customers during an outage as follows:

$$\frac{1}{\mu}\frac{dR_L}{dT_s} = (1 - r_l)\mathrm{e}^{-\mu T_s} \tag{4.14}$$

Hence,

$$R_L = 1 - \frac{1}{\mu}\frac{dR_L}{dT_s} \tag{4.15}$$

As a result,

$$\frac{1}{\mu}\frac{dR_L}{dT_s} = -R_L + 1 \tag{4.16}$$

where t_{down}= $1/\mu$, shows how much restrictions there is locally for a shift in local resilience by attaching new energy storage systems near the load or it is indicating in order to obtain the same local system resiliency, how much more or less energy storage devices have to be available.

Reference [9] defined a multiple-component resiliency metric for power distribution system based on the network topology as:

$$\Re_\tau = \sum_{j=1}^{\eta} V_j \lambda'(i,j) \tag{4.17}$$

where η is the number of metric components. V is equal to:

$$V = \begin{bmatrix} A_{fc} & B_D & C_{C_B} & D_{lG} & E_{C_n} & F_{\Lambda_2} \end{bmatrix}^T \tag{4.18}$$

In (4.18) $A, B, C, \ldots, F$ are the obtained weights to indicate the importance of its corresponding measure. $\lambda(i,j)$ is an element of $\vec{\Re}_\tau \vec{\Re}_\tau^T$ and given by:

$$\lambda'(i,j) = \frac{\lambda(i,j) - \min_{i=1}^{\eta}(\lambda(i,j))}{\max_{i=1}^{\eta}(\lambda(i,j)) - \min_{i=1}^{\eta}(\lambda(i,j))} \tag{4.19}$$

$$\vec{\Re}_\tau \vec{\Re}_\tau^T = \begin{matrix} f_c \\ D \\ C_B \\ l_G \\ C_n \\ \Lambda_2 \end{matrix} \overset{\begin{matrix} f_c & D & C_B & l_G & C_n & \Lambda_2 \end{matrix}}{\begin{pmatrix} 1 & a & b & c & d & e \\ 1/a & 1 & f & g & h & i \\ 1/b & 1/f & 1 & j & k & l \\ 1/c & 1/g & 1/j & 1 & m & n \\ 1/d & 1/h & 1/k & 1/m & 1 & o \\ 1/e & 1/i & 1/l & 1/n & 1/o & 1 \end{pmatrix}} \tag{4.20}$$

where $a, b, c, \ldots, o$ are weight coefficients in the interval (0, 1] [4].

$$\vec{\Re}_\tau = \begin{bmatrix} f_c & D_G & l_G & C_B & C_n & \Lambda_2 \end{bmatrix} \tag{4.21}$$

Assume the power distribution system demonstrated by a graph $H = (M, S, V)$ comprising of M nodes, a set of section (edges) S with each element connected from node x to node y with a corresponding weight V.

In (4.21) D_G (the optimal (shortest) path between the farthest nodes) calculated as:

$$D_G = \frac{2E}{|N|(|N| - 1)} \tag{4.22}$$

l_G represents the length of the graph and obtained by (4.23).

$$l_G = \frac{\sum_{i \neq j} d(n_i, n_j)}{N(N-1)} \tag{4.23}$$

C_B is the betweenness centrality of the graph and calculated as:

$$C_B(i) = \sum_{n_i \neq n \neq n_l} \frac{n_k \to n_l, n_i}{n_k \to n_l} \tag{4.24}$$

where $n_k \to n_l, n_i$ is 1 if the optimal path between the node n_k to n_l passes through n_i and 0 if n_k to n_l does not pass through n_i. The phrase $n_k \to n_l$ is to show the optimal (shortest) path between the nodes n_k and n_l [9].

C_n in (4.21) shows the clustering factor of the power distribution system and calculated by (4.25) [9].

$$C_n = \frac{2e_n}{k_n(k_n - 1)} \tag{4.25}$$

Algebraic connectivity of the power distribution network is indicated as Λ_2 and is calculated by (4.26).

$$\Lambda_2 = \text{eig } L^2_{(i,j)} \tag{4.26}$$

where Laplacian Matrix is obtained as [9]:

$$L_{(i,j)} = \begin{cases} \deg(n_i) & \text{if } i = j \\ -1 & \text{if } i \neq j \text{ and } n_i \text{ is adjacent to } n_j \\ 0 & \text{otherwise} \end{cases} \tag{4.27}$$

There are more metrics which can be used for defining resiliency of a power system and (4.21) is only one combination.

In Ref. [10] six metrics that can measure the operational resiliency of microgrids have been identified based on graph theory and Choquet integral. This definition is based on three main assumptions:

- The number of paths between supply and load nodes affects the resiliency.
- Increasing the ratio of power supply resources to system loads improves (increases) the system resiliency.
- The increase in the number of switches in the system will increase the system resiliency, while the increase in the number of switching actions required to connect critical loads to the power supply will reduce the system resiliency.

For the six resiliency metrics defined in Ref. [10] it is assumed the power distribution system is equivalent to a graph that has n nodes and their nodes are connected to one another by e branches. In this equalization, the buses and lines of the power distribution system are demonstrated with nodes and branches, respectively.

Branch Number Impact (BNI) This measure is equal to the ratio of the total number of joined branches for each *RIWL* in a *PN* to the number of all *CLs* [10].

$$BNI_q = \frac{\sum_{k=1}^{N_q} \frac{\text{Nodes in } \mathit{RIWL} \text{ for } k\text{th } \mathit{PN}}{\text{Number of } \mathit{CLs} \text{ in } k\text{th } \mathit{PN}}}{N_q} \tag{4.28}$$

Overlapping Branches (OB) This metric is equal to the total number of joint branches in each *PCWL* in a *PN* [10].

$$OB_q = \frac{\sum_{k=1}^{N_q} \text{common branches in } k\text{th } PN}{N_q} \tag{4.29}$$

Switching Actions (SA) This measure represents the total number of switching operations (change in the state of switches, i.e. closed to open and vice versa) needed to connect all the *CL*s to sources through different *FN*s [10].

The Number of Resources (NoR) It is equal to the ratio of the total number of possible resources utilized to supply all *CL*s to the number of all *CL*s in each *PN* [10].

$$NoR_q = \frac{\sum_{k=1}^{N_q} \frac{\text{Resurces supplying all } \mathit{CLs} \text{ in } k\text{th } \mathit{PN}}{\text{Number of } \mathit{CLs} \text{ in } k\text{th } \mathit{PN}}}{N_q} \tag{4.30}$$

Route Abundance (RA) This is the ratio of the total number of routes that is possible for all *CL*s joining to all resources to the total number of *CL*s in each *FN* [10].

$$RA_q = \frac{\text{Routes joning all } CLs \text{ to all resources in } q\text{th } FN}{\text{Number of } CLs \text{ in } q\text{th } FN} \tag{4.31}$$

The Probability of Accessibility and Penalty Factor (PoA & PF) This metric is based on two factors: the probability of availability of the source, and the losses in distribution or penalty factor *PoA & PF* for a *FN* is calculated by (4.32) [10].

$$PoA \,\&\, PF_q = \frac{\sum_{k=1}^{N_q} PoA \times PF \text{ for } k\text{th } PN}{N_q} \tag{4.32}$$

Figure 4.3 shows the framework that is used in Ref. [10] for quantifying resiliency in a power distribution system using graph theory and Choquet integral computation.

Reference [11] is introduced four metrics as (4.33) to measure the resiliency of power grid under extreme events.

$$\Im = \{K, LOLP, EDNS, G\} \tag{4.33}$$

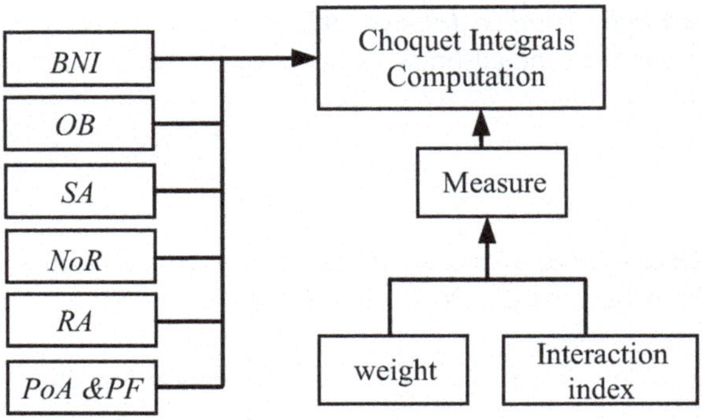

Fig. 4.3 Flowchart of quantifying resiliency in power distribution system [10]

In (4.33) K demonstrates the expected number of lines are on outage due to the inordinate event and is calculated as:

$$K = \int_0^\infty kf(k)dk \tag{4.34}$$

$$f = P_d(k|\Upsilon) \tag{4.35}$$

where, P_d refers to the conditional probability of outage of k lines in Υ [11].

LOLP and *EDNS* which are known reliability indices [12] are modified in Ref. [11] and defined as survivability following extreme events.

$$LOLP = \sum_{e_i \in S_e} P_{e_i} \tag{4.36}$$

$$EDNS = \sum_{e_i \in S_e} P_{e_i} C_{e_i} \tag{4.37}$$

where C_{e_i} is obtained using optimal power flow (OPF) [11].

Parameter G in (4.33) measures the complexity of grid restoration. It must be mentioned, the power system restoration process after an extreme event, depending on the kind and amount of intensity of the disaster and the extent of damage to the critical infrastructures of the system, may take several hours to several days. The grid recovery index is expressed as (4.38) [11].

$$G = \sum_{i=1}^{5} w_i \eta_i \quad \text{where} \sum_{i=1}^{5} w_i = 1 \tag{4.38}$$

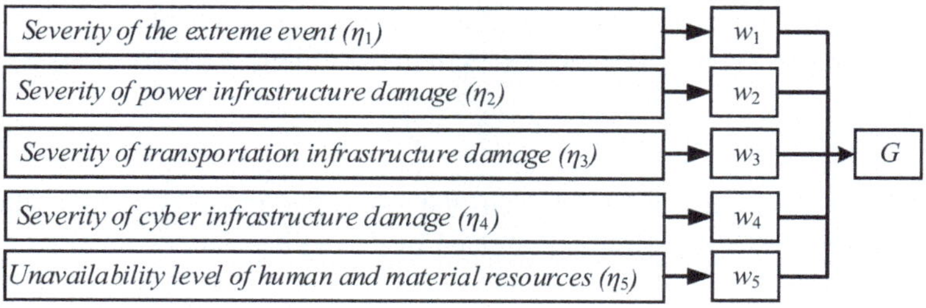

Fig. 4.4 Grid recovery index factors [11]

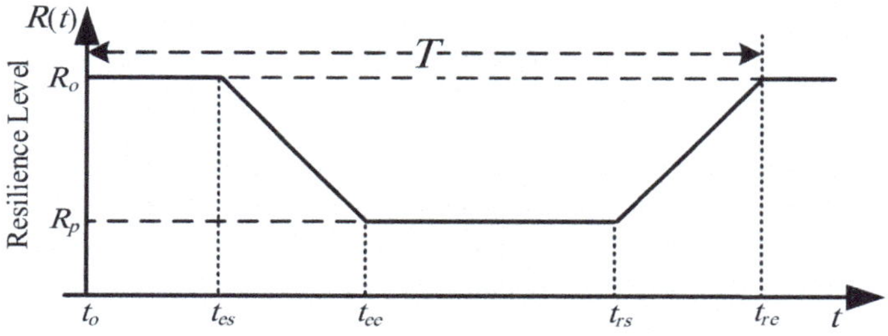

Fig. 4.5 The resilience level of the power system under a natural disaster

where, w_i and η_i are described in Fig. 4.4.

In accordance with Refs. [13, 14], the resilience level of the power system under a natural disaster can be plotted as in Fig. 4.5 in five-time intervals. The first stage, which covers the time interval [t_o t_{es}], shows the system's resilience level before the horrible event. The devastating event starts at time t_{es} and continues until the time t_{ee}. During this period, the level of resilience of the system gradually decreases from the initial value R_o to its minimum value, i.e. R_p (the gradient depends on the structure and capabilities of the power network). The preparation is then followed to start the recovery process at the fastest possible time interval, i.e. [t_{ee} t_{rs}]. With the onset of restoration process at time t_{rs}, the level of resilience of the power system gradually returns to its original state (the desired value before the disaster, R_o). The time after t_{re} is the state of resilience after the completion system recovery process.

Reference [14] has considered a set of network performance indices as a benchmark for measuring the resilience level of the power system against extreme events and called it as $\Phi\Lambda E\Pi$.

In this definition, Φ represents the number of lines that are tripped per hours (during the extreme event occurrence) and calculated by (4.39):

$$\Phi = \frac{R_p - R_o}{t_{ee} - t_{es}} \tag{4.39}$$

The parameter Λ refers to the amount of power system resilience level reduction due to the occurrence of a malicious event (number of lines tripped) and is equal to:

$$\Lambda = R_p - R_o \tag{4.40}$$

The time duration that it takes to start the restoration/recovery process after the occurrence of an extreme event represented by E and is equal to:

$$E = t_{rs} - t_{ee} \tag{4.41}$$

After the start of the recovery/restoration process, the number of lines that are retrieved per hour is shown using Π and is equal to:

$$\Pi = \frac{R_o - R_P}{t_{re} - t_{rs}} \tag{4.42}$$

In addition to the $\Phi\Lambda E\Pi$ metric, for calculating the lines that were in service from the beginning of the disaster to the end of the recovery/restoration process (the lines that have not experienced the outage), a criterion called the *Area* is defined and, in accordance with Fig. 4.5, is equal to [14]:

$$Area = \int_{t_{es}}^{t_{re}} R(t)dt = \frac{\Lambda \times (t_{ee} - t_{es})}{2} + (R_p \times (t_{rs} - t_{es})) + \frac{\Lambda \times (t_{rs} - t_{re})}{2} \tag{4.43}$$

Similarly, by plotting the variations of resilience level of the critical infrastructures under a natural disaster, we can also calculate the $\Phi\Lambda E\Pi$ and *Area* metrics for them [14].

Reference [15] has improved the power system resiliency based on the grid reconfiguration. In this regard, three metrics have been suggested for quantitative evaluation of power system resiliency.

$$\Psi = [\Psi^{\lambda}_{i,n,d,t}, \Psi^{\mu}_{i,n,d,t}, \Psi^{\partial}_{i,n,d,t}] \tag{4.44}$$

In (4.44) when the *i*th plan of the network reconfiguration is considered at time t, $\Psi^{\lambda}_{i,n,d,t}$ calculated as:

$$\Psi^{\lambda}_{i,n,d,t} = \frac{\sum_{i \in I} \sum_{n \in N} P^{t|\varepsilon}_{d_n,i}}{P^T_d} \tag{4.45}$$

The term $\Psi^{\mu}_{i,n,d,t}$ in (4.44) is equal to:

$$\Psi^{\mu}_{i,n,d,t} = \sum_{i \in I} \sum_{n \in N} C_{dn}(P^{t|\varepsilon}_{d_n,i+1} - P^{t|\varepsilon}_{d_n,i}) \tag{4.46}$$

The last parameter of Ψ in (4.44) is $\Psi^{\partial}_{i,n,d,t}$ which is calculated as:

$$\Psi^{\partial}_{i,n,d,t} = \sum_{i \in I} \sum_{n \in N} \frac{P^{t|\varepsilon}_{d_n,i} - P^{t_d|\varepsilon}_{d_n}}{P^{T}_{d} - P^{t_d|\varepsilon}_{d_n}} \times 100 \tag{4.47}$$

4.3 Conclusion

In this chapter, quantitative metrics that were proposed in the literature to assess the resilience of power systems were explained. Researchers have proposed different metrics for the resiliency of power grid in various viewpoints such customer perspective and power distribution level. Physical structure and network topology, severity and type of the threat, system performance under malicious event, restoration/recovery time after the disaster, network reliability indices, number of critical infrastructures such as transformers, storage resources, distributed energy resources etc., are effective in the assessment of the power network resiliency. In a general viewpoint, resilience metrics may be classified in three categories such simulation-based methods (whose are based on the performance of the system), analytical methods (whose are based on the probability and reliability indices), and statistical analysis of historic outage data. It should be noted that the power system planner can use one or a several numbers of the metrics for an accurate measurement of the resilience of the power system for a specific event with a known severity considering the purpose of system planning.

References

1. J.P. Watson, R. Guttromson, C. Silva Monory, R. Jeffers, K. Jones, J. Ellison et al., in *Conceptual Framework for Developing Resilience Metrics for the Electricity, Oil, and Gas Sectors in the United States, USA* (2015)
2. A. Kwasinski, Quantitative model and metrics of electrical grids' resilience evaluated at a power distribution level. Energies **9**, 93 (2016)
3. Critical Infrastructure Security and Resilience, https://obamawhitehouse.archives.gov/the-press-office/2013/02/12/presidential-policy-directive-critical-infrastructure-security-and-resil (2013)
4. A. Kwasinski, Field technical surveys: an essential tool for improving critical infrastructure and lifeline systems resiliency to disasters, in *IEEE Global Humanitarian Technology Conference (GHTC 2014)*, pp. 78–85 (2014)
5. Measurement frameworks and metrics for resilient networks and services: challenges and recommendations, European Network and Information Security Agency (ENISA) (2010)
6. K. Zhao, A. Kumar, T.P. Harrison, J. Yen, Analyzing the resilience of complex supply network topologies against random and targeted disruptions. IEEE Syst. J. **5**, 28–39 (2011)
7. M. Ouyang, L. Duenas Osorio, Time-dependent resilience assessment and improvement of urban infrastructure systems. Chaos Interdiscip. J. Nonlinear Sci. **22**, 033122 (2012)

8. A. Kwasinski, Local energy storage as a decoupling mechanism for interdependent infrastructures, in *IEEE International Systems Conference*, pp. 435–441 (2011)
9. S. Chanda, A.K. Srivastava, Defining and enabling resiliency of electric distribution systems with multiple microgrids. IEEE Trans. Smart Grid **7**, 2859–2868 (2016)
10. P. Bajpai, S. Chanda, A.K. Srivastava, A novel metric to quantify and enable resilient distribution system using graph theory and choquet integral. IEEE Trans. Smart Grid, no. 99, p. 1 (2016)
11. X. Liu, M. Shahidehpour, Z. Li, X. Liu, Y. Cao, Z. Bie, Microgrids for enhancing the power grid resilience in extreme conditions. IEEE Trans. Smart Grid **8**, 589–597 (2017)
12. R. Billinton, W. Li, in *Basic Concepts of Power System Reliability Evaluation, Reliability Assessment of Electric Power Systems Using Monte Carlo Methods*, ed. by R. Billinton, W. Li (Springer US, Boston, MA, 1994), pp. 9–31
13. M. Panteli, D.N. Trakas, P. Mancarella, N.D. Hatziargyriou, Power systems resilience assessment: hardening and smart operational enhancement strategies. Proc. IEEE **105**, 1202–1213 (2017)
14. M. Panteli, P. Mancarella, D.N. Trakas, E. Kyriakides, N.D. Hatziargyriou, Metrics and quantification of operational and infrastructure resilience in power systems. IEEE Trans. Power Syst. **32**, 4732–4742 (2017)
15. P. Dehghanian, S. Aslan, P. Dehghanian, Quantifying power system resiliency improvement using network reconfiguration, in *IEEE 60th International Midwest Symposium on Circuits and Systems (MWSCAS)*, pp. 1364–1367 (2017)

Part II
Microgrids and Optimal Operations of Resilience Systems

Chapter 5
Resilience Thorough Microgrids

Shahram Mojtahedzadeh, Sajad Najafi Ravadanegh, Mahmoudreza Haghifam and Naser Mahdavi Tabatabaei

Abstract Currently the number of Microgrids (MGs) is continuously increased in distribution network. In this view, the future advanced distribution network can be regarded as clusters of MGs. Hence the MGs is the building blocks of smart distribution networks. There are many technical, economic and social reason for MGs implementation. One of the main advantages of MGs is the ability to encounter with abnormal conditions in the network such as occurrence of natural disasters with island operation capability. Based on the above discussion, the problem of optimal planning of distribution network based-on MGs is an interesting topic. In this chapter the optimal MG-based smart distribution grid planning problem is formulated and tested on a planning area. While the natural disasters are low probability and high impact phenomena, there are not enough historical data to extract an accurate component failure model. In this chapter the initial geographical area of MGs is supposed as input data in a large scale Greenfield study area and based on the resiliency constraints and index, the optimal configuration of total distribution network including MGs is determined. The distribution network configuration is planned such that all MGs meet the predefined requirement based on definition and supply the predefined critical loads within each MGs. In this work the optimal size and site of network elements and its configuration is determined by a

S. Mojtahedzadeh
Department of Electrical Engineering, Azarshahr Branch, Islamic Azad University, Azarshahr, Iran
e-mail: mojtahed.s@gmail.com

S. Najafi Ravadanegh (✉)
Smart Distribution Grid Research Lab, Azarbaijan Shahid Madani University, Tabriz, Iran
e-mail: s.najafi@azaruniv.ac.ir

M. Haghifam
Electric Transmission and Distribution Research Lab, Faculty of Electrical and Computer Engineering, Tarbiat Modares University, Tehran, Iran
e-mail: haghifam@modares.ac.ir

N. Mahdavi Tabatabaei
Department of Electrical Engineering, Seraj Higher Education Institute, Tabriz, Iran
e-mail: n.m.tabatabaei@gmail.com

© Springer Nature Switzerland AG 2019

N. Mahdavi Tabatabaei et al. (eds.), *Power Systems Resilience*, Power Systems,
https://doi.org/10.1007/978-3-319-94442-5_5

multi-objective optimization algorithm. The effect of the natural disasters on resilient MG-based distribution network planning, the geographical data for disasters is modelled to give a geographical map that joins the spatial risk index with distribution network component location. The main goal of this work is to propose a framework for optimal MG-based resilient distribution networks.

Keywords Critical load · Greenfield · Microgrid · Natural disasters Optimal configuration · Resilience · Resiliency constraint · Resiliency index Spatial risk index

Nomenclatures

B_{di}	Binary decision variable for DGs
$(\cos\varphi_{ave})_{ic}$	Power factor of *c*th cluster connected to *i*th DG
$(\cos\varphi)_{ick}$	Power factor of *k*th load of *c*th cluster supplied by *i*th DG
$C_{Dept.}$	Depreciation value
$C_{DGi,fuel}$	Fuel price for DG unit
$C_{DGi}^{Levelized}$	Levelized cost of energy
$C_{DGi,O\&M}^{Var.}$	O&M fixed cost
$C_{DERi,O\&M}^{Var.}$	O&M variable cost
C_{DGi}^{Cap}	Capital cost for DG
C_i	Current temperature
$C_{LVF,ic}^{Cap}$	LV feeder capital cost from *i*th DG to *c*th Cluster
$C_{LVF,kic}^{Cap}$	LV feeder cost for *k*th load of *c*th cluster connected to *i*th DG
C_{tax}	Tax rate
C_v	Voltage temperature
Co_{cost}^{colony}	Colonies cost coefficient
D_{rate}	Discount rate
$Dist_{\max}$	Max distance or max feeder length (m)
$Dist_{ck}$	Distance between *k*th load and *c*th cluster (m)
$Dist_{ic}$	Distance between *i*th DG and *c*th Cluster (m)
F	Cost function
$f_{DGi,cap.}$	Capacity factor
f_{ELC}	Energy loss cost factor (USD/kWh)
$f_{rec.}^{Cap}$	Capital recovery factor
$F_{DG_{MG}}$	Cost of DG in MG
$F_{LVF_{MG}}$	Cost of LV feeder in MG
$H_{DGi,rate}$	Heat rate
$i_{LOSS,ck}$	Loss index for feeder from *c*th cluster to *k*th load point
$i_{LOSS,ic}$	Loss index for feeder from *i*th DG to *c*th cluster
$I(s_i)$	PV current
I_{ms}	Max current
I_{sc}	Short circuit current

k	Weibull PDF shape factor
l	Distance
l_t	Life time
L_{DGi}	Connected load to DG (kW)
LF_{ave}	Average annual load factor
$MD_{\max}$	Max elec. distance (m.kW)
N	PV number
N_c	Number of countries
N_{DG}	Number of DG units
N_i	Number of clusters
N_I	Number of imperialists
N_L	Total number of loads
N_{Li}	Number of load blocks supplied by *i*th DG
ND	Number of decades
P	Turbine output power
P_r	Rated power turbine
P_k	Block demand (kW)
P_{DGi}	Rated capacity of DG (kW)
PF_c	Customer power factor
PF_D	DG power factor
R	Line resistance (Ω)
R_{rev}	Revolution rate
S	Demand
s	Scale factor
s_i	Solar irradiance
s_{ave}	Average radiation
T	Time period (years)
T'	Ambient air temperature
T_n	Nominal operating cell temperature
T_{cell}	Cell temperature
V	LV line voltage (kV)
$V(s_i)$	PV voltage
$V_{\max}$	Maximum permitted voltage drop value
V_{ms}	Max voltage
V_{oc}	Open circuit voltage
x	x coordinate of certain load or source point
X	Line reactance (Ω)
y	y coordinate of certain load or source point
μ	MG's number
γ	Mean of demand
σ	Variance of demand
v	Wind speed
η	Efficiency
α	Alpha PDF shape parameter

β Beta PDF shape parameter
ρ Specific resistance
λ Failure rate

5.1 Introduction

Previous blackouts have highlighted the weakness of the interconnected power system, which is complicated, fragile and vulnerable to natural disaster and other attacks. The ability of the distribution system to efficiently withstand low-probability, high-impact events; while enabling a quick recovery and restoration to the normal state is interpreted as resiliency. In [1], Distribution System Resiliency is discussed as a process of modernizing the grid that needs considerable time, effort and innovation. It outlines a distribution resiliency roadmap, identifying critical pieces of such an effort. As distribution networks are still vulnerable to extreme weather events, it is necessary to design new restoration techniques to reduce the outage time to improve the network resiliency. Traditional restoration techniques aim to restore as much load as possible in areas where service is lost.

By transforming the power system, electricity can be delivered where and when needed with high reliability [2]. Distribution system restoration, which aims to restore loads after a fault by altering the topological structure of the distribution network, has been extensively studied in the literature using various methodologies, including expert systems [3, 4], fuzzy logic [5, 6], multi-agent systems [7, 8], heuristic search [9], and optimization [10]. In [11], spanning tree search algorithm is used for reliability analysis of smart distribution systems with Distribution System Restoration technique and remote control capability. Switching sequence is decided by a set of rules. In [12], an optimization method is proposed to restore loads after a fault by changing the topological structure of the distribution network while meeting electrical and operational constraints.

Other benefits of the smart grid include enhanced cyber-security, integration of renewable energy resources, demand response and the integration of electric vehicles onto the grid for achieving grid resilience is increasing system flexibility and robustness [13]. As protecting the power distribution grids from catastrophic natural events is quite complicated and uneconomical, quick restoration ability after disasters is considered as the suitable solution. However, during extreme events, the main substations may fail and the distribution network cannot be supplied, and this may cause widespread outages. In these cases, traditional restoration methods cannot recover power supply in a short time. The integration of distributed generations and micro-grids provides new ways to supply critical loads and enable faster recovery of system. According to the IEEE Standard 1547.4, splitting a distribution system into multiple MGs can improve the operation and reliability of the distribution network. The self-sufficiency of the micro-grids is critical when the main network is down due to a widespread outage. This ability can provide a faster system recovery [14]. Microgrids could be formed in natural disasters by using available emergency generators through expanding their supply coverage beyond their designed service areas; or renewable energy resources such as wind/PV supply units that are gaining popularity to the environmental concerns [15, 16]. This ability, known as islanding, is arguably the most important characteristic of a microgrid. Since natural disasters cannot be avoided, it is important to enhance the capacity of systems to resist and recover from these events [17]. The use of microgrids can increase the power system's defences against natural disasters. By having active components such as distributed generation and energy storage at the distribution level, microgrids provide more flexibility, shorten the distance between generation and load and reduce the susceptibility of the conventional centralized grid and control architecture [18]. Microgrids manage distributed generation units, energy storage and other resources to increase the availability of generation, thereby improving resiliency. During a blackout, local generation within microgrids can be used to start a local restoration procedure [19]. Recently research on distribution network restoration with the presence of micro-grids is increased. In [20] the self-healing ability of a distribution network by dividing the system into multiple micro-grids is investigated [14], simulated the system black start for micro-grids to ensure the power quality, stability and robustness during the restoration. In [12], a restoration method based on micro-grids is used. The approach in [12] aims to restore max number of loads and min number of switching operations.

5.2 Formulation on Smart Resilient Distribution Network Planning Based on MGs

Distribution network planning is a difficult problem to solve, because we must satisfy so many constraints. In recent years, because of increasing penetration of distributed generations that change network topology and structure, engineers are faced with problems related to that change the facts we believed in before. More than this, network optimization costs are so high. Then for an optimal planning, these important factors must be considered in the planning process. Due to the description above, clustering existing grid to multiple micro-grids is a new solution for planning problem which can postpone the network upgrade to benefit technical and economic advantages. In this chapter, we present a framework to plan a distribution system based on micro-grids, form micro-grids, and determine the optimal service area, distributed energy resources combination, DG location, sizing, MV/LV substation's size and its location in each microgrid. Designed cost function is (F_t):

$$F_t = \sum_{\mu=1}^{N_{MG}} \left(F_{DG_{MG(\mu)}} + LVF_{MG(\mu)} \right) \tag{5.1}$$

This function consists of two parts, one is related to energy sources costs (fixed and variable) and other is related to low voltage feeders cost. The first term represents the cost of both dispatchable units (such as CHP, FCH and MT) and non-dispatchable units (such as wind and PV) within each MGs. The second term is the cost of LV network feeders that connect the LV loads to energy resources in each MGs. Decision variables are energy resources' size, location and type, and MV/LV transformers size and location and micro-grid's service area.

5.3 Planning Model Details

To start required data (geographical, technical, and load forecasting data, irradiation and wind speed data) must be collected. After this step, the Greenfield area must be divided into load blocks that their load gravity center is considered to be at their centroid and they may have different demands. Then, candidate places for distributed generators must be specified. By preparing decision variables, some candidate places may be specified to install DGs, and then some of the load points will be assigned to each source by K-Means clustering method. Proposed algorithm steps are:

5.3.1 Load Assignment

For load allocation, voltage magnitude constraints are considered. Product of a load point demand and its distance to a DG is calculated by:

$$MD = \text{Max Load} \times \text{Distance} \tag{5.2}$$

Dist = distance between a load block and a DG;

If constraints below are satisfied, then the corresponding candidate location will be kept:

$$\begin{array}{l} P_k \cdot Dist_{ik} < MD_{\max} \\ \begin{pmatrix} P_k = V_k I_k & \text{if } V_k \approx 1\,\text{pu} \Rightarrow P_k \propto I_k \\ R_{ik} = \rho \frac{l_{ik}}{A} \Rightarrow & R_{ik} = kl_{ik} \Rightarrow R_{ik} \propto Dist_{ik} \end{pmatrix}. \\ Dist_{ik} < Dist_{max} \end{array} \tag{5.3}$$

where

$$Dist_{ik} = \sqrt{(x_i - x_k)^2 + (y_i - y_k)^2} \tag{5.4}$$

$$MD_{\max} = V_{\max} \cdot \left(\frac{10^3 V^2 \cos \varphi}{R \cos \varphi + X \sin \varphi} \right) \tag{5.5}$$

Equations (5.3) and (5.5) determine the voltage inequality constraint and maximum rated voltage drop. To determine the real distance GIS data can be used or the method presented in [21] can be used to calculate the real distance between DG_i and $load_k$. Then by sorting distance between each load block and DGs, loads points will be assigned to the nearest DG considering following constraints:

$$\sum_{k=1}^{N_{Li}} P_k < LF_{aveDG_i} . P_{\max DG_i} \tag{5.6}$$

i belongs to set of selected DG sites and LF_{ave} is:

$$LF_{ave} = \frac{\text{total annual energy}}{\text{annual peak load} \times 8760} \tag{5.7}$$

To ensure that all load points are supplied, constraint below must be satisfied:

$$\sum_{i=1}^{N_{DG}} N_{Li} = N_L \tag{5.8}$$

Also, to avoid having not supplied load points a penalty is defined.

5.3.2 K-Means Method

K-Means clustering method is used to grouping load points into "N" groups. Clustering is done by minimizing Euclidean distances between load points and the corresponding centroid. By using K-Means, N_{Li} blocks which are connected to a DG (ith DG), will be clustered to N_i groups related to ith DG and N_i is:

$$N_i = ceil \frac{\sum_{j=1}^{N_{Li}} P_j}{\sqrt{3} V I_{\max} \cos \varphi_{DGi}} \tag{5.9}$$

5.3.3 Cost of Feeders

The following equation is used to calculate LV feeder's costs considering its variable and capital costs (costs of feeder from DG to cluster centroid (between ith DG and cth cluster) and from cluster centroid to load point (between cth cluster and kth load block):

$$\begin{aligned} F_{LVF_{MG(\mu)}} &= \sum_{i=1}^{N_{DG}} \sum_{c=1}^{N_i} (C_{LVF,ic}^{Cap} Dist_{ic} + \left(\frac{R_{ic} i_{LOSS,ic} f_{ELC}}{3 \times 10^3 V^2 (\cos \varphi_{ave})_{ic}^2} \right) \cdot 8760T \\ &+ \sum_{k=1}^{N_{L,j,i}} (C_{LVF,kic}^{Cap} Dist_{ck} + \left(\frac{R_{jk} i_{LOSS,ck} f_{ELC}}{3 \times 10^3 V^2 (\cos \varphi)_{ick}^2} \right) \cdot 8760T)) \cdot CE \cdot B_{di} \end{aligned} \tag{5.10}$$

$$i_{LOSS,ic} = \left(\sum_{k=1}^{N_{L,c,i}} P_k \right)^2 .Dist_{ic} \tag{5.11}$$

$$i_{LOSS,ck} = P_k^2 .Dist_{ck} \tag{5.12}$$

$$(\cos \varphi_{ave})_{ic} = \frac{\sum_{k=1}^{N_{L,c,i}} P_k}{\sum_{k=1}^{N_{L,c,i}} \frac{P_k}{(\cos \varphi_{ave})_{ick}}} \tag{5.13}$$

In cost modeling, all costs are in USD and currency exchange factor (CE) is used. As mentioned before, to calculate distance between cth cluster and kth load block GIS data or method presented in [21] is used.

5.3.4 Cost of DGs

To determine capacity of each DG, the sum of the demands assigned to them is needed:

$$L_{DGi} = \sum_{k=1}^{N_{L,i}} \frac{P_k}{LF_{ave}} \tag{5.14}$$

Fixed and variable costs for different types of DGs are considered in related cost function and covers capital, fuel, operations, performance and maintenance costs (USD/kWh), $C_{DGi}^{Levelized}$ [22]:

$$F_{DG_{MG(\mu)}} = \sum_{i=1}^{N_{DG}} ((C_{DGi}^{Levelized} .P_{DGi}.CE).T.8760).B_{di} \tag{5.15}$$

$$C_{DGi}^{Levelized} = \frac{C_{DGi}^{Cap} f_{rec.}^{Cap} \cdot (1 - C_{tax} C_{Dep.})}{8760 f_{DGi,cap.} \cdot (1 - C_{tax})} + \frac{C_{DGi,O\&M}^{Fixed}}{8760 f_{DGi,cap.}} + \frac{C_{DGi,O\&M}^{Var.}}{10^3} + \frac{C_{DGi,fuel} H_{DGi,rate}}{10^6} \tag{5.16}$$

$$f_{rec.}^{Cap} = \frac{D_{rate} \cdot (1 + D_{rate})}{(1 + D_{rate})^{l_t} - 1} \tag{5.17}$$

5.3.5 *Partitioning and Forming of Low-Voltage Network into Autonomous Microgrids*

After all steps before, it's time to cluster and form the distribution network to community of MGs. It is assumed that all micro-grids will have a main dispatchable DG to supply critical loads reliably. Main DG types are CHP or Fuel Cell. After energy resource placement and load allocation, according to main units, according to the distance between main units and other types of DGs, microgrids are formed by a dispatchable and some non-dispatchable units. The service area of each low voltage MG will be determined by these set of sources and their loads according to predefined criteria. Then load gravity center coordinates of each MG are determined to place the microgrid PCC bus (MV/LV transformer).

5.3.6 *Uncertainty Modeling*

Renewable energy resources have intermittent nature which can affect microgrid's performance. The best way to deal with this problem is to adapt the system to some operating scenarios and make it less sensitive to the variations. To speed up calculation a proper scenario reduction method must be applied to decrease the number of generated scenarios due to uncertainties. In this chapter, uncertain parameters are wind turbines and PV panels outputs and also demand. To represent errors in load forecasting methods, wind speed and irradiation predictions, an appropriate probability density function is used for each input variable:

- Normal PDF (probability density function) for load modeling

$$f(s) = \frac{1}{\sqrt{2\pi\sigma}} e^{\frac{(s-\gamma)^2}{2\sigma^2}} \tag{5.18}$$

The load in each point is modeled with a normal PDF with γ (mean of demand) and σ (variance of demand) is equal to small percentage of the base load.

- Weibull PDF for wind speed, which depends on wind speed and is presented by

$$f(\upsilon) = \frac{k}{s}\left(\frac{\upsilon}{s}\right)e^{-\left(\frac{\upsilon}{s}\right)^k}, \quad 0 \leq \upsilon \leq \infty \tag{5.19}$$

and the transformation of wind speed to wind turbine output power is given by:

$$P = \begin{cases} 0 & \text{if } \upsilon \leq \upsilon_{in} \text{ or } \upsilon \geq \upsilon_{out} \\ \frac{\upsilon - \upsilon_{in}}{\upsilon_{rated} - \upsilon_{in}} P_r & \text{if } \upsilon_{in} \leq \upsilon \leq \upsilon_{rated} \\ P & \text{else} \end{cases} \tag{5.20}$$

- Beta PDF for solar irradiation modeling is given in (5.21)

$$f(s) = \begin{cases} \frac{\Gamma(\alpha+\beta)}{\Gamma(\alpha)+\Gamma(\beta)} s_i^{\alpha-1}(1-s_i)^{\beta-1} & \text{if } 0 \leq s_i \leq 1, 0 \leq \alpha, \beta \\ 0 & \text{else} \end{cases} \tag{5.21}$$

The output power of PV panel is calculated by (5.22)

$$P(s_i) = N.\eta.V(s_i).I(s_i) \tag{5.22}$$

$$\eta = \frac{V_{rms} I_{rms}}{V_{oc} I_{sc}} \tag{5.23}$$

Finally, the PV voltage (depends on radiation) is

$$V(s_i) = V_{oc} - C_v T_{cell} \tag{5.24}$$

PV current can be obtained from

$$I(s_i) = s_{ave}[I_{sc} + C_i(T_{cell} - 25)] \tag{5.25}$$

$$T_{cell} = T' + s_{i,ave}\left(\frac{T_n - 20}{0.8}\right) \tag{5.26}$$

Then $\pm 10\%$ deviation is considered for error in predicted peak values. After generating scenarios, the number of them is reduced by the simultaneous backward reduction method [23].

5.3.7 Resiliency Modeling

As was detailed in [24, 25], power grids in a certain area may be subject to extreme events that could affect them by damaging infrastructure components or facilities. Also they may be affected by less severe events such as economic crisis and capital investment reduction that could lead to aging infrastructure components. The focus in this chapter is on extreme events like natural disasters. Extreme events can be considered as hazards for the operation of power systems, which can be defined based on [26] as "a potentially damaging physical event, phenomenon and/or human activity, which may cause operations or service disruptions." Extreme events not only affect grids' operations when they are active, but also in their aftermath.

As Fig. 5.1 represents, the period of time from one extreme event and the next one can be divided into four phases. These phases briefly are [24]:

- During the event: This phase lasts few minutes or longer like few days. This phase lasts from the time the first signs of the event are being noticed to the time the first repairs are made, Δt_1.
- Immediate aftermath: This phase may last from a few hours to a few weeks. This phase lasts from the time that the restoration and repair activities initiate to when they are mostly completed, Δt_2.
- Intermediate aftermath: This phase lasts from a few weeks to several months. The focus during this phase is to studying the effect of the event on the power grid by performance metrics evolution and documenting damage.
- Long-term aftermath: This phase usually last from a few months to several years. The focus in this phase is to preparing system for the future event by planning and modifying the existing infrastructure and operating processes according to third phase output.

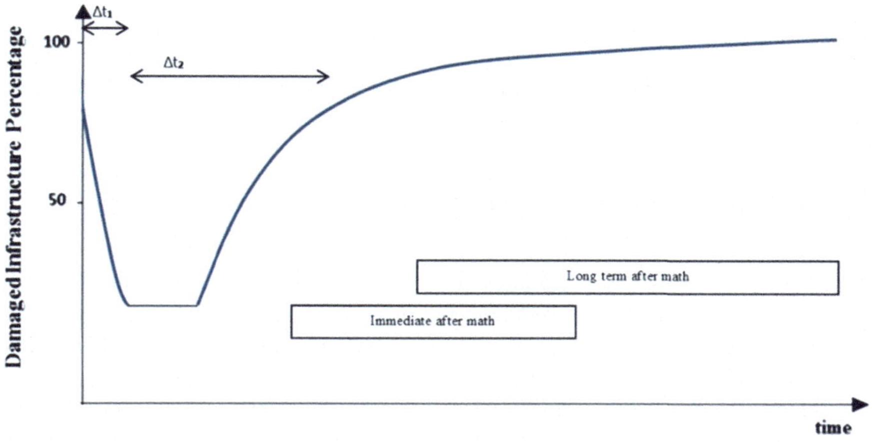

Fig. 5.1 Phases of an extreme event and resiliency curve

In this research, definition used for resilience is “the ability to prepare for and adapt to changing conditions and withstand and recover rapidly from disruptions” [27]. That is, resilience is an attribute with four components:

- Withstanding capability; i.e., the ability to operate during an extreme event.
- Recovery speed; the time needed to recover with respect to a disruption.
- Preparation/planning capacity; the ability to implement measures to decrease the effect of future events on networks’ performance.
- Adaptation capability; the ability to operate and manage network to react to conditions that could affect its performance.

Based on this definition for resilience, the used metric for resilience in this framework, R_I, is defined as [5–26]:

$$RI = \frac{T_{Up}}{T_{Up} + T_{Down}} \tag{5.27}$$

This equation presents a measure of resiliency through the dependence on the up (T_{Up}) and down times (T_{Down}). The service recovery time is represented by T_{Down}, which is mostly influenced by human driven activities and hardware related aspects. T_{Up} depends on the withstanding capability of a network to during an event. Its value is related network design and hardware characteristics. The sum of up and down time is the total evaluating time T.

Then an index is designed for MGs which calculate MG’s feeders’ length to total feeder length of network. This parameter is:

$$R_{MG} = \frac{MG_{Line}}{Total_{Line}} \tag{5.28}$$

Then its average value is obtained for all MGs and then is applied to problem.

$$R_{MG_{Av}} = \frac{\sum_{mg=1}^{N_{MG}} R_{MG}}{N_{MG}} \tag{5.29}$$

In addition, for resilience improvement network hardening costs are considered in cost function.

5.4 Component Fragility Modelling

Fragility modelling is a key concept in planning of distribution networks against natural disasters. In this case to decide the hardening degree of the network components with respect to the severity of the disaster it is important to use the fragility curve for each component to show that a network component is damaged in encountering to a specific natural event. There is different fragility curve for different event that should be used for the current event. In this chapter the fragility curve for distribution network poles and feeder against hurricane wind speed is applied to model the degree of fragility with respect to hurricane in a specific are of the planning zone. The fragility curve for distribution network poles and feeder are given by (5.30) and (5.31), respectively [26, 27]. In the above equations the parameter λ define the component failure frequency with respect to hurricane and v wind is speed.

$$\lambda_{Pole}(v) = 1 \times 10^{-4}\, e^{0.421v} \tag{5.30}$$

$$\lambda_{Line}(v) = 8 \times 10^{-12}\, v^{5.173} \tag{5.31}$$

Figures 5.2 and 5.3 are a graphical representation of Eqs. (5.30) and (5.31).

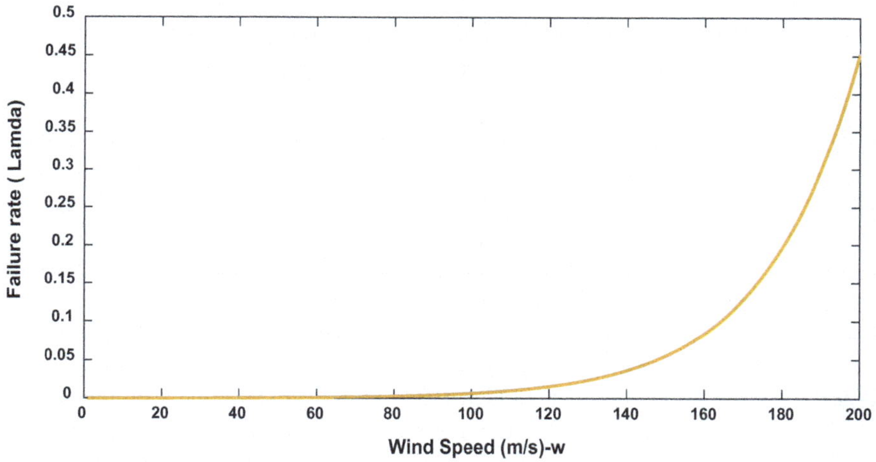

Fig. 5.2 Distribution poles failure rate with respect to v

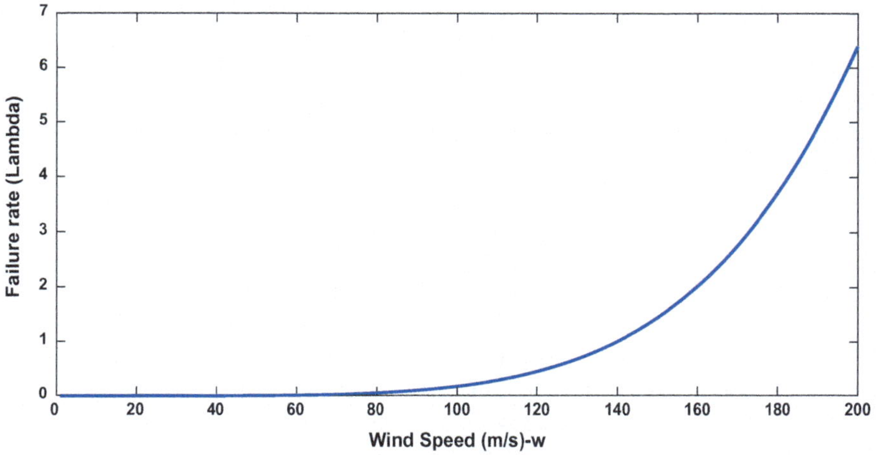

Fig. 5.3 Distribution conductor's failure rate with respect to W

5.5 Joint Geospatial Map Construction for Natural Disasters

Geospatial map for the hurricane wind speed pattern in the study zone is another important concept that joints the geographic map of the natural disasters to the fragility curve. The goal of this concept is to model a relation between the severities of the natural disaster to the fragility curve of the network component. In this chapter a geographical map is illustrated the show the planning zone with different hurricane speed.

5.6 Simulation and Results for Optimal Resilient Planning of Smart Distribution Network Based on MG

In this chapter, the autonomous multiple microgrid forming is studied but without loss of generality, it can be applied for autonomous and non-autonomous cases. MGs' borders are determined using some predefined geographical, economic and electrical constraints, considering uncertainty sources in the network. The proposed method is applied to a certain Greenfield (the area considered in [23]). After preparing data, we divide the proposed area into small load blocks (50 $\times$ 50 m). The total forecasted load is 3778.72 kW with 0.9 power factor and candidate locations for each type of DGs are specified [23]. Four types of DG units are used, Wind Turbine (WT), Photovoltaic Cell (PV), Fuel Cell (FC) and Combined Heat and Power (CHP). To increase the reliability of critical customers, candidate locations for dispatchable DGs are selected to be near these loads and due to this criterion, all microgrids at least have one dispatchable DG. In Fig. 5.4, considered area with its blocks and their load density is presented. It is assumed that system is a smart grid, all needed infrastructures are ready, and customers are engaging demand response or load control programs. The ICA algorithm is used to solve the optimization problem. Parameters used in proposed method are described in Table 5.1. The parameters related to the levelized cost of energy are prepared from [27]. To test the proposed method, planning considering uncertainties and resiliency constraints is evaluated.

5.6.1 Planning Considering Resiliency Constraints

In this test all constraints in previous cases and also the resiliency and hardening criteria are considered. In Fig. 5.5, the area which may has destructive storms is presented. Then according to geographical structure, large elliptic in Fig. 5.6, determine the critical area and red elliptic determine the area which needs under grounded feeders. In other parts, reinforced poles will be used (for hardening the

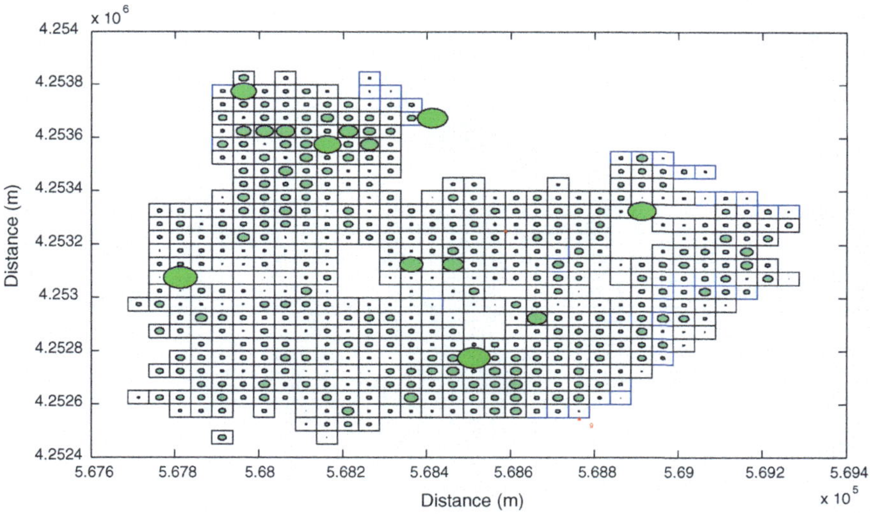

Fig. 5.4 Considered green field area with load density

Table 5.1 Proposed framework's parameters

Parameters		
1	Planning time period	8 years
2	Network nominal voltage	0.4 kV
3	f_{ELC}	800 IRR/kWh
4	C_{TAX}	0.392
5	D_{rate}	0.07
6	CE	30,000
7	N_c	30
8	AC	1.1
9	N_I	3
10	R_{rev}	0.2
11	Co_{cost}^{colony}	0.02
12	l_t	30 years
13	ND	150
14	PF_c	0.9
15	PF_D	0.8
16	Feeder conductor impedance	$0.2116e^{-3}+j0.08e^{-3}$ Ω/m
17	Line length	400 m
18	Max allowed current	200 A
19	Failure rate	2 failure/year/km
20	Average repair time	5 h

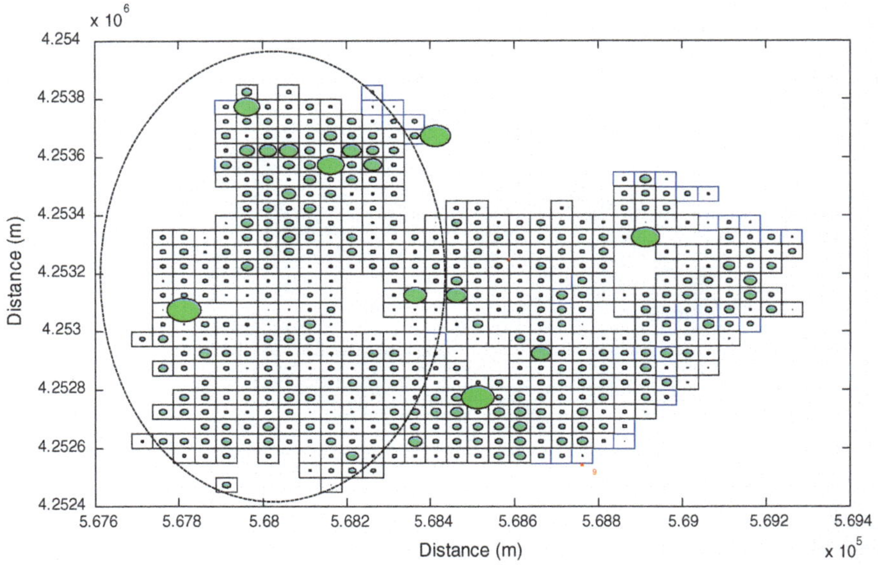

Fig. 5.5 Critical area

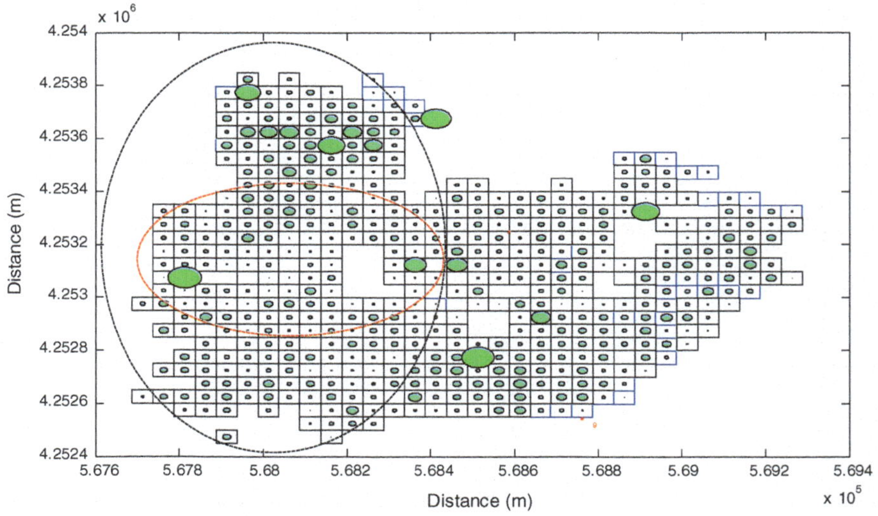

Fig. 5.6 Area which needs under grounded feeders

network). Notice that all of the costs related to hardening is added to objective function to be optimized. Then results are given by Fig. 5.7 and Table 5.2 (number of micro-grids and their specifications). The data related to unavailability time and resiliency index is represented by Table 5.3. We can see that in the second planning

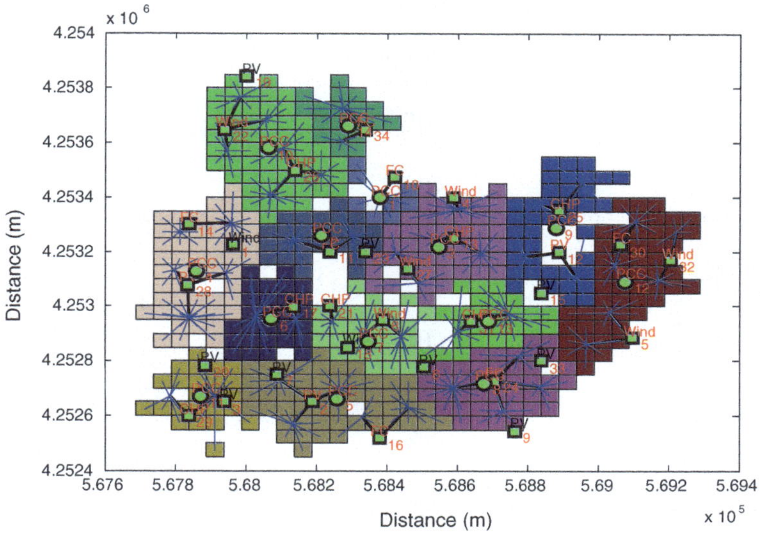

Fig. 5.7 Clustering of low voltage grid into microgrids (resiliency constraints)

Table 5.2 Clustering of LV network into multiple MGs considering resiliency constraints

MG	DGs	DG number	DG type	Capacity (kW)	PCC Trans. (kVA)
1	Non-Disp.	–	–	–	100
	Disp.	10	FC	75	
2	Non-Disp.	–	–	–	250
	Disp.	11	FC	175	
3	Non-Disp.	4 27	WT WT	125 175	200
	Disp.	13	CHP	150	
4	Non-Disp.	28	PV	350	315
	Disp.	14	FC	200	
5	Non-Disp.	2	PV	300	250
	Disp.	16	FC	175	
6	Non-Disp.	–	–	–	315
	Disp.	17	CHP	200	
7	Non-Disp.	8	WT	150	100
	Disp.	21	CHP	50	
8	Non-Disp.	–	–	–	630
	Disp.	24	FC	500	
9	Non-Disp.	12	PV	175	250
	Disp.	25	CHP	200	

(continued)

Table 5.2 (continued)

MG	DGs	DG number	DG type	Capacity (kW)	PCC Trans. (kVA)
10	Non-Disp.	22	WT	325	630
	Disp.	26	CHP	400	
11	Non-Disp.	–	–	–	315
	Disp.	29	CHP	225	
12	Non-Disp.	5 32	WT WT	175 150	200
	Disp.	30	FC	150	
13	Non-Disp.	–	–	–	315
	Disp.	31	CHP	225	
14	Non-Disp.	–	–	–	315
	Disp.	34	FC	250	

Table 5.3 Unavailability time and resiliency index

MG	1	2	3	4	5	6	7	8	9	10	11	12	13	14
R_{MG} ($\times 10^{-4}$)	450	923	2121	3437	4837	5152	5960	6611	7565	8834	9476	11,104	11,500	12,130
T_D ($\times 10^{-1}$)	4.65	0	0	0	14.1	0	0	6.74	9.87	13.13	6.65	16.84	0	6.52
T_D average	0.5607													

case, average down time for network is lower than previous case because of considered algorithm and hardening criteria. Using undergrounded feeder make the down time zero.

5.7 Conclusion

In this chapter the problem of optimal planning of distribution network based-on MGs in a green field area is investigated. The optimal MG-based smart distribution grid planning problem is formulated using a cost based and resilient-oriented method. The boundaries of low-voltage distribution MGs are determined considering network adequacy and security constraints. Besides the optimal size and location of network components is determined using a multi-objective optimization algorithm. Failure rate of main components in distribution network such as distribution poles and conductors are modelled as a function of hurricane wind speed. A new metric is defined to show the degree of planned network resiliency with respect to conventional planning. The network configuration is designed and compare with normal planning case. Finally, the effect of resilient planning of distribution network on the number of MGs is evaluated. It is shown that with resilient planning the number of MG is increased in a given network.

References

1. G. Davis, A.F. Snyder, J. Mader, The future of distribution system resiliency, in *Clemson University Power Systems Conference*, pp. 1–8 (2014)
2. R. Galvin, K. Yeager, J. Stuller, *Perfect Power: How the Microgrid Revolution Will Unleash Cleaner, Greener, and More Abundant Energy* (McGraw-Hill, New York, NY, 2009)
3. C.C. Liu, S.J. Lee, S.S. Venkata, An expert system operational aid for restoration and loss reduction of distribution systems. IEEE Trans. Power Syst. **3**(2), 619–626 (1988)
4. C.S. Chen, C.H. Lin, H.Y. Tsai, A rule-based expert system with colored petri net models for distribution system service restoration. IEEE Trans. Power Syst. **17**(4), 1073–1080 (2002)
5. S.I. Lim, S.J. Lee, M.S. Choi, D.J. Lim, B.N. Ha, Service restoration methodology for multiple fault case in distribution systems. IEEE Trans. Power Syst. **21**(4), 1638–1644 (2006)
6. S.J. Lee, S.I. Lim, B.S. Ahn, Service restoration of primary distribution systems based on fuzzy evaluation of multi-criteria. IEEE Trans. Power Syst. **13**(3), 1156–1163 (1998)
7. J.M. Solanki, S. Khushalani, N.N. Schulz, A multi-agent solution to distribution systems restoration. IEEE Trans. Power Syst. **22**(3), 1026–1034 (2007)
8. C.P. Nguyen, A.J. Flueck, Agent based restoration with distributed energy storage support in smart grids. IEEE Trans. Smart Grid **3**(2), 1029–1038 (2012)
9. A.L. Morelato, A. Monticelli, Heuristic search approach to distribution system restoration. IEEE Trans. Power Del. **4**(4), 2235–2241 (1989)
10. S. Khushalani, J.M. Solanki, N.N. Schulz, Optimized restoration of unbalanced distribution systems. IEEE Trans. Power Del. **22**(2), 624–630 (2007)
11. Y. Xu, C.C. Liu, H. Gao, Reliability analysis of distribution systems considering service restoration, in *IEEE PES Conference Innovative Smart Grid Technologies*, pp. 1–5 (2015)
12. J. Li, X.Y. Ma, C.C. Liu, K.P. Schneider, Distribution system restoration with microgrids using spanning tree search. IEEE Trans. Power Syst. **29**(6), 3021–3029 (2014)
13. Executive Office of the President, in *Economic Benefits of Increasing Electric Grid Resilience to Weather Outages* (2013)
14. C.L. Moreira, F.O. Resende, J.P. Lopes, Using low voltage microgrids for service restoration. IEEE Trans. Power Syst. **22**(1), 395–403 (2007)
15. C. Abbey, D. Cornforth, N. Hatziargyriou, K. Hirose, A. Kwasinski, E. Kyriakides, G. Platt, L. Reyes, S. Suryanarayanan, Powering through the storm. IEEE Power Energy Mag. **12**(3), 67–76 (2014)
16. C. Chen, J. Wang, F. Qiu, D. Zhao, Resilient distribution system by microgrids formation after natural disasters. IEEE Trans. Smart Grid **7**, 958–966 (2015)
17. H. Zhou, J.A. Wang, J. Wan, H. Jia, Resilience to natural hazards: a geographic perspective. Nat. Hazards **53**, 21–41 (2010)
18. A. Kwasinski, V. Krishnamurthy, S. Junseok, R. Sharma, Availability evaluation of micro-grids for resistant power supply during natural disasters. IEEE Tran. Smart Grid **3**, 2007–2018 (2012)
19. C. Gouveia, C. Leal Moreira, J.A. Pecas Lopes, D. Varajao, R. Esteves Araujo, Microgrid Service restoration: the role of plugged-in electric vehicles. IEEE Ind. Electron. Mag. **7**, 26–41 (2013)
20. Z. Wang, J. Wang, Self-healing resilient distribution systems based on sectionlization into microgrids. IEEE Trans. Power Syst. Press **30**(6), 3139–3149 (2015)

21. S.N. Ravadanegh, S.H. Hosseinian, M. Abedi, A. Vahidnia, S. Abachezadeh, A framework for optimal planning in large distribution networks. IEEE Trans. Power Syst. **24**(2), 1019–1028 (2009)
22. Levelized Cost Calculations, http://en.openei.org/apps/TCDB/levelized_cost_calculations.html
23. Sh Mojtahedzadeh, S. Najafi Ravadanegh, M.R. Haghifam, Optimal multiple microgrids based forming of greenfield distribution network under uncertainty. IET Renew. Power Gener. **11**(7), 1059–1068 (2017)
24. A. Kwasinski, Field technical surveys: an essential tool for improving critical infrastructure and lifeline systems resiliency to disasters, in *IEEE Global Humanitarian Technology Conference*, San Jose, CA, USA, pp. 1–7, October 2014
25. A. Kwasinski, Quantitative model and metrics of electrical grids' resilience evaluated at a power distribution level. Energies **9**(2), 93 (2016)
26. A.F. Mensah, Resilience assessment of electric grids and distributed wind generation under hurricane hazards, Ph.D. Thesis, Rice University, May 2015
27. Presidential Policy Directive—Critical Infrastructure Security and Resilience, www.whitehouse.gov/the-press-office/2013/02/12/presidential-policy-directive-critical-infrastructure security-and-resil. Accessed on 25 May 2015

Chapter 6
Optimal Scheduling of Networked-Microgrids to Resiliency Enhancement Under Uncertainty

Pouya Salyani, Sajad Najafi Ravadanegh and Naser Mahdavi Tabatabaei

Abstract By increasing the penetration of distributed energy resources (DERs) and developing the microgrids (MGs) due to its implication for the existing circumstance of power system, Networked-Microgrid (NMG) approach is becoming an important issue in smart distribution grids. On the other hand, low probability but extreme events like natural disasters or the fuel interruption can threaten the grid security. For this end resiliency problem is discussed in this paper for the MMG structure of distribution network. The main objective of this chapter is to globally reach the minimum cost operation of microgrids in normal condition while meeting the adequacy in resiliency operation mode. The dispatchable unit status, energy storage and adjustable loads scheduling and the energy trade status between microgrids in each hour are evaluated with stochastic modeling of load and renewable power generations. It is of important because these microgrids have energy exchange between each other under control of energy management system (EMS) in addition of energy drawing from main grid. This can prepare an option to count on DER capacity of other microgrids in resiliency mode of system operation when the main grid is disconnected.

Keywords Distribution network · Grid security · Networked-microgrids
Resiliency enhancement · Resiliency operation · Uncertainty

Nomenclatures

Ω_{mg}	Set of microgrids in the MMG
$\Omega_{d,i}$	Set of dispatchable units in microgrid i

P. Salyani · S. Najafi Ravadanegh (✉)
Smart Distribution Grid Research Lab, Azarbaijan Shahid Madani University, Tabriz, Iran
e-mail: s.najafi@azaruniv.edu

P. Salyani
e-mail: pouya1370salyani@yahoo.com

N. Mahdavi Tabatabaei
Department of Electrical Engineering, Seraj Higher Education Institute, Tabriz, Iran
e-mail: n.m.tabatabaei@gmail.com

© Springer Nature Switzerland AG 2019
N. Mahdavi Tabatabaei et al. (eds.), *Power Systems Resilience*, Power Systems,
https://doi.org/10.1007/978-3-319-94442-5_6

$\Omega_{sl,i}$	Set of shiftable loads in microgrid i
$\Omega_{cl,i}$	Set of curtalable loads in microgrid i
$\Omega_{nd,i}$	Set of nondispatchable units in microgrid i
μ	Mean value of load P_L
σ	Standard deviation of load P_L
v	Wind speed
$P_{r,WT}$	Wind turbine nominal power
v_{cut-in}	Wind turbine cut in speed
$v_{cut-off}$	Wind turbine cut off speed
P_{pv}	PV output power
P_{STC}	PV power at standard condition
G_{ING}	Solar irradiation
G_{STC}	Solar irradiation at standard condition
T_c	PV standard temperature
T_r	PV real temperature
$x_{i,j,t}$	Binary decision variable that defines if the dispatchable unit j in microgrid i operates in hour t or not
$P^d_{i,j,t}$	Generated active power of dispatchable unit j in microgrid i and in hour t (MW)
$P^{nd}_{i,q,t}$	Generated power of nondispatchable unit q in microgrid i and in hour t (MW)
$P^g_{i,t}$	Active power transacted from main grid to microgrid i in hour t (MW)
$\zeta_{i,t}$	Binary decision variable that defines whether microgrid i buys energy from other microgrids or not
$P^{Tr}_{i'i,t}$	Active power transferred from microgrid i' to microgrid i in hour t (MW)
$\xi_{i,t}$	Binary decision variable that defines whether microgrid i sells its surplus energy to other microgrids or not
$P^{Tr}_{ii',t}$	Active power transferred from microgrid i to microgrid i' in hour t (MW)
$s_{i,t}$	Binary decision variable that defines whether if the battery storage microgrid i is in charge state or not
$v_{i,t}$	Binary decision variable that defines whether if the battery storage microgrid i is in discharge state or not
$P^{st}_{i,t}$	Active power of battery storage in microgrid i and in hour t (MW)
$z_{i,c,t}$	Binary decision variable that is 1 if the curtailable load c is shed in microgrid i and in hour t
$D^{cl}_{i,c,t}$	Consumption of curtailable load c in microgrid i and in hour t (MW)
$\omega_{i,s,t}$	Binary decision variable that is 1 if the shiftable load s is shed in microgrid i and in hour t

$D^{sl}_{i,s,t}$	Consumption of shiftable load s in microgrid i and in hour t (MW)
$C_{i,t}$	Available capacity of battery storage in microgrid i and in hour t (MWh)
$\alpha_{i,s}, \beta_{i,s}$	Optimum start and end time of shiftable load s consumption in microgrid i (hr)
$FP_{i,j,t}$	Fuel price of dispatchable unit j in microgrid i and in hour t ($/MWh)
$FP_{i,j,t}$	Market price in hour t ($/MWh)
f^{sh}_{s}	Defined penalty in order of shifting the load s within specified time horizon ($/hr)
f^{cr}_{c}	Defined penalty in order of curtailing the load c within specified time horizon ($/hr)
$D_{i,t}$	Certain demand in microgrid i and in hour t
$Pd^{\min}_{i,j}, Pd^{\max}_{i,j}$	Power limitation of dispatchable unit j in microgrid i (MW)
$RU^{d}_{i,j}, RD^{d}_{i,j}$	Ramp up and ramp down rates for dispatchable unit j in microgrid i
$Pg^{\max}_{i}$	Maximum permissible power to be transferred in microgrid i (MW)
$OT_{i,j}, DT_{i,j}$	Up and down time of dispatchable unit j in microgrid i
$P^{ch,\min}_{i}, P^{ch,\max}_{i}$	Minimum and maximum charge state power of storage in microgrid i (MW)
$P^{dch,\min}_{i}, P^{dch,\max}_{i}$	Minimum and maximum discharge state power of storage in microgrid i (MW)
$C^{\min}_{i}, C^{\max}_{i}$	Minimum and maximum capacity of storage in microgrid i (MWh)
MCT_{i}, MDT_{i}	Minimum charging and discharging time of storage of microgrid i (hr)
$E^{sh}_{i,s}, E^{cr}_{i,c}$	Total consumption energy of adjustable loads in microgrid i (MWh)
$\tau_{i,s}, \delta_{i,s}$	Defined start and end times of shiftable load s consumption in microgrid i (hr)
$D^{sh,\min}_{i,s}, D^{sh,\max}_{i,s}$	Demand limitation of shiftable load s in microgrid i (MW)
$D^{cr,\min}_{i,c}, D^{cr,\max}_{i,c}$	Demand limitation of curtailable load c in microgrid i (MW)

6.1 Introduction

Networked-microgrids are cutting-edged technologies which have great capabilities to improve the resiliency of existing power distribution networks. This brand new technologies play a vital role in making highly reliable and secure situations for local loads especially during natural disasters and emergency conditions and prevent consumers from facing with unnecessary interruptions and load shedding. In other words, the important purpose of this chapter is to providing optimal scheduling for microgrids (MGs) in different operational modes which ensures better conditions for resilient microgrids. On the other hand, the sharp fluctuations and uncertainties in load demands and renewable power generations enforce researchers to handle the power system operation problems in microgrids environment with stochastic methods. Therefore, in the chapter, a scenario-based heuristic algorithm is introduced to provide more effective solution to the proposed problem. The suggested optimization problem is divided into normal and emergency conditions. In normal situation, all MGs operate in a regular condition and dispatch of power between MGs and utility grid is done in order to satisfy the requirements of local loads. In this condition, MGs have a great coordination with together through flowing power and data in safe environment. In emergency condition, there can be some natural or man-made disasters in the network and prevent an efficient delivering of electricity through distribution lines. Although, MGs may encounter with critical situation in providing power for their own loads, a new proposed energy management system (EMS) for the presented NMGs structure can ensure a trustworthy and effective condition for both MGs owners and consumers. One of the significant differences in the structures of normal and emergency conditions has to do with the connection of MGs to the utility grid. While, all MGs have access to the utility grid in whole time of normal situation, in other condition, due to some fault events, MGs cannot benefit from main distribution grid, but due to the perfect advantages of the presented EMS, the disconnection affects the optimal operation of MGs in low level which plays a key role in enhancing the resiliency of power systems. At last, in order to have a proper focus on the correctness of results.

Going toward, power system is facing the penetration of distributed energy resources (DERs) in its all sectors. The integration of these DERs can result in formation of microgrids which leads to more decentralization of power grid. To facilitate and regulate the incorporation of DERs in distribution network, microgrid concept is a best choice. Microgrids have the capability of improving the voltage profile, loss reduction, reliability enhancement and service procurement with a cheaper price relative to the grid price [1–6]. More over reduction in the air pollution by central power plants and firing of fossil fuels [7], flattening the load profile by applying the demand response mechanism [8], making the network more resilient [9] and reducing the transportation cost can be accounted as the benefits of microgrids.

One level higher than microgrid concept is the idea of several microgrids known as networked-microgrid [10–12]. A main problem for a single microgrid is that in isolated mode and when it is disconnected from the main grid, just load shedding has to be carried out to resolve the shortage power of local sources if there is. Of course when the penetration of DERs is high, there is not any significant problem for the microgrid itself. To overcome this issue, microgrids can be linked together to have energy exchange with each other [13, 14] or in other word construct a networked-microgrid.

In this context, [15] gives the approach of braking the distribution network down into number of interconnected microgrids in order to have a better control on its operation under significant penetration of DERs. In [16], it is aimed to design a networked-microgrid including electricity market and distribution system design. A multiobjective approach is employed to maximize the utility of microgrids, power grid and independent system operator simultaneously. In [17], the total cost of power grid compromising multiple areas of wind power generation is minimized. Optimal operation of networked-microgrids concerning the probability density function of DERs such as wind turbines (WTs) and photovoltaic cells (PVs) is discussed in [18]. This economic dispatch problem involves cost generation cost of each microgrid plus the cost of buying and selling energy to the main grid.

Authors in [19] have proposed a control strategy to have a proper operation for NMG in which main grid operator is considered as the higher decision maker. In [20], the problem of congestion in distribution network is studied by managing the microgrids of NMG equipped with electric vehicle (EV) and using smart charging strategy for EV. Networked-microgrid operation in its unsymmetrical condition is discussed in [21]. Because asymmetric line impedance between Distributed Generators (DGs) impacts the power sharing between them. In [12] networked-microgrid control system (MMCS) is introduced to make a connection between microgrid concept and smart grid concept and to have a management on areas containing ICTs. Cooperation between distribution company (DisCo) and microgrids is modeled by [22] in which applying a bi-level optimization in this paper, the profit maximization of DisCo and cost minimization of each microgrid is fulfilled but there is not any energy exchange between microgrids.

Transaction of energy among microgrids in a networked-microgrid is investigated in [23], so that by defining the price offered by each DG and the energy price offered by main grid in every time slot, the overall operation cost of networked-microgrid is optimally determined in a probabilistic manner.

Energy trading between islanded microgrids is allowed in [24] to economically service the loads within the networked-microgrid. This cost of each microgrid contains the inner operation cost of DERs and cost of transferring energy from a known microgrid to another one. Cost of energy transferring between each microgrid in this paper is determined using Lagrange relaxation method. Therefor optimal buying and selling prices of microgrids are obtained based on their inner generation and system topology.

In [25], with the purpose of optimal operation of networked-microgrid, the problem is solved in two layer. The first layer relates to the local optimization of each microgrid cost. Global optimization is discussed in second layer where the amount of energy transacted between each microgrid are determined based on their buying and selling price. This procedure is followed in [26] where the energy surplus/shortage is obtained for each microgrid and then internal trading between microgrids is assessed. In [27], first the optimal power flow is used to get the surplus/shortage of each microgrid by satisfying the network constraints.

With respect to these results, market layer model is used in energy transaction in which naive auction mechanism is applied to determine the amount of sold or bought energy of each microgrid in the 24-h horizon and to maximize the utility of microgrids. The control and management of mentioned power mismatch in networked-microgrid is performed by multi-agent system (MAS) [28, 29]. In [30], energy exchange among autonomous microgrids is addressed and prospect theory as a game theory is implemented to study the energy trading behavior of microgrids.

In case of resiliency state, the network encounter with a high extent events and it is disconnected from the main grid. Therefor if the scaffolding of network is based on a smart grid, it is capable of self-healing and supplying its interrupted loads. In other word, microgrids are able to operate in islanding mode. Resilience operation of microgrids is discussed in [31, 32]. Also in [33], author has involved the adjustable loads in the microgrid resiliency assessment and the total operation cost of microgrid is minimized through the Benders decomposition and by reducing the power mismatch in the resilience mode of operation.

The main contributions of the work can be categorized as follow:

- A grid of microgrids or networked-microgrids structure is taken into account for studying MG optimal daily scheduling to resolve the prevalent drawbacks of conventional structures of microgrids.
- The proposed EMS can optimally operate in different MG operation. The microgrids not only can fulfil the optimal operation for local resources through sharing their information with central controller but also provide reliable and economic conditions for MGs to address their own economic and environmental conditions efficiently.
- Two different problem formulations and scheduling horizons are defined for normal and resiliency modes. In the case of any natural disaster event, the microgrid is switched to the emergency operation to maximise the service reliability to local loads.

As mentioned before the networked-microgrid structure for smart distribution grid can enhance both of reliability in normal operation mode and resiliency in case of natural disaster mode comparing with a conventional distribution grid or a single entity smart microgrid. What is investigated in this paper is the study of resiliency for networked-microgrid which are connected to each other and there is energy transaction among them. Furthermore, they can be supplied from the main grid in the normal operation of networked-microgrid. Section 6.2 explains the uncertainty modeling and Sect. 6.3 defines problem and associated objective function. Section 6.4 gives the results obtained from simulation and finally in last section the conclusion is presented.

6.2 Probabilistic Model of the Input Data

In this section for network input data such as load, wind speed and solar irradiation a probabilistic model is provided.

6.2.1 Probabilistic Model of Load

The increasing in penetration of renewable and intermittent energy sources for electrical energy increases the uncertainty in the operation of distribution networks. In deterministic calculations of a system's operational state the set of input parameters are determined. The probabilistic analysis in power systems is a very powerful tool operation and planning studies. In probabilistic calculations based on real numbers input parameters and state variables are described by probability distribution function (PDF) and these data can be described by cumulative distribution function (CDF). Obtained results from probabilistic analysis also are presented in PDF and CDF forms. In this work, PDF is used for input data such as load and MGs small scale units' output power. Moreover, the results such as bought and sold powers by microgrids, cost of energy transactions are described in form of PDF or CDF. Time and climate are two factors for deterministic component of the demand variation whereas stochastic component is an independent random variation. The behavioral patterns of different energy consumers lead to a variable demand in each bus. These variations can be calculated by statistical analysis. Consequently, demand varies continually with a high degree of uncertainty. The load can be described by a PDF. Several random variables have been used to model the demand, e.g. uniform, Weibull, beta and normal PDF. In this work, load demand is modelled as a normal distribution with mean value μ and standard deviation σ.

$$f(P_L) = \frac{1}{\sigma\sqrt{2\pi}} \exp(-\frac{(P_L - \mu)^2}{2\sigma^2}) \tag{6.1}$$

A random variable with a Gaussian distribution is said to be normally distributed and is called a normal deviate. The generation of renewable energy resources depend on the availability of the primary energy resource such as wind speed, solar irradiation. In this section the probabilistic model of load and renewable energy resources is discussed.

6.2.2 Probabilistic Model of Wind Power

The generated power of wind turbine depends on the wind speed. Wind speed changes every time of year, that shows the importance of a probability model.

Based on published report the Weibull distribution is suitable to model the probability density distribution of wind speed for long-term planning purposes. Weibull PDF is as follows:

$$f_v(v) = \begin{cases} \frac{\rho}{\alpha} \times (\frac{v}{\rho})^{\rho-1} \times \exp((\frac{v}{\alpha}) \times \rho) & v \geq 0 \\ 0 & \text{otherwise} \end{cases} \tag{6.2}$$

If the wind speed is generated by (6.2), the real power generated by wind turbine (WT) can be given by (6.3).

$$P_{WT}(v) = \begin{cases} \mathrm{P}_{r,WT} \times (\frac{v^2 - v^2_{cut-in}}{v^2_r - v^2_{cut-in}}) & v_{cut-in} \leq v \leq v_r \\ \mathrm{P}_{r,WT} & v_r \leq v \leq v_{cut-out} \\ 0 & \text{otherwise} \end{cases} \tag{6.3}$$

6.2.3 Probabilistic Model of Solar Power

For power generation in photovoltaic system the solar radiation and air temperature are two important parameters. Like the wind speed, these parameters are variable in any time of year. In this chapter, irradiance and air temperature are modeled by normal distribution function. The standard condition for PV is in G_{ING}= 1000 W/m^2, T_r= 25 °C cell temperature and the actual operating condition may differ from the base case. The output power of the PV module can be evaluated by:

$$P_{pv} = P_{STC} \times \frac{G_{ING}}{G_{STC}} \times (1 + k(T_c - T_r)) \tag{6.4}$$

6.3 Resilience Modeling of Networked-Microgrids

Suppose a networked-microgrid community compose of cooperative microgrids such as Fig. 6.1. These microgrids have coordinated operation and are controlled by central EMS. They have externally energy trade with main grid and internally with each other. However, for either of main grid and microgrids their own 24-h energy price is determined to sell the produced power. Furthermore the management of energy inside the microgrid is assumed to be done locally by μEMS. The μEMS effectively controls and makes the balance between supply and demand in the microgrid by adjusting output power of DERs and consumed power of loads. A microgrid studied in this paper can involve dispatchable units like microturbines (MTs) or CHP, renewable resources and adjustable loads.

With respect to abovementioned assumptions and explanations about the attended networked-microgrid, the decision maker aims to optimize the overall cost

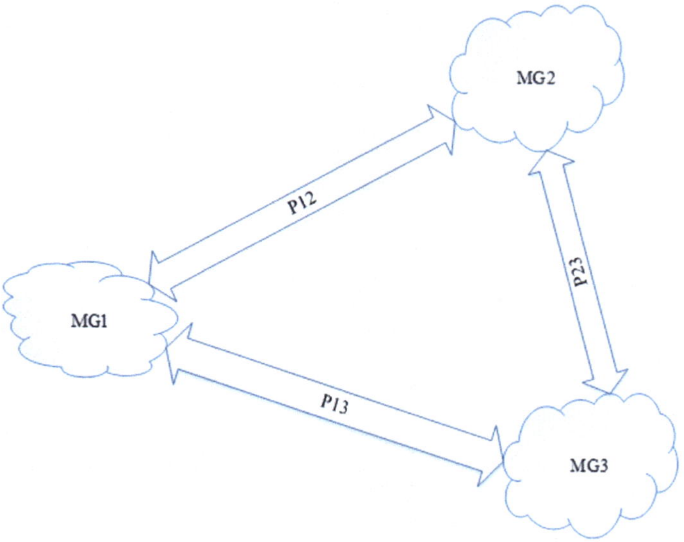

Fig. 6.1 Networked-microgrid schema

of operation in normal condition of the system. But on the other hand, it is aimed to take into account the resiliency of networked-microgrid in the operation cost in that the intended networked-microgrid is subject to high duration interruptions. This can be achieved by optimally scheduling of microgrids and determine the state of switches between each microgrid, so that normal operation cost should be minimized whereas networked-microgrid meets the power balance in resiliency state. The cost function for 24-h normal operation condition of a networked-microgrid is represented below which is addressed in [31] and the related constraints are given in (6.5–6.27).

$$\begin{aligned} C_{oper} = & \sum_{t=1}^{N_h} \sum_{i \in \Omega_{mg}} \sum_{d \in \Omega_{d,i}} x_{i,j,t} . P_{i,j,t}^{d} . FP_{i,j,t} + \sum_{t=1}^{N_h} \sum_{i \in \Omega_{mg}} P_{i,t}^{g} . MP_t \\ & + \sum_{t=1}^{N_h} \sum_{i \in \Omega_{mg}} \zeta_{i,t} . \sum_{i' \in \Omega_{mg}} \gamma_{i'i,t} . P_{i'i,t}^{Tr} . MP_{i',t} + \sum_{i \in \Omega_{mg}} \sum_{s \in \Omega_{sl,i}} f_s^{sh} . \Delta h_{i,s}^{sh} \\ & + \sum_{i \in \Omega_{mg}} \sum_{c \in \Omega_{cl,i}} f_c^{cr} . \sum_{t=1}^{N_h} z_{i,c,t} \end{aligned} \tag{6.5}$$

Subject to

$$\sum_{j\in\Omega_{d,i}} x_{i,t}.P^{d}_{i,j,t} + \sum_{q\in\Omega_{nd,i}} P^{nd}_{i,q,t} + P^{g}_{i,t} + \zeta_{i,t}.\sum_{i'\in\Omega_{mg}} \gamma_{i'i,t}.P^{Tr}_{i'i,t} + (s_{i,t} - v_{i,t}).P^{st}_{i,t}$$
$$= D_{i,t} + \sum_{s\in\Omega_{sl,i}} \omega_{i,s.t}.D^{sl}_{i,s,t} + \sum_{c\in\Omega_{cl,i}} z_{i,c.t}.D^{cl}_{i,c,t} + \xi_{i,t}.\sum_{i'\in\Omega_{mg}} \gamma_{ii',t}.P^{Tr}_{ii',t} \tag{6.6}$$

$$x_{i,j,t}.Pd^{\min}_{\mathrm{i,j}} \le P^{d}_{i,j,t} \le x_{i,j,t}.Pd^{\max}_{\mathrm{i,j}} \tag{6.7}$$

$$x_{i,j,t}.P^{d}_{i,j,t} - x_{i,j,t-1}.P^{d}_{i,j,t-1} \le RU^{d}_{i,j} \tag{6.8}$$

$$x_{i,j,t-1}.P^{d}_{i,j,t-1} - x_{i,j,t}.P^{d}_{i,j,t} \le RD^{d}_{i,j} \tag{6.9}$$

$$x_{i,j,t} - x_{i,j,t-1} \le \sum_{k=1}^{\phi_{i,j}} x_{i,j,t+k}, \quad \phi_{i,j} = OT_{i,j} - 1 \tag{6.10}$$

$$x_{i,j,t-1} - x_{i,j,t} \le \sum_{k=1}^{\varphi_{i,j}} \left(1 - x_{i,j,t+k}\right), \quad \varphi_{i,j} = DT_{i,j} - 1 \tag{6.11}$$

$$-Pg^{\max}_{i} \le P^{g}_{i,t} \le Pg^{\max}_{i} \tag{6.12}$$

$$P^{st}_{i,t} \le s_{i,t}.P^{ch,\max}_{i} + v_{i,t}.P^{dch,\max}_{i} \tag{6.13}$$

$$P^{st}_{i,t} \ge s_{i,t}.P^{ch,\min}_{i} + v_{i,t}.P^{dch,\min}_{i} \tag{6.14}$$

$$C_{i,t} = C_{i,t-1} + (s_{i,t} - v_{i,t}).P^{st}_{i,t} \tag{6.15}$$

$$C^{\min}_{i} \le C_{i,t} \le C^{\max}_{i} \tag{6.16}$$

$$s_{i,t} - v_{i,t} \le \frac{1}{B_i}\sum_{k=1}^{B_i} s_{i,t+k}, \quad B_i = MCT_i - 1 \tag{6.17}$$

$$v_{i,t} - s_{i,t} \le \frac{1}{G_i}\sum_{k=1}^{G_i} v_{i,t+k}, \quad G_i = MDT_i - 1 \tag{6.18}$$

$$D^{\mathrm{sh},\min}_{i,s} \le D^{sh}_{i,s,t} \le D^{\mathrm{sh},\max}_{i,s} \tag{6.19}$$

$$\sum_{t=1}^{N_h} D^{sh}_{i,s,t} = E^{sh}_{i,s} \tag{6.20}$$

$$D_{i,c}^{\text{cr,min}} \leq D_{i,c,t}^{cr} \leq D_{i,c}^{\text{cr,max}} \tag{6.21}$$

$$\sum_{t=1}^{N_h} D_{i,c,t}^{cr} = E_{i,c}^{cr} \tag{6.22}$$

$$\zeta_{i,t}.\gamma_{i'i,t} + \xi_{i',t}.\gamma_{ii',t} \leq 1 \tag{6.23}$$

$$\xi_{i,t}.\gamma_{i'i,t} + \zeta_{i',t}.\gamma_{ii',t} \leq 1 \tag{6.24}$$

$$\gamma_{ii',t}.P_{ii',t}^{Tr} + \gamma_{i'i,t}.P_{i'i,t}^{Tr} = 0 \tag{6.25}$$

$$\zeta_{i,t} + \xi_{i,t} \leq 1 \tag{6.26}$$

$$-\gamma_{ii',t}.Ptr_{i'i}^{\max} \leq P_{ii',t}^{Tr} \leq \gamma_{ii',t}.Ptr_{i'i}^{\max} \tag{6.27}$$

As it is seen the cost function in (6.1) includes five terms. The first term is operation cost of dispatchable units in each microgrid depending on the fuel price and hourly dispatched power. Cost of power purchased from main grid is included in second term. A microgrid can sell its power in the same price MP_t to the upstream network. The third term is about the cost of power bought from other microgrids. If it is decided to buy the power in hour t, the decision variable $\zeta_{i,t}$ is 1. Binary decision variable $\gamma_{i'i,t}$ defines the that the power is transferred from microgrid i' to microgrid i. The forth term denotes the inconvenience cost which depends on the shifted time interval of energy consumption for shiftable loads. This shifted time for each adjustable load is represented by $\Delta h_{i,s}^{sl}$ that is calculated as follows.

$$\Delta h_{i,s}^{sl} = \alpha_{i,s} - \tau_{i,s} + \delta_{i,s} - \beta_{i,s} \tag{6.28}$$

Regarding to the sensitivity of each shiftable load, a penalty factor $f_{i,s}^{sh}$ is defined to exhibit the flexibility of these customers to operating outside their specified time interval. Higher values of $f_{i,s}^{sh}$ indicate that to shift the operating interval for one hour, higher cost is imposed to the microgrid respect to other loads. The last term relates to the curtailable loads. Apart from shiftable loads, a penalty factor is considered for these loads based on their sensitivity. Sum of hours in a day that the load is curtailed determines the cost that must be paid to the customers.

The power balance for microgrid i must be satisfied through Eq. (6.6) in each hour. In addition of produced and consumed power within the microgrid and also the power purchased from main grid, the imported or exported power is considered in this equation too. If this microgrid is in sell state, $\xi_{i,t}$ is 1, otherwise it is 0. Power limitation for dispatchable units is considered in (6.7). Inequalities (6.8) and (6.9) ensure the ramp up and ramp down rate for these units. Minimum up time and down time limitation for these units is given in (6.10) and (6.11).

Due to some reasons like thermal limit of feeders, the power delivered from main grid to microgrids is restricted to $Pg_i^{\max}$ in (6.12). Energy storage has limitation in charged or discharged power (6.13) and (6.14) and its stored energy or released energy cannot exceed the maximum capacity (6.15) and (6.16). Minimum time for energy storage to be charged or discharged is represented by (6.17) and (6.18).

Both the shiftable and curtailable loads are counted as adjustable loads. Restriction in demand of adjustable loads and the amount of their consumed energy over the 24 h is shown in (6.19–6.22). In other word it is assumed that these customers have a specific amount of consumption over a day. According to inequalities (6.23–6.26), a microgrid is not able to simultaneously sell and buy the energy and can just be in one of these states. Also (6.27) indicates that the amount of power transferred from microgrid i' is same as the power received by microgrid i.

Now the main problem is that how to schedule the MMG in a way that by going into resiliency mode, the power mismatch reaches to zero. Resiliency analysis for MMG mostly depends on the time of incident in which the main grid supply disconnects and MMG must rely on its renewable generation. When the MMG goes to resiliency mode of operation, all the loads must be supplied and a zero power mismatch should be attained in each hour. This is represented in (6.29) and the constraints associated with resiliency mode are shown in (6.30–6.35).

$$\begin{aligned}&\sum_{j\in\Omega_d} x_{i,t}^r.P_{i,j,t}^{d,r}+\sum_{q\in\Omega_{nd}} P_{i,q,t}^{nd,r}+\zeta_{i,t}^r.\sum_{i'\in\Omega_{mg}}\gamma_{i'i,t}^r.P_{i'i,t}^{Tr,r}+\left(s_{i,t}^r-v_{i,t}^r\right).P_{i,t}^{st,r}\\&=D_{i,t}+\sum_{a\in\Omega_{sl}}\omega_{i,a.t}^r.D_{i,a,t}^{sl,r}+\sum_{c\in\Omega_{cl}} z_{i,c.t}^r.D_{i,c,t}^{cl,r}+\xi_{i,t}^r.\sum_{i'\in\Omega_{mg}}\gamma_{i'i,t}^r.P_{ii',t}^{Tr,r}\end{aligned} \tag{6.29}$$

$$x_{i,t}^r=x_{i,t},\quad \forall t=(h,h+rd) \tag{6.30}$$

$$s_{i,t}^r=s_{i,t},\quad \forall t=(h,h+rd) \tag{6.31}$$

$$v_{i,t}^r=v_{i,t},\quad \forall t=(h,h+rd) \tag{6.32}$$

$$\zeta_{i,t}^r=\zeta_{i,t},\quad \forall t=(h,h+rd) \tag{6.33}$$

$$\xi_{i,t}^r=\xi_{i,t},\quad \forall t=(h,h+rd) \tag{6.34}$$

$$\gamma_{ii',t}^r=\gamma_{i'i,t}^r=\gamma_{ii',t},\quad \forall t=(h,h+rd) \tag{6.35}$$

To include the resiliency state in the main objective function, we can model the Eqs. (6.29–6.35) as the other constraints for C_{oper}. In other word, not only the technical constraints in normal operation mode of the MMG such as power balance or limitations for energy storage must be fulfilled, but the MMG must be resilient to have the ability of servicing the loads and recovering after disruptive events.

More over by occurrence of a disruptive events like disconnection in main grid, it is not possible for some equipment to change their state due to their technical limitations. Dispatchable units cannot change their mode of operation to be on or off (6.30) and energy storage must maintain its charging or discharging state (6.31) and (6.32). Also the MMG must be optimally scheduled, so that microgrids be able to support each other and fill the gap of supply shortage during the grid disconnection. By occurrence of such an unexpected event, it is assumed that microgrids are not permissible to change their buy-sell state (6.33–6.35).

To put it more simply, if it is determined for microgrid i to buy energy from microgrid i' in hour t, then among the resiliency mode, microgrid i cannot change its decision to buy energy from other microgrids. However the amount of power bought from microgrid i' can be variable in either normal condition or resilient condition. It depends on required energy of ith microgrid and surplus power of microgrid i.

6.4 Numerical Results

A networked-microgrid test constructed of three microgrid is employed in order to validate the proposed approach which its required data are represented below. The problem is solved by GAMS and the solver BARON. The required data for dispatchable units belong to each microgrid are given in Table 6.1. Table 6.2 gives the data about shiftable loads in each microgrid. There are totally 4 shiftable loads in microgrid 1, 3 shiftable loads in microgrid 2 and just 1 existing in the third microgrid.

About curtailable loads, it must be noted that for all microgrids, it is assumed that for each microgrid there is just one curtailable load with hourly demand limited between 1.7 and 2 MW and overall required energy of 42 MW. Figure 6.1 depicts the profile of certain demand of each microgrid over the 24-h of day and the expected produced power by non-dispatchable units in each microgrid is given in Fig. 6.2a–c.

Table 6.1 Required data for dispatchable units in microgrids

MG	Unit	Operation cost ($/MWh)	Produced power limitation (MW)	Min up/down time (h)	Ramp up/down rate (MW/h)
1	1	27.7	1–5	3	2.5
1	2	39.1	1–5	3	2.5
2	1	27.7	1–5	3	2.5
2	2	27.7	1–5	3	2.5
2	3	27.7	1–5	3	2.5
2	4	27.7	1–5	3	2.5
2	5	39.1	0.8–3	1	3
2	6	39.1	0.8–3	1	3
3	1	39.1	0.8–3	1	3

Table 6.2 Shiftable loads required data

MG	Min demand (MW)	Max demand (MW)	Required energy (MWh)	Start time (h)	End time (h)	Min up time (h)
1	0	0.8	4	11	15	1
1	0.07	1	5	15	19	1
1	0.02	1.2	3.6	16	18	1
1	0.02	0.8	4.8	14	19	1
2	0.01	0.6	3	13	18	1
2	0.09	1.1	4	11	17	1
2	0.08	1	3	14	20	1
3	0.06	0.4	2.7	12	18	1

Figure 6.3 represents the market energy price and Table 6.3 shows the energy selling price for each microgrid over the 24-h of the day. These prices are supposed to be determined in the day-ahead market. Maximum permissible power transaction between main grid and microgrids is 10 MW. About energy storage and for microgrids 1 and 2, its maximum capacity is considered to be 12 MWh with maximum power limited to 0.4–2 MW. But for microgrid 3 these quantities are 18 MWh and 0.4–3 MW, respectively. However, both the minimum charge and discharge time are assumed to be 5 h for all the energy storage systems (Fig. 6.4).

Furthermore, in the case of resiliency mode of operation, the distribution network disconnects from the main grid in hour 13 and goes into resilience mode through a period of about 8 h by construction of an island. Figure 6.4 shows the power profile of energy storage systems in microgrid 1 based on their charging and discharging state. It can be observed that for every three systems, up to about hour 10 or 12, they have been charged and after that they have gone into discharge state.

This is because of that minimum charging and discharging time for these units is 5 h and maximum power that can be provided or absorbed is 5 MW. Hence they are charged till the start time of resiliency and then based on the load amount of microgrid in the resiliency mode, it is discharged. In other word, if the energy storage be fully charged before the hour 13, its available power to be delivered is 10 MW, while the resiliency period is 7 h and this 10 MW cannot response to the total period. Therefor it is necessary to manage this energy and discharge it in the specific hours that microgrid incurs higher demand. Of course, by pay attention to the certain demand, it can be seen that in the period of about 14–16, the demand has an increase in microgrids and its impact is considered in results.

The profile of produced energy for dispatchable units for MG 2 and in the normal operation mode is demonstrated in Fig. 6.5. The microgrids are encountered to a light load in the initial hours of the day and because of that most of the associated units are in off state. Instead the microgrids are supplied by the existing WT units. It must be mentioned that dispatchable units directly affect the total operation cost and as the units are in off state, it is economical for ISO.

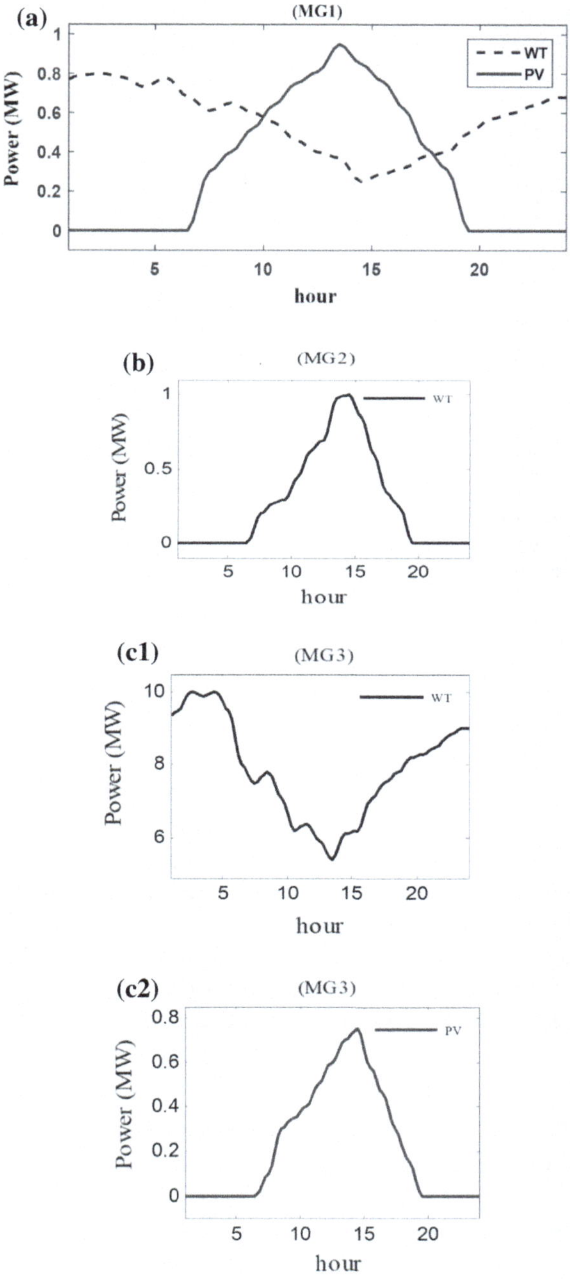

Fig. 6.2 Expected power profile for WT and PV operated in each microgrid (**a–c2**)

For the periods that microgrids are confronted with heavy load, the dispatchable units operate in full capacity. These units should be optimally scheduled so that in the resilience mode of network the power mismatch could be minimized. Due to

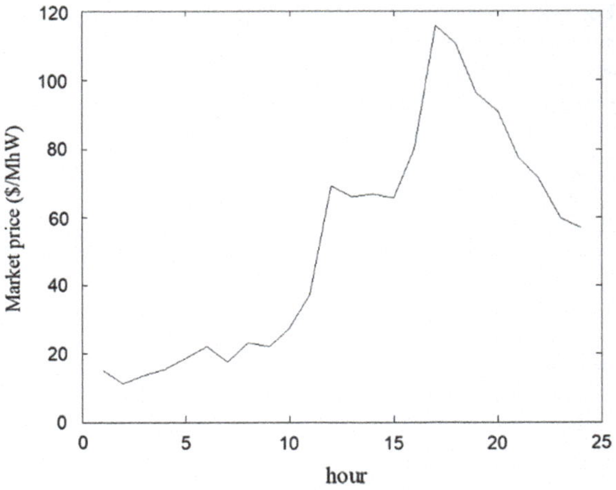

Fig. 6.3 Market price over 24-h

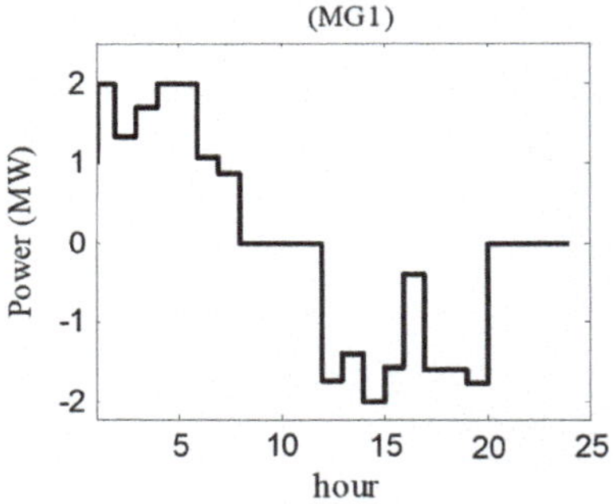

Fig. 6.4 Energy profile of battery storage for each microgrid

limitation in dispatch states of these units (minimum on and down time) and also their ramp limitation which impacts on the generated power in the next hours, they are scheduled till hour 13 by taking into account the resiliency period.

As it is apparent in these plots (Fig. 6.5), close to the resiliency start time the units in each microgrid are in on state and operate close to their specified maximum

Table 6.3 Energy selling price for each microgrid

MG	Time horizon (1–12) h											
1	14	10.4	12.7	15	17.9	21.5	22.12	23.9	30.59	32.78	33.08	60.6
2	14.7	10.2	12	14.8	17.5	21.1	24.32	27.9	29.19	31.98	34.08	58.13
3	13	10	12.2	14.5	17.8	20.9	30	40.9	50.19	51.78	55.08	70.13
MG	Time horizon (13–24) h											
1	70.2	62.17	80.36	80.69	55	50.14	98	89.1	61	60	56.6	49
2	69.9	61.7	75.66	82.69	55.13	51.14	100.6	85.5	65.5	58.3	55	48
3	63.9	70.27	78.36	85.4	56	52.1	105	90.3	55.5	50	53.2	39

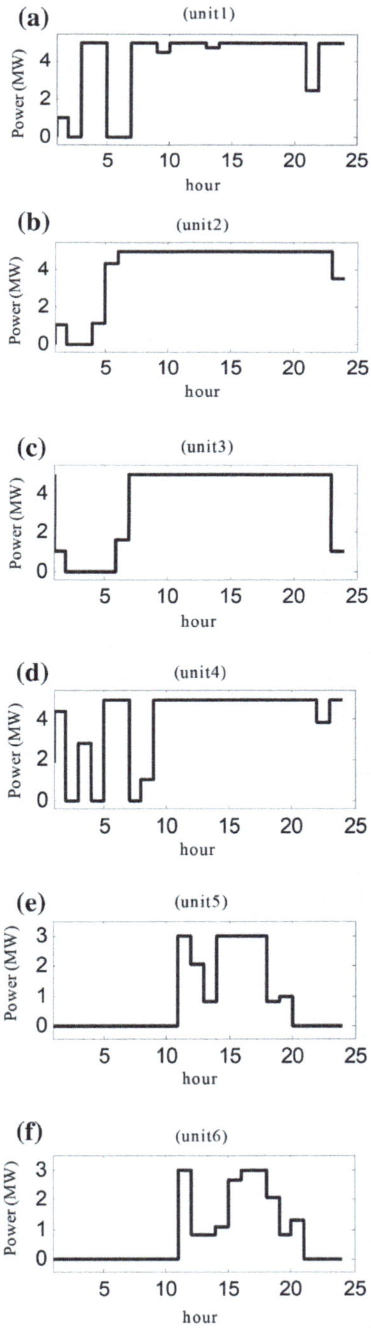

Fig. 6.5 Energy profile of dispatchable units in MG2 (**a**–**f**)

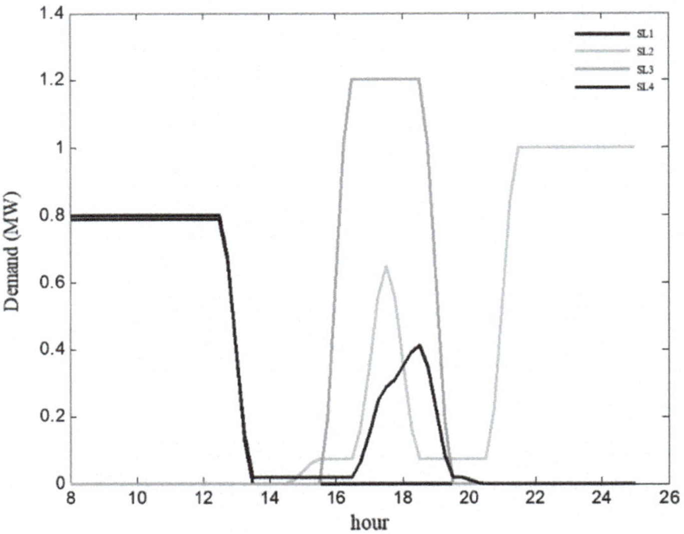

Fig. 6.6 Demand profile of shiftable loads in MG1

capacity. Its reason is that when the network shifts to the resiliency mode, it is assumed that system operator is not able to change the state of dispatchable units and their state must maintain constant through this period. In addition, by going into resilience mode, they cannot have a sudden increase in their produced power and so near the hour 12–13, they have an almost full capacity operation.

Table 6.4 represents the resilient operation of dispatchable units within the microgrids. This table points out that in resilience mode of the MMG, all the dispatchable units have dispatched and operated in their full capacity. The main reason is that in resilience mode, the MMG has been disconnected from the main grid which would be able to supply the microgrids in case of power shortage condition.

Table 6.4 Operation of dispatchable units in resilience mode of the MMG

MG	Unit	Resiliency period (13–20)							
1	1	5	5	5	5	5	5	5	5
1	2	5	5	5	5	5	5	5	5
1	3	3	3	3	3	3	3	3	3
1	4	3	3	3	3	3	3	3	3
2	1	5	5	5	5	5	5	5	5
2	2	5	5	5	5	5	5	5	5
2	3	5	5	5	5	5	5	5	5
2	4	5	5	5	5	5	5	5	5
2	5	3	3	3	3	3	3	3	3
2	6	3	3	3	3	3	3	3	3
3	1	3	3	3	3	3	3	3	3

Table 6.5 Start and end time of shiftable loads

MG	Optimal start time				Optimal end time			
	SL1	SL2	SL3	SL3	SL1	SL2	SL3	SL4
1	8	15	16	8	15	24	18	19
2	12	10	14	–	18	17	21	–
3	11	–	–	–	18	–	–	–

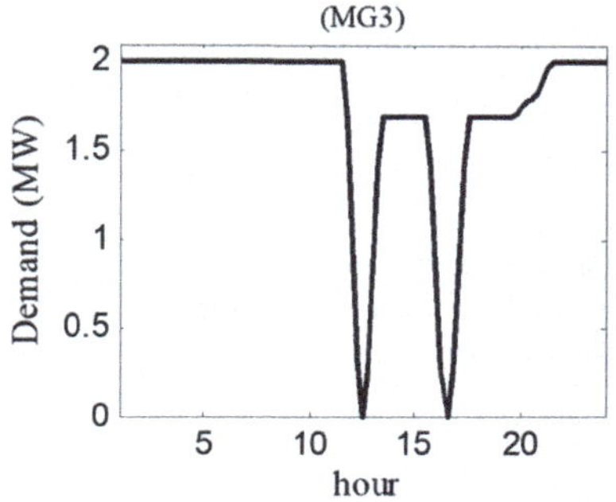

Fig. 6.7 Demand profile of curtailable loads in microgrid 3

The optimal start and end time of each shiftable load is obtained that is shown in Table 6.5. It can be seen that major of these customers have faced a shift in their start time of consumption, especially in microgrid 1.

To elaborate the reason, in resilience mode of MMG all the existing dispatchable units have been dispatched and have increased their production to the maximum level, albeit its high incurred operation cost. Energy storage devices that were fully charged up to the resilience start time, are not capable of supplying the whole demand and the energy provided by WT or PV is insufficient. Furthermore, the transacted energy is highly dependent on the other microgrids scheduling. Therefor there is not any way except to shift and curtail the adjustable loads.

Demand profile of shiftable loads within their specified consumption period are depicted in Fig. 6.6 for microgrid 1. Out of this time period these loads have not any consumption and because of that just demand within this period is provided. In some hours the loads are curtailed or they have a low demand to fulfil power balance in the resilience mode of microgrids. Also Table 6.6 shows the up and down states of shiftable loads over the 24-h of the day. Figure 6.7 represents the consumption profile of curtailable loads over the 24-h of the day for MG 3. Considering this figure, in hours 13, 14, 15, 16 and 17 these loads have interrupted which imposed a total penalty of 600 $ to the DisCo.

Energy transaction among microgrid 1 and the main grid is represented in Fig. 6.8. There is not any energy transfer from microgrids to the upstream main grid (energy selling to main grid), but in some specific hours, main grid has sold its power to the microgrids. In some hours, this energy exchange is becoming zero. In these hours, it has been beneficial for microgrid to be supplied by its existing units or compensate its shortage power by buying from other microgrids if it is available.

Table 6.6 Up/Down state of shiftable loads over the specified period in microgrids

MG	SL	Time horizon (8–24 h)																
1	1	1	1	1	1	1	0	0	0	0	0	0	0	0	0	0	0	0
1	2	0	0	0	0	0	0	0	1	1	1	1	1	1	1	1	1	1
1	3	0	0	0	0	0	0	0	1	1	1	0	0	0	0	0	0	0
1	4	1	1	1	1	1	1	1	1	1	1	1	1	0	0	0	0	0
2	1	0	0	0	0	1	1	1	1	1	1	1	0	0	0	0	0	0
2	2	0	0	1	1	1	1	1	1	1	1	0	0	0	0	0	0	0
2	3	0	0	0	0	0	0	1	1	1	1	1	1	1	1	0	0	0
3	1	0	0	0	1	1	1	1	1	1	1	1	0	0	0	0	0	0

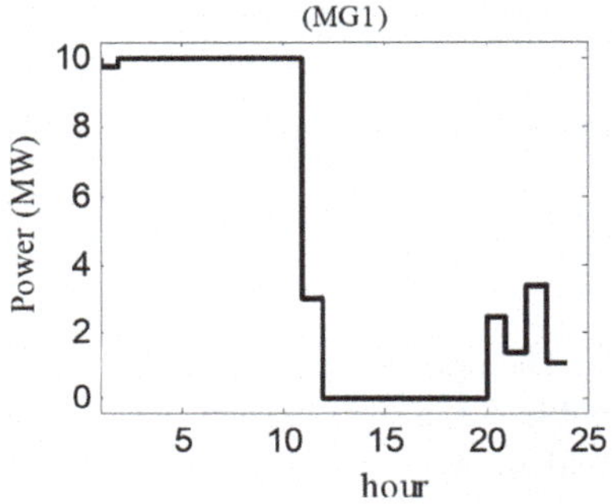

Fig. 6.8 Profile of energy transaction among main grid and microgrids

Table 6.7 Buy-sell states of microgrids

MG	Time horizon (1–12 h)												
1	1	1	1	1	0	0	1	1	1	0	1	1	1
2	1	0	1	1	1	1	1	1	1	0	1	1	0
3	0	0	0	0	0	0	0	0	0	1	0	0	0
MG	Time horizon (13–24 h)												
1	1	0	0	0	0	1	0	0	1	0	0	1	1
2	0	0	0	0	0	1	0	0	0	0	0	1	0
3	0	1	1	1	1	1	1	1	1	1	1	1	0

Table 6.7 gives the buy-sell state of microgrids through the 24 h. These states are determined based on the shortage and surplus power of each microgrid and their price of selling energy too. Table 6.8 shows the amount of energy transaction between these microgrids in the resilience mode of MMG respectively. Considering abovementioned results, total operation cost of MMG in its normal mode is 32,847 $.

Table 6.8 Amount of power transacted among microgrids in resiliency mode of operation (MW)

MG	Resiliency period (h)							
	13	14	15	16	17	18	19	20
1–3	0	0	3.85	2.81	0	0	0	0
2–1	0	0	0	0	0	0	0	0.5
2–3	3.61	5.16	2.39	0.83	1.01	3.64	4.72	3.61

6.5 Conclusion

What was investigated in this work is the resiliency assessment of distribution networks with networked-microgrids. The problem of how to schedule these microgrids to reach the minimum 24-h operation cost while difficulties associated with resiliency condition of networked-microgrids be overcame. The main difficulty for the system operator is the power mismatch occurrence for microgrids. On the other hand, system operator has an extra alternative relative to a single microgrid network that is the energy exchange facility which can play a key role in the power shortage problem of microgrids and also in the resilience operation mode of network.

As it was observed, optimal dispatch of units, charge and discharge state of energy storage systems and consumption profile of adjustable loads could be obtained in order the minimum operation cost be attained in addition of fulfilling the power balance in each microgrids through the resiliency mode of network. However overall operation cost of MMG could be reduced by providing energy transaction among microgrids in some hours and this transaction helped to increase the self-healing feature of microgrids when the MMG runs into a high extent event and shifts to resiliency mode of operation.

References

1. S.A. Arefifar, Y.A.R.I. Mohamed, T.H.M. El-Fouly, Optimum microgrid design for enhancing reliability and supply-security. IEEE Trans. Smart Grid **4**(3), 1567–1575 (2013)
2. A. Ameli, S. Bahrami, F. Khazaeli, M.R. Haghifam, A multiobjective particle swarm optimization for sizing and placement of DGs from DG owner's and distribution company's viewpoints. IEEE Trans. Power Deliv. **29**(4), 1831–1840 (2014)
3. W. Caisheng, M.H. Nehrir, Analytical approaches for optimal placement of distributed generation sources in power systems. IEEE Trans. Power Syst. **19**(4), 2068–2076 (2004)
4. Y. Wang, K.T. Tan, X.Y. Peng, P.L. So, Coordinated control of distributed energy-storage systems for voltage regulation in distribution networks. IEEE Trans. Power Delivery **31**(3), 1132–1141 (2016)
5. M.E. Khodayar, M. Barati, M. Shahidehpour, Integration of high reliability distribution system in microgrid operation. IEEE Trans. Smart Grid **3**(4), 1997–2006 (2013)
6. A.K. Basu, S. Chowdhury, Impact of strategic deployment of CHP-based DERs on microgrid reliability. IEEE Trans. Power Deliv. **25**(3), 1697–1705 (2010)
7. M.H. Moradi, M. Eskandari, S.M. Hosseinian, Operational strategy optimization in an optimal sized smart microgrid. IEEE Trans. Smart Grid **6**(3), 1087–1095 (2015)

8. S.A. Pourmousavi, M.H. Nehrir, Real-time central demand response for primary frequency regulation in microgrids. IEEE Trans. Smart Grid **3**(4), 1988–1996 (2012)
9. L. Che, M. Shahidehpour, D.C. Microgrids, Economic operation and enhancement of resilience by hierarchical control. IEEE Trans. Smart Grid **5**(5), 2517–2526 (2014)
10. N. Nikmehr, S. Najafi Ravadanegh, Probabilistic optimal scheduling of networked-microgrids considering time-based demand response programs under uncertainty. Appl. Energy **198**, 267–279 (2017)
11. N. Nikmehr, S. Najafi Ravadanegh, Solving probabilistic load flow in smart distribution grids using heuristic methods. J. Renew. Sustain. Energy **7**(4), 043138 (2015)
12. N. Nikmehr, S. Najafi Ravadanegh, Heuristic probabilistic power flow algorithm for microgrids operation and planning. IET Gener. Transm. Distrib. **9**(11), 985–995 (2015)
13. G.S. Kasbekar, S. Sarkar, *Pricing Games Among Interconnected Microgrids* (IEEE Power and Energy Society, New Jersey, 2012), pp. 1–8
14. M. Fathi, H. Bevrani, Statistical cooperative power dispatching in interconnected microgrids. IEEE Trans. Sustain. Energy **4**(3), 586–593 (2013)
15. E.J. Ng, R.A. EL-Shatshat, Multi-microgrid control systems (MMCS), in *IEEE Power and Energy Society General Meeting*, RI, USA, 2010
16. W.Y. Chiu, H. Sun, H.V. Poor, A multiobjective approach to multimicrogrid system design. IEEE Trans. Smart Grid **6**(5), 2263–2272 (2015)
17. Y. Cao, M. He, Z. Wang, T. Jiang, J. Zhang, Multiple resource expansion planning in smart grids with high penetration of renewable generation, in *Smart Grid Communications (SmartGridComm)*, Tainan, Taiwan (2012), pp. 564–569
18. N. Nikmehr, S. Najafi Ravadanegh, Optimal power dispatch of multi-microgrids at future smart distribution grids. IEEE Trans. Smart Grid **6**(4), 1648–1657 (2015)
19. M.A. Sofla, R. King, Control method for multi-microgrid systems in smart grid environment —stability, optimization and smart demand participation, in *Innovative Smart Grid Technologies (ISGT)*, Washington, DC, USA (2012)
20. G. Del Rosario Calaf, M. Cruz Zambrano, C. Corchero, R. Gumara Ferret, Distribution network congestion management by means of electric vehicle smart charging within a multi-microgrid environment, in *Electric Vehicle Conference (IEVC)*, Florence, Italy (2014), pp. 1–8
21. B. Dag, M.T. Aydemir, M.S. Smiai, Modelling and analysis of unsymmetrical multi-microgrid operation of active distribution networks, in *Power Engineering, Energy and Electrical Drives (POWERENG)*, Istanbul, Turkey, 13–17 May 2013
22. M. Marzband, N. Parhizi, M. Savaghebi, J.M. Guerrero, Distributed smart decision-making for a multimicrogrid system based on a hierarchical interactive architecture. IEEE Trans. Energy Convers. **31**(2), 637–648 (2016)
23. H. Keshtkar, J. Solanki, S. Khushalani Solanki, Analyzing multi-microgrid with stochastic uncertainties including optimal PV allocation, in *Smart Cities and Green ICT Systems (SMARTGREENS)*, Lisbon, Portugal (2015)
24. P. Li, X. Guan, J. Wu, D. Wang, An integrated energy exchange scheduling and pricing strategy for multi-microgrid system, in *IEEE Region 10 Conference (TENCON 2013)*, Xi'an, China (2013)
25. N.O. Song, J.H. Lee, H.M. Kim, Y.H. Im, J.Y. Lee, Optimal energy management of multi-microgrids with sequentially coordinated operations. Energies **8**(8), 8371–8390 (2015)
26. V.H. Bui, A. Hussain, H.M. Kim, Demand bidding and real-time pricing-based optimal operation of multi-microgrids. Int. J. Smart Home **10**(4), 193–208 (2016)
27. B. Kim, S. Bae, H. Kim, Optimal energy scheduling and transaction mechanism for multiple microgrids. Energies **10**(4), 1–17 (2017)
28. Q. Wang, P. Zhang, Energy management system for multi-microgrid, in *International Conference on Electricity Distribution (CICED 2014)*, Shenzhen (2014)
29. A.L. Dimeas, N.D. Hatziargyriou, Operation of a multiagent system for microgrid control. IEEE Trans. Power Syst. **20**(3), 1447–1455 (2005)

30. D. Gregoratti, J. Matamoros, Distributed energy trading: the multiple-microgrid case. IEEE Trans. Ind. Electron. **62**(4), 2551–2559 (2015)
31. V. Meyer, C. Myres, N. Bakshi, The vulnerabilities of the power-grid system: renewable microgrids as an alternative source of energy. J. Bus. Cont. Emerg. Plan. **4**(2), 142–153 (2010)
32. C. Gouveia, J. Moreira, C.L. Moreira, J.A. Pecas, Lopes, coordinating storage and demand response for microgrid emergency operation. IEEE Trans. Smart Grid **4**(4), 1898–1908 (2013)
33. A. Khodaei, Resiliency-oriented microgrid optimal scheduling. IEEE Trans. Smart **5**(4), 1585–1591 (2014)

Chapter 7
Resilient Optimal Power Flow with Evolutionary Computation Methods: Short Survey

Basar Baydar, Haluk Gozde, M. Cengiz Taplamacioglu and A. Osman Kucuk

Abstract Economic issues of power systems are formulated as optimization problems to enhance reliable operation and safe security of the real-time and hierarchical systems including complex control structures. The optimization problems have been formulated as combination of objective functions and constraints which Optimal Power Flow (OPF) must be increased to combine security constraints. The OPF problem is basically a network analysis challenge and the main objective of this challenge is to plan and to predict the undesirable situations that may arise by adding various assumptions to the account. This challenge can be solved using well-known numerical approaches, however these include derivatives and the solution of them is relatively difficult. However, the Evolutionary Computation (EC) based optimization algorithms provide more easy solutions for the OPF. In this chapter, the algorithms that contain the heuristic methods used on EC based algorithms and their applications on OPF are described.

Keywords Artificial intelligence · Evolutionary computation · Optimal power flow · Optimization · Reliability · Resilient · Search algorithm

---

B. Baydar (✉)
Baskent Electricity Distribution Company, Enerjisa, Ankara, Turkey
e-mail: basarbaydar@gmail.com

H. Gozde
Electronic Engineering Department, Military Academy,
National Defense University, Ankara, Turkey
e-mail: halukgozde@gmail.com

M. C. Taplamacioglu · A. O. Kucuk
Electrical and Electronics Engineering Department, Gazi University,
Ankara, Turkey
e-mail: taplam@gazi.edu.tr

A. O. Kucuk
e-mail: aliosmankucuk@gazi.edu.tr

© Springer Nature Switzerland AG 2019
N. Mahdavi Tabatabaei et al. (eds.), *Power Systems Resilience*, Power Systems,
https://doi.org/10.1007/978-3-319-94442-5_7

7.1 Introduction

It is utmost importance that the electrical infrastructure, which is responsible for providing electricity to the most important services of electricity and modern society, which has an important place in the functioning and development of today's world, is to be reliable and at the same time to operate safely. In recent years, the need for electric energy has increased day by day in parallel with technological developments, but the fact that the raw energy sources can not be actuated in the same way has made it necessary to benefit from the energy resources in the best way. In order to ensure that the power plants are dispersed in different regions and that optimum operation of the energy systems is ensured, interconnection networks of different power systems are formed. Nowadays, electricity energy exchanges between some countries have caused interconnected networks of these countries to connect with each other. Thus, the problems that arise during the planning and operation of networks, which are increasing in their qualities and sizes in the face of electrical engineers, have become increasingly complex and the use of computers has become compulsory in order to solve the emerging problems.

An electrical power system should not only be safe and reliable, but at the same time economically optimum and efficient. In other words, it does not reduce electricity generation costs and transmission losses the most. In general, several economic, operational or environmental objectives and constraints are met. There are two important concepts for an electric power system to operate satisfactorily and safely: these are reliability and safety concepts. The concept of reliability refers to the possibility that the power system continues to be successful and satisfactory in the long run [1]. The concept of safety is the ability to tolerate unexpected conditions without interruption of the risk level and supply of the power system. Security can also be defined as the ability of the supply to be protected in the most general sense [1].

Reliable operation of the electrical system is a real-time system that includes a hierarchical and complex control structure, with reliable accountability for the balance between electricity generation and consumption. While economic issues are included in the top layer of this system, power flow transfer is also formulated as optimization problems that should be optimal. This optimization problem can be formulated as different objective functions, constraint functions, or a combination thereof. The conventional objective function is minimization of production cost.

It is within the scope of the system security analysis work safely. It examines possible scenarios and includes corrective and preventive actions to alleviate them. If a system is said to be running in "N − 1" safety conditions under current operating conditions, the system can withstand a single chance without losing its ability to obtain full load and without violating any limits [1]. Safety and optimality are normally competing requirements. So it would not be appropriate to treat them separately [2]. Instead, the Optimal Power Flow (OPF) must be increased to combine security constraints.

In this section, the power system and its components will be described briefly. Then the Power Flow will be defined, and why the power flow is needed and power flow equations will be given together with defining the aim and purpose. OPF will be defined by making a transition from OPF to constraints and objective functions applied to power flow equations. In applying OPF, it will be explained both classical and intuitive methods, classical methods will be briefly described and heuristic methods will be emphasized. The algorithms that contain the heuristic methods used on OPF will be described and explained together with the examples in the literature. At the same time, the structure of the algorithms will be examined by mentioning the algorithms that are never used on OPF.

7.2 Electrical Power System

An electrical power system is a grid that consists of generation, distribution and transmission system. It completely uses the form of energy (like as diesel and coal etc.) and converts it into the electrical energy. The electrical power system consists of devices connected to the system such as a synchronous generator, motors, transformers, circuit breakers, conductors, relays etc. [3].

Power plants, transformers, transmission lines, sub-stations, distribution lines and transformers are the six major components of the power system. A power plant generates the power which is increased through the step-up transformer for transmission.

The transmission line conducts the power to the various sub-stations. Through sub-station, the power is conducted to the distribution transformer that step-down the power to the appropriate value which is suitable for the consumers. Finally, the appropriate level of electric energy could be presented to the consumers. All of these processes require a resilient and comprehensive planning, and a secure and reliable operation.

7.2.1 Purposes of Electrical Power System Planning

A power system needs an appropriate planning in order to evaluate and meet the future growing needs of the system in advance. The following issues should be considered in this plan [4];

- The cost of production, equipment, fuel and workforce need to be at minimum.
- The quality of the source needs to be proper.
- The growth in the system should be able to done without reducing the source's quality.
- The safety of the employee and system integration need to be ensured.

This purpose could be achieved with,

- Proper current and voltage ratios,
- Proper standards for illustration and design,
- Space between circuits, insulation and grounding compatibility, safety fuses, etc.

7.3 Power (Load) Flow

The growth and complexity of the electrical energy systems have led to the necessity of detailed studies in the planning stages. Inefficient planning and operation of a network leads to cost loss. With the development of the modern industry, renewed efforts are being made to generate electricity. The use of renewable resources such as wind and solar energy is also on the agenda. The acceleration and development of the electric industry is increasing parallel to the development of mathematics and the computer industry. The main way to solve any problem in a mixed system is to work on an analog or mathematical model. The main information obtained in load flow analysis studies is the amplitude, phase angle and active and reactive forces flowing in each line, as well as the optimal operation of existing power systems as well as the planning of future developments in systems. Previously power system analyzes and therefore load flow analyzes were carried out with AC. This process was quite annoying and time consuming. As a result of the rapid improvements in computers, previously used analysis methods had to leave their place to computer analysis methods. The speed, reliability and high accuracy of today's computers have led them to quickly become the most used tool in the analysis of power systems, especially load flow analysis. Numerical methods of analysis have come to the forefront as computers have begun to be used in the analysis of power systems [5].

7.3.1 Why the Power (Load) Flow Studies Are Needed?

First, the power flow problem emerged when different network configurations had to be planned for the expected loads for the future. Also, this problem then became an operational requirement for engineers and businesses to instantly follow the values of voltage and currents in the network.

7.3.2 Power (Load) Flow Problem Definition

The following definition of the load flow challenge is the most general definition, since it covers the simplest and most common cases in practice. The problem of load flow is to find the voltages and power flow in the electrical grid.

It has been accepted that the buses in the network are divided into two groups, the cargoes and the production lines to which the cargoes and the plants are connected. The active power (*P*) and reactive power (*Q*) in load buses, how the active power requirement of the network is distributed between the power plants and the voltage amplitudes of the plant are known. Therefore, these values are used as the data of the problem. What remains is how the complex tension in each bus is distributed to the reactor's reactive productions and to the lines drawn. These are the unknowns of the problem.

It is not necessary to know the other specifications of the plant and loads in the load flow study. These are represented as currents in the buses. In this way, the problem of the load flow is reduced to dissolve the circuit, which is known and unknown from the bus and lines of the network, having the node currents and voltages. The conditions that must be solved are the active and reactive powers and the voltage magnitudes of some buses. After the solution is obtained, these conditions determine whether the phase differences between voltage and lines, transformers, synchronous generators, voltage levels and voltages at each point of the system have been met. The load flow itself is the first step in other short circuit and stability operations than the benefits it provides.

7.3.3 Power Flow Analysis Data and Slack Bus

Y_{BUS} and Z_{BUS} matrices can be used in load flow analysis. Since Z_{BUS} matrix is more suitable for short circuit analysis, Y_{BUS} matrix is used in load flow analysis. By moving from the single line diagram of the system and taking into account the series impedances and shunt admittances of the transmission lines, the Y_{BUS} matrix can be obtained.

For each analysis, the working conditions must always be defined and the active power entering the grid in all other buses except one bus should be identified. The power drawn by the load is the negative power input to the system. The other input powers are the positive and negative powers coming from the generators and the system. Furthermore, the amplitude of reactive power or voltage flowing into the system must be defined in each of these buses.

It can be obtained solutions of the load flow problem via personal computers. A precise solution can be obtained within the calculation precision limits. At this point it is necessary to make an assumption to the above definition of the load flow. Even though it can be predicted very closely in practice, it is impossible to know fully the active production of all the plants in the network. That's why the line losses are unknown. Therefore, it is necessary to make one of the active bus powers unknown and to achieve this at the end of the solution. For this a generation bus is chosen and this bus is called the slack bus.

It is not compulsory to choose the release bus from the transmission lines. The active power variable and value of the oscillating bus equals the difference between the active generation of the other plants and the sum of the active loads and the active losses. It is useful to give a number to the oscillation bus when the buses in the network are numbered and to take the voltage there as a phase reference of the other voltages, although it is not obligatory for the solution.

The choice of the swing bus greatly affects the convergence in some cases. As a general rule, the oscillating bus is selected from the electrical center of the craft or from the buses to which it is connected by several lines. These solutions are completely empirical.

7.3.4 *Definition of Power Flow Problem*

Power flow problem requires knowledge of four variables in every k bus in the system, where,

P_k Real power (W)
Q_k Reactive power (VAr)
V_k Voltage magnitude (pu)
θ_k Phase angle (radians)

Solving the power flow problem, it is enough to know two of variables. Load flow application is to solve the other two variables. It should be defined three different buses based on fixed system frequency and stable voltage acceptance (Table 7.1).

Voltage Controlled Bus (Generator or PV Bus) Total P_k real power is specified. The voltage magnitude V_k is conserved at a certain value by the provision of reactive power. This type of bus usually corresponds to a generator or from static parallel capacitors; the synchronous compensators are a constant voltage source with a reactive power supply (auxiliary power stations).

Non-voltage Controlled Bus (Load or PQ Bus) In this bus, the total $P_k + jQ_k$ power is indicated. In physical power systems this corresponds to a center of

Table 7.1 Bus type variables [5]

Bus type	Known variables	Unknown variables
Generator (PV)	$\|V_i\|$, P_i	Q_i, δ_i
Load (PQ)	P_i, Q_i	$\|V_i\|$, δ_i
Slack (Reference)	$\|V_i\|$, δ_i	P_i, Q_i

gravity, such as a city or an industrial center. It is assumed here that small changes in the buckle voltage do not affect the P_k and Q_k forces.

Slack (Reference) Bus Since the system losses are not known, the advantage of load current calculation arises. The total power given for this reason is not mentioned in every bus. In general, one of the suitable voltage control buses is selected as the free bus. Free break voltage is taken as reference.

It is derived from nonlinear loop power equations for two unknown variables at each node of the load flow system. The iteration method is applied to this group of linear equations.

System data such as bus power ratings, network connections, admittance and impedance are read. Initial voltages are assigned to all buses for basic load flow. When *P*, *V* are set to 1, and *P*, *Q* are set to 1 + *j*0.

The iteration is terminated when the bus voltage and angles provide determined power and production. This requirement is acceptable for all buses when the power dispute is smaller than a tolerance value and the voltage increases are smaller than a defined error value. When this result is achieved, the power requirements are calculated for all the buses. Then line power flows, losses and system totals are calculated.

7.3.5 Methods of Power Flow Solution

Methods used in power flow solution in electric power networks are divided into two;

- Direct Methods
- Iterative Methods.

7.3.5.1 Direct Methods

In order to find the voltages of the busbar in the power systems with current information, it is necessary to take the inverse of the [*Y*] admittance matrix in Eq. (7.1):

$$[V] = [Y]^{-1} \times [I] = [Z] \times [I] \tag{7.1}$$

7.3.5.2 Iterative Methods

Among the many iterative solution methods, well-known Gauss-Seidel and Newton-Raphson methods are usually used in Power Flow applications.

7.4 Optimal Power (Load) Flow

OPF is described as the sharing of production to the generators in the system and the optimal power exchange between the buses, without exceeding the physical limits of the equipment used in the power systems and operating limits. It is the minimization of the energy production cost of the energy system by providing the objective equality and inequality constraints of the OPF problem. These constraints can include;

- Active output powers of generator buses (except slack bus),
- Voltage amplitude values of generator buses,
- Transformer tap changing level values,
- Shunt capacity values.

Particularly, OPF is the one of the four main elements of power system analysis as depicted in Fig. 7.1.

7.4.1 Original Optimal Power Flow

OPF was firstly introduced by Carpentier in 1962 [6]. Generally, the OPF is a nonlinear and non-convex problem including an optimal flow which must be optimized (maximized or minimized), a set of equality and inequality constraints which must be satisfied, and a problem solving method [7, 8].

In other words, OPF optimizes a given optimal flow controlling power flow within an electrical system without violating power flow constraints or operational limits [9, 10]. In fact, it determines the optimal operation state for the system. Unlike the conventional power flow, OPF works with an under-constrained network [11]. Also, it can provide a useful support to the operator to overcome many difficulties in the planning, operation and control of power networks [12].

A variety of extended OPF versions has been reported in the related literature so far. Some of them are as follows:

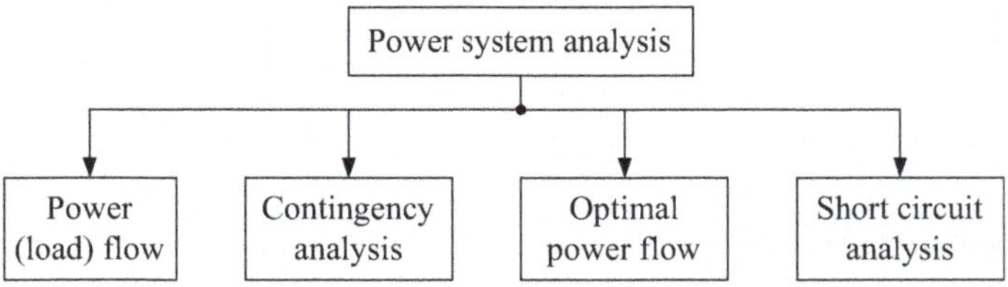

Fig. 7.1 Main elements of power flow analysis

7.4.1.1 Static OPF

This type of OPF method optimizes optimal flow under various constraints at a certain time of interest. In other words, it can only handle one load level at a specific time [13]. It can be named classic OPF method.

7.4.1.2 Dynamic OPF

This type of OPF method is a prolonged version of the static OPF and determines the optimal operating point over a time horizon. That is, it covers multiple time periods [14, 15]. The time periods can be days, weeks, months or years. The static OPF is run for each period taking into account the changes in these periods.

7.4.1.3 Transient Stability-Constrained OPF

This type of OPF method considers static and dynamic constraints of the power network during the optimization process simultaneously [16]. Under this condition, the system can withstand severe contingencies [17]. The static OPF method is run in these conditions.

7.4.1.4 Security-Constrained OPF

This is another extended version of the OPF that involves constraints arising from the operation of the system under a set of postulated contingencies as similar to the transient stability-constrained OPF.

7.4.1.5 Deterministic OPF

This widely used type of OPF method does not consider stochastic factors. The static OPF method can be named deterministic OPF.

7.4.1.6 Stochastic OPF

This type of OPF method considers uncertainties in power system parameters [18–20]. That is, it regards the uncertainty as a section of the constraints and objective models. Hence, the optimization process as well as the final OPF results can be affected uncertain factors [21]. For example, the solar and wind uncertainties are applied to the power system analysis with the stochastic OPF method. These uncertainties can be taken into account by using probability distribution functions such as Weibull, Rayleight or the other probability distribution functions.

7.4.1.7 Probabilistic OPF

It estimates the possibility distribution functions of dependent variables based on the possibility distributions of loads and another uncertain factors through using Monte-Carlo Simulation [22], Cumulant method [23], Point-Estimate Method (PEM) [24], customized Gaussian mixture model [25], and etc. In this type of OPF method, the uncertain factors do not affect the final results of the analysis [21].

7.4.1.8 AC OPF

This is associated with the AC power networks and is based on the natural power flow characteristics of the power system [26]. Consequently, the results obtained by this type of OPF method are more accurate for AC power systems [27, 28].

7.4.1.9 DC OPF

This type of OPF method does not consider the reactive power and transmission losses [26]. It can be used to DC analysis of the power system. Also it can be applied to the HVDC systems.

7.4.1.10 Mixed AC/DC OPF

It is associated with OPF analysis in both AC and DC grids [29, 30].

7.4.2 Purpose of Optimal Power Flow

Before creating an OPF system, the data needed to perform OPF should be considered. The first goal of a comprehensive OPF is to meet the load demand for a power system while protecting the safety of the system, least costly. The second objective of the OPF is to determine the marginal cost data. Some general objectives of OPF can be defined as follows.

Active Power Goals;

- Economic Distribution (minimum cost and loss, MW generation and transmission losses)
- Environmental Distribution (CO_2 emission)
- Maximum Power Transfer.

Reactive Power Goals;

- Minimize MVAR losses.

7.4.3 *Optimal Power Flow Problem*

OPF problem is expressed as a limited and non-linear optimization problem and is formulated as below;

- $f(x, u) = 0$ (objective function)
- $g(x, u) = 0$ (equality constraints)
- $h(x, u) \leq 0$ (inequality constraints)

where,

- $f(x, u)$ is the desired function to find the minimum value.
- $g(x, u)$ represents power flow equations.
- $h(x, u)$ represents security limit values.

and x and u denote the state and control variables, respectively.

State variables in the energy system are an active output power of slack bus, the voltage amplitude values of load buses, and the reactive output powers of the generator buses as depicted in Eq. (7.2).

$$x = \left[P_{swing}, V_L, Q_g\right] \tag{7.2}$$

Control variables in energy systems are the active output powers of the generator buses except for the swing bus, the voltage amplitude values of the generator buses, the transformer tap changing values and the shunt capacity values as represented in Eq. (7.3).

$$u = \left[P_g, V_g, T, Q_c\right] \tag{7.3}$$

To deliver optimal actual power, the aim function f is the whole cost of production like results in Eq. (7.4):

$$F_{\cos t} = \sum_{i=1}^{N_g} \left(\alpha_i + \beta_i.P_{gi} + \gamma_i.P_{gi}^2\right) \tag{7.4}$$

where,

N_g total number of generators in the system
P_{gi} active power of generators in the system
α_i, β_i, γ_i generator fuel cost coefficients.

In the equations below, active and reactive power at *k* bus taken from the system are given in Eqs. (7.5) and (7.6).

$$P_k = 0 = V_k \sum_{m=1}^{N} [V_m.[g_{km}.\cos(\delta_k - \delta_m) + b_{km}.\sin(\delta_k - \delta_m)]] - P_{GK} + P_{LK} \quad (7.5)$$

$$Q_k = 0 = V_k \sum_{m=1}^{N} [V_m.[g_{km}.\sin(\delta_k - \delta_m) - b_{km}.\cos(\delta_k - \delta_m)]] - Q_{GK} + Q_{LK} \quad (7.6)$$

The required generator active power, generator reactive power, busbar voltage amplitude, transformer step value and shunt capacity limit values are shown in the following equations:

- The active power generated by the generator must be between the specified min and max production capacity values.

$$P_{Gk}^{\min} \leq P_{Gk} \leq P_{Gk}^{\max} \quad (7.7)$$

- The reactive power transmitted by the generator to the system must be between the specified min and max production capacity values.

$$P_{Gk}^{\min} \leq P_{Gk} \leq P_{Gk}^{\max} \quad (7.8)$$

- The value of the voltage at the *k* bus must be between the min and max voltage values specified for that bus.

$$V_k^{\min} \leq V_k \leq V_k^{\max} \quad (7.9)$$

- The phase angle of the k bus must be between the min and max values specified for each bus.

$$\phi_k^{\min} \leq \phi_k \leq \phi_k^{\max} \quad (7.10)$$

- The power carried in the transmission line must not exceed the max power carrying capacity of the transmission line.

$$S_{km} \leq S_{km}^{\max} \quad (7.11)$$

- Transformer step rates should be between the min and max values specified for each transformer.

$$T_k^{\min} \leq T_k \leq T_k^{\max} \tag{7.12}$$

- The shunt capacitors to be energized must be within the specified limits.

$$Q_{Ck}^{\min} \leq Q_{Ck} \leq Q_{Ck}^{\max} \tag{7.13}$$

7.4.4 Solution Methods of Optimal Power Flow Problem

The solution methods of OPF problem are generally divided into two as the traditional methods and the artificial intelligence based methods as represented in Fig. 7.2.

7.4.4.1 Traditional Methods

Traditional methods are called deterministic optimization methods and generally involve the methods depicted in Fig. 7.3. These methods are based on mathematical programming. While excellent progress has been made in classical methods, the following disadvantages can be presented [31];

Disadvantages:

- Required linearization
- Required differentiability
- May get stuck at local optima
- Poor convergence
- Weak in handling qualitative constraints
- If number of variables are large, become too slowly
- In a single simulation run, can find only a single optimized solution.

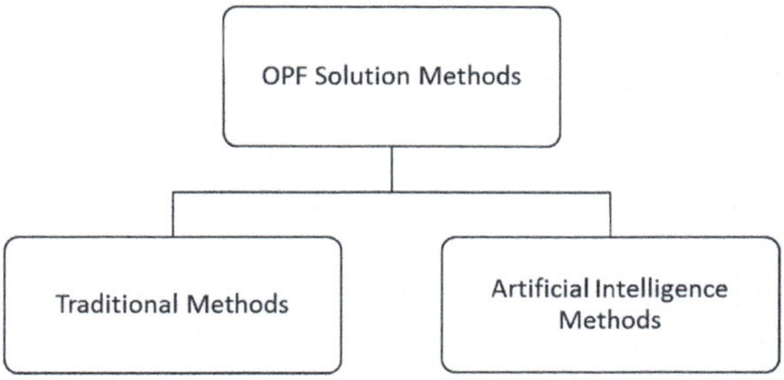

Fig. 7.2 Classification of OPF solution methods [64]

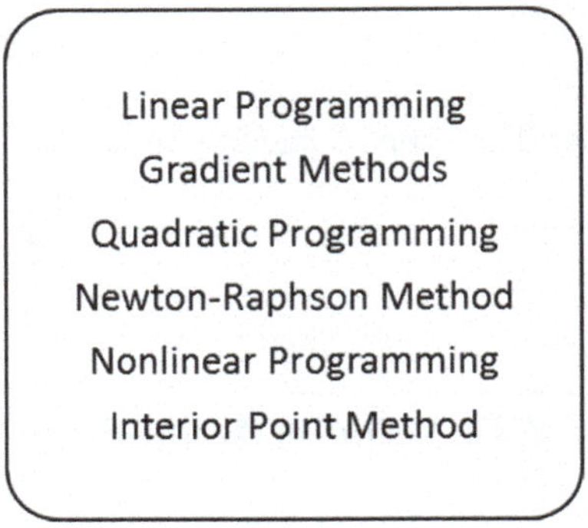

Fig. 7.3 Traditional solution methods

7.4.4.2 Artificial Intelligence (AI) Based Methods

Artificial Intelligence (AI) which is excited the human being since ancient Egypt can be defined in general as the intelligence which the machines have. Although, the philosophers have explored the thought structure of human intelligence throughout history, the first applicable scientific studies about AI has started by Alan Mathison Turing at USA in 1943 because of the crypto analysis requirements in World War 2. On the other hand, it could only be named as artificial intelligence term at first at Dartmouth College in USA in 1956 [32].

At first glance, it can be considered that the intelligence can only belong to the humans and animals because the intelligence is needed some advanced tools such as eyes for sensing, nerve system for transmitting the sensed knowledge, brain for evaluation and interpretation of these knowledge and mouth or arms for presenting the results and these could only be belonged to the humans and animals. But, modeling these tools as software and hardware has offered an opportunity to realize artificial intelligence to the scientists. These have called intelligent agents.

The intelligent agents which are perceived their environment and are improved their behavior that maximizes its success chance are the basic block of AI as depicted in Fig. 7.4. The intelligent agent can sense its environment by using the vision and language processing tools like as seeing and hearing by eyes and ears of human being. It can process and interpret the sensed knowledge by using machine learning tools. After that, it can develop its appropriate behavior by its expert systems for succeed. Finally, it can transform this behavior to the work by using its speaking or robotic tools. From this point of view, the AI can be defined as the imitation of the human's or animal's intellectual behavior.

Recently, the AI is historically divided into three groups in general: (1) traditional statistical methods, (2) traditional symbolic AI, and (3) computational intelligence as represented in Fig. 7.5 [33, 34]. The first group includes well-known algebraic statistical methods such as regression analysis. They have deterministic approaches only depended upon the given inputs. Traditional symbolic AI has classic logic structures such as basic logic gates and combinational or sequential logic circuits.

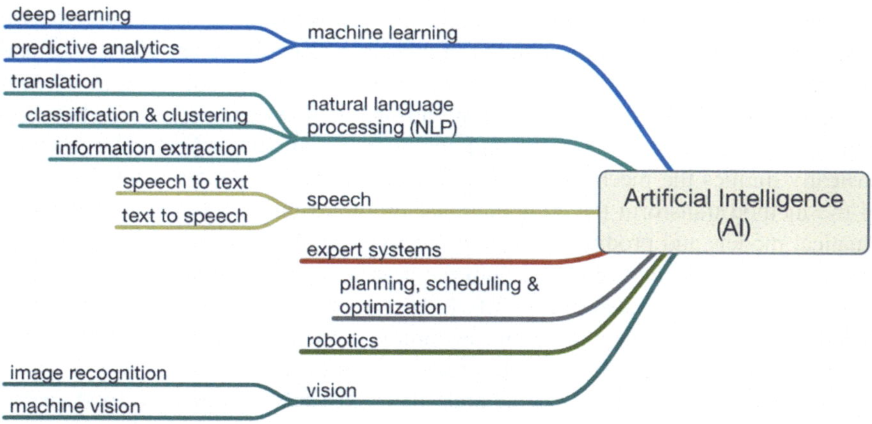

Fig. 7.4 The intelligent agent of AI [65]

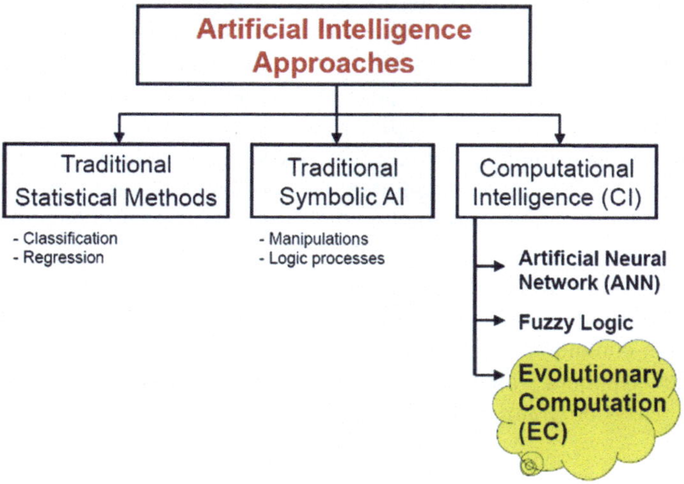

Fig. 7.5 Three approaches of AI

The last group of AI approaches is called computational intelligence (CI) and involves some heuristic methods apart from the other two approaches. Particularly, the artificial neural networks can reach to the results by generalizing among the little information about the problem [35]. Fuzzy logic uses linguistic expression to account for the intermediate values as different from the classic logic [36]. Evolutionary computation (EC) which is the area of CI using idea of biological evolution is also one of the heuristic methods of AI. The underlying behind this method is natural selection: *given a population of individuals the environmental pressure causes natural selection (survival of the fittest) and this causes a rise in the fitness of the population* [37]. This clearly means "optimization" in mathematics.

The EC methods are which is based on 1950s generally used in order to optimize computational problems and separated in two such as evolutionary algorithms and swarm intelligence according to their search principle [38]. The concept of evolutionary algorithms has introduced in terms of developing genetic algorithm optimization method by Holland and his students in 1960s [39]. This method basically mimics the mechanisms of evolution and natural genetics. First versions of the method transform the computational problems into the binary coded mathematical models and produce the results by applying the genetic operators such as selection, crossover and mutation onto the binary coded candidates to minimize a fitness function as depicted in Fig. 7.6. In later versions of the method such as real coded genetic algorithm, differential evolution, differential search algorithm etc., these operators can be applied to the real numbers [40].

The swarm intelligent based methods have basically inspired by food search behavior of the bird flocks, fish schools or ant colonies. In these type optimization algorithms, these decentralized, collective and self-organized behaviors are modeled mathematically and then, their results are updated iteratively by minimizing a fitness function produced according to the nature of the problem. The intelligent agents (birds, fishes, ants, bacteria etc.) interact locally with one another and with their environment. This interaction finds out complex global behavior and swarm intelligence to realize the target purpose [41]. After its introduction by Beni and Wang in 1989 in the context of cellular robotic systems [42], a lot of different algorithms such as ant colony optimization algorithm, particle swarm optimization algorithm, artificial bee colony algorithm, cuckoo search algorithm etc. have been improved by the researchers inspired by different behaviors of different animals or living organisms so far.

The actual factor in the emergence of these optimization algorithms is that the classical optimization techniques have limited in practical applications which includes non-continuous and non-differentiable natural functions. The swarm intelligence techniques which have completely algebraic update functions and also, short and easy program code can cope with these challenges successfully.

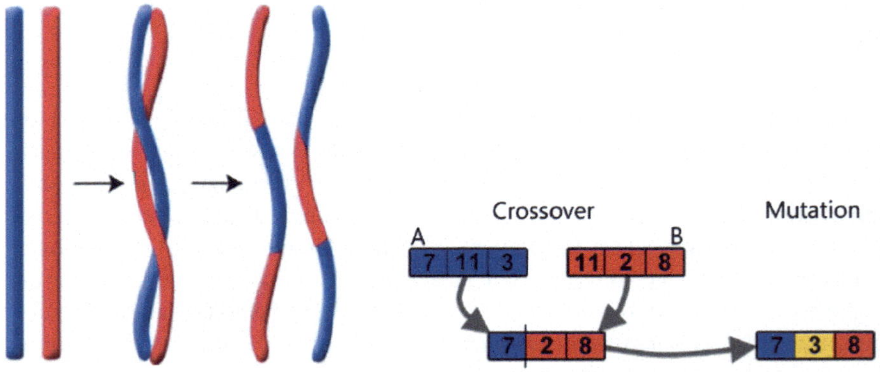

Fig. 7.6 Genetic operators [40]

Additionally, their small memory needs and pure algebraic contents allow easy implementation in simple microprocessors depend on a process. The swarm intelligence based optimization algorithms have been used in different scientific disciplines and applications so far, and one of them is OPF problem [43]. The advantages and disadvantages of these methods are:

Advantages:

- No possibilities for problem area
- Completely applicable
- Low development and application costs
- Easy to incorporate other optimization methods
- Solutions are interpretable
- It is to provide so many alternative answers.

Disadvantages:

- Not guarantee for optimal solution within finite time
- Poor theoretical principle
- Need parameter regulation
- Often computationally expensive or/and slow.

7.4.5 Comparison of Evolutionary Computation Based Algorithms and Classical Optimization Methods

- Adaptation to mathematical formulation changes in classical optimization methods is rather difficult, but in the EC based algorithm the parameters can be easily modified and adapted.
- While there is a need to understand mathematical expressions in classical optimization methods, there is no need to understand such expressions in EC based algorithms.
- In terms of probing solution search, EC based algorithms are easier to solve than many alternative points, while starting from a single point to search for classical methods.
- As a further advantage, no assumptions need to be made in EC based algorithms while classical methods use a number of additional data such as derivatives of the objective function.

7.5 OPF and Evolutionary Computation Methods

In this section, the some used and still unused EC based optimization methods for OPF are briefly investigated from literature.

7.5.1 Most Used Evolutionary Computation Based Optimization Algorithms for OPF Problem

Most used EC based optimization algorithms determined from the literature for OPF problem can be listed as follows (Table 7.2).

7.5.2 Literature Overview for Most Used Evolutionary Computation Based Algorithms

The most commonly used EC algorithms in the literature can be summarized in Table 7.3.

Table 7.2 Most used EC based optimization algorithms [66]

Particle swarm optimization algorithm
Artificial bee colony algorithm
Ant colony optimization
Electromagnetism-like algorithm
Cuckoo search optimization algorithm
Flower pollination optimization algorithm
Harmony search optimization algorithm
Bat optimization algorithm
Firefly optimization algorithm
Spider optimization algorithm
Collective animal behavior algorithm
Colonel selection optimization algorithm
Cultural optimization algorithm
Genetic expressing programming
Genetic optimization algorithm
Genetic programming
Evolutionary programming
Evolutionary strategy
Differential evolution optimization algorithm
Differential search optimization algorithm
Grammatical evolution optimization algorithm
Learning classifier optimization algorithm
Gene expression programming
Non dominated-sorting genetic algorithm
Strength-Pareto evolution algorithm
Simulated annealing
Tabu search algorithm

Table 7.3 The brief literature study of most used EC based optimization algorithms

References	Publication name	Publication date	Used algorithms	Compared algorithms
[67]	Improved genetic algorithms for optimal power flow under both normal and contingent operations states	1997	Improved genetic algorithms	Gradient based conventional method
[68]	Optimal power flow using particle swarm optimization	2001	Particle swarm optimization algorithms	Evolutionary programming Genetic algorithms
[69]	Optimal power flow using Tabu search algorithm	2001	Tabu search algorithms	Evolutionary programming Nonlinear programming
[70]	Optimal power flow of the Algerian electrical network using an ant colony optimization method	2005	Ant colony optimization algorithms	Genetic algorithms
[71]	Modified differential evolution algorithm for optimal power flow with non-smooth cost functions	2007	Improved differential evolution	Genetic algorithms Evolutionary programming Particle swarm optimization Simulated annealing Tabu search, differential evolution
[72]	Optimal power flow using differential evolution algorithm	2008	Differential evolution algorithm	–
[73]	An improved particle swarm optimization algorithm for optimal power flow	2009	Improved particle swarm optimization algorithm	Particle swarm optimization Genetic algorithms
[74]	Optimal power flow by a fuzzy based hybrid particle swarm optimization approach	2009	Fuzzy based hybrid particle swarm optimization	Particle swarm optimization (Local random search) Particle swarm optimization Evolution programming

(continued)

Table 7.3 (continued)

References	Publication name	Publication date	Used algorithms	Compared algorithms
[75]	Particle swarm optimization applied to optimal power flow solution	2009	Particle swarm optimization algorithms	Matpower
[76]	A solution to the optimal power flow using artificial bee colony algorithm	2010	Artificial bee colony algorithm	Particle swarm optimization Genetic algorithms
[77]	Application of biogeography-based optimization to solve different optimal power flow problems	2011	Biogeography-based optimization algorithm	Particle swarm optimization Genetic algorithms Evolutionary programming Differential evolution Gradient method
[78]	Application of particle swarm optimization to optimal power systems	2011	Particle swarm optimization algorithm (using loss minimization)	Interior point algorithm
[79]	Optimal power flow using gravitational search algorithm	2011	Gravitational search algorithms	Other methods in literature
[80]	A solution to multi-objective optimal power flow using hybrid cultural-based bees algorithm	2012	Cultural-based bees algorithm	Artificial bee colony algorithm Particle swarm optimization Genetic algorithms
[81]	Optimal power flow with emission controlled using firefly algorithm	2013	Firefly algorithms	Genetic algorithms Particle swarm optimization algorithms
[82]	Temperature dependent optimal power flow using GBest-guided artificial bee colony algorithm	2014	GBest-guided artificial bee colony algorithm	Gravitational search algorithms

(continued)

Table 7.3 (continued)

References	Publication name	Publication date	Used algorithms	Compared algorithms
[83]	Optimal power flow using glowworm swarm optimization	2015	Glowworm swarm optimization Particle swarm optimization	Glowworm swarm optimization Particle swarm optimization
[84]	Optimal power flow using moth swarm algorithm	2016	Moth swarm algorithms	Modified particle swarm optimization Modified differential evolution Moth flame optimization Flower pollination algorithm

7.5.3 Some Unused EC Based Algorithms for Optimal Power Flow

- *Population Based Incremental Learning Algorithms (PBIL)*

A Population Based Incremental Learning (PBIL) algorithm proposed firstly Baluja in 1994 [44]. It is a method that combines the mechanisms of a general genetic algorithm with simple competitive learning. In dynamic environments, for improving adaptability, a PBIL-specific combinatorial memory scheme has been explored, which stores the best solutions and relevant environmental information in memory [45].

- *Tabu Search Continuous Optimization (TSC)*

Tabu Search Continuous Optimization algorithm (TSC) is a meta-heuristic originally developed by Glover in 1989 [46, 47]. In this algorithm in general, at a local minimum the cue performs to accept some non-cue points from that point to allow the search to search for new areas of the area.

- *Firework Algorithm*

In the sky at night, inspired by fireworks explosions, the fireworks algorithm (FWA) was proposed in 2010 to Verlag Berlin Heidelberg, in swarm intelligence algorithms, through the investigation of the fact that fireworks explosion is like to the way a singular researches for optimal solution [48].

- *Raindrop Optimization*

The Raindrop Optimization Algorithm was created by the treatment of rain droplets and proposed by Shah-Hosseini in 2009. When the rain falls to the ground, it normally flows downwards from the normal due to gravity; the landscape selects the optimum path to the lowest point [49].

- *Bayesian Optimization Algorithms*

The Bayesian Optimization Algorithm (BOA) was proposed by Pelikan, Goldberg, and Cantu-Paz. They had associated the opinion of using probabilistic models for guiding optimization and the methods to learn and sample Bayesian networks. BOA builds a Bayesian network for the set of promising solutions for learning an adequate decomposition of the problem. By sampling the built network, new candidate solutions are generated [50].

- *The Wind Driven Optimization Algorithm*

The Wind Based Optimization (WDO) technique is an iterative and it is also heuristic global optimization algorithm with application potentials for search area constraints for multi-domain and multi-mode problems and also, was created and maintained by Zikri Bayraktar in 2010 [51].

- *Normalized Normal Constraint Algorithm*

The Normalized Constraint (NC) technique generates a set of equally spaced solutions on a Pareto frontier for multi-objective optimization challenge and also proposed first, by A. Mesac, A. Ismail-Yahya and C. A. Matsson in 2003 [52].

- *Binary Bat Algorithm*

Bat algorithm (BA), mimicking the behavior of echo detection bat to perform global optimization which is one recently proposed heuristic algorithms and was proposed by Yang [52].

- *Hill Climbing Optimization Algorithm*

It is one of the search algorithms used in computer science. It gets the name from the hills in the graph where the search is made. Simply looking at the lowest point in a graph, the movement in the graph is actually similar to the climbing of the hill and was searched by Davis, in 1991 [53].

- *Perception Learning Algorithm*

The Perception Learning Algorithm was essentially built up by Fr. Rosenblatt in 1950s and is used to predict the outcome of new future data using observed data. The algorithm first takes any vector in space and looks at whether it provides complete separation for all data. If it does, it scrolls the vector and repeats it until it is available for all the data. If it is not available with any vector it will continue forever [54].

- *Self-organizing Map Algorithm*

Kohonen, named for the first time is developed by Finnish scientists map the Kohonen (Kohonen map). The name given to these networks operate in two different ways, like all other artificial neural networks [55].

- *Adaptive Random Search Optimization Algorithm*

Adaptive random search algorithm, known as the Global Optimization and Stochastic Optimization is a set of general approaches. Adaptive random search optimization technique has been revised by Kregting and White and assert an approach named Adaptive Random Search Optimization Algortihm [56]. It does not require the derivative financial instruments in the search field to navigate directly to a search method [57].

- *Stochastic Hill Climbing Optimization Algorithm*

The Stochastic Hill Climbing algorithm is a Stochastic Optimization algorithm and is a Local Optimization algorithm (contrasted to Global Optimization). Because it does not require derivatives of the search field, it is a direct search technique. Stochastic Hill Climbing is an extension of deterministic hill climbing algorithms such as Simple Hill Climbing (first-best neighbor), Steepest-Ascent Hill Climbing (best neighbor), and a parent of approaches such as Parallel Hill Climbing and Random-Restart Hill Climbing [57].

- *Iterated Local Search Optimization Algorithm*

Iterative Local Search (ILS) is a search algorithm that produces a series of solutions generated by an embedded heuristic and also was presented in 2007, by Ruiz and Stutzle [58].

- *Guided Local Search Optimization Algorithm*

Guided Local Search (GLS) is a meta-heuristic method proposed to solve combinatorial optimization problems. It is a high level strategy, applies an efficient penalty-based approach to interact with the local improvement procedure [59].

- *Extremal Optimization Algorithm*

A new heuristic approach that combines the modularity and community fitness and uses extremal optimization algorithm (EO) as an underlying method has been proposed Bak and Sneppen [60].

- *Cross-Entropy Optimization Algorithm*

Cross-Entropy Optimization Method is used to find a way out difficult estimation and optimization problems used. The Kullback-Leibler (or cross entropy) is a versatile intuitive tool based on the most downsizing of works and this method was given in de Boer et al. in 2005 [61].

- *Negative Selection Optimization Algorithm*

Negative Selection Algorithm (NSA) in Artificial Immune Systems (AIS) is the main method. It was inspired by self and non-self-discrimination process observed in the Mammalian Immune System (MIS) and was proposed by Forrest in 1994 [62].

- *Continuous Scatter Search Optimization Algorithm*

Continuous Scatter Search Memory is a population-based approach based on spatial combination ideas that will be empowered to exploit designs and was first introduced in Glover in 1977 [49].

- *Variable Neighborhood Search Optimization Algorithm*

It was discovered in 1997 by Mladenovi'c and Hansen. The main idea of this meta-intuitive is the systematic change of neighbors when search regions are used [63].

Other optimization algorithms can be listed as below;

- Swine Flow Optimization Algorithm
- Lloyd's Algorithm
- Huffman Algorithm
- Global Neighborhood Algorithm
- Scatter Search Optimization Algorithm
- Alternating Conditional Expectation Algorithm
- Immune Network Optimization Algorithm
- Dendritic Cell Optimization Algorithm
- Non-dominated Sorting Genetic Optimization Algorithm
- Univariate Marginal Distribution Optimization Algorithm
- Greedy Randomized Adaptive Search Optimization Algorithm
- Generative Algorithms
- Active-set Algorithm
- Charged System Search Optimization Algorithm.

References

1. M. Zima, M. Bockarjova, *Operation, Monitoring and Control Technology of Power Systems*, Lecture Notes 227-0528-00 (ITET ETH, Zurich, 2007)
2. B. Stott, O. Alsac, A.J. Monticelli, Security analysis and optimization. IEEE Invited Pap. **75** (12), 1623–1644 (1986)
3. http://circuitglobe.com/power-system.html
4. P. Schavemaker, L. Van der Sluis, *Electrical Power System Essentials* (Wiley, 2017)
5. http://www.yildiz.edu.tr/~inan/LU_Hesap/Yuk_Akisi_Genel_Bilgi.doc
6. J. Carpentier, Contribution to the study of the economic dispatching. Bull La Societe Fr Des Electr **3**, 431–447 (1962)

7. B. Ghaddar, J. Marecek, M. Mevissen, Optimal power flow as a polynomial optimization problem. IEEE Trans. Power Syst. **31**, 539–546 (2016)
8. J. Lin, V.O.K. Li, K.C. Leung, A.Y.S. Lam, Optimal power flow with power flow routers. IEEE Trans. Power Syst., 1–13 (2016)
9. S. Frank, I. Steponavice, S. Rebennack, Optimal power flow: a bibliographic survey-I: formulations and deterministic methods. Energy Syst. **3**, 221–258 (2012)
10. K.C. Almeida, A. Kocholik, Solving III-posed optimal power flow problems via Fritz-John optimality conditions. IEEE Trans. Power Syst. **1–10** (2016)
11. A. Vaccaro, C.A. Canizares, A knowledge-based framework for power flow and optimal power flow analyses, IEEE Trans. Smart Grid **1–11** (2016)
12. V. Radziukynas, I. Radziukyniene, *Optimization Methods Application to Optimal Power Flow in Electric Power Systems* (Springer, Berlin Heidelberg, 2009)
13. A.M. Shaheen, R.A. El-Sehiemy, S.M. Farrag, Solving multi-objective optimal power flow problem via forced initialised differential evolution algorithm, **10**, 1634–1647 (2016)
14. S. Gill, I. Kockar, G.W. Ault, Dynamic optimal power flow for active distribution networks. IEEE Trans. Power Syst. **29**, 121–131 (2014)
15. T. Niknam, M.R. Narimani, M. Jabbari, Dynamic optimal power flow using hybrid particle swarm optimization and simulated annealing. Int. Trans. Electr. Energy Syst. **23**, 975–1001 (2013)
16. Y. Xu, J. Ma, Z.Y. Dong, D.J. Hill, Robust transient stability-constrained optimal power flow with uncertain dynamic loads. IEEE Trans. Smart Grid, 1–11 (2016)
17. Y. Xu, Z.Y. Dong, Z. Xu, R. Zhang, K.P. Wong, in *Power System Transient Stability Constrained Optimal Power Flow: A Comprehensive Review*, IEEE Power Energy Society General Meeting, San Diego, CA (2012), pp. 1–7
18. M. Perninge, C. Hamon, A stochastic optimal power flow problem with stability constraints—part II: the optimization problem. IEEE Trans. Power Syst. **28**, 1849–1857 (2013)
19. A. Vaccaro, C. Canizares, An Affine arithmetic-based framework for uncertain power flow and optimal power flow studies. IEEE Trans. Power Syst. **1–15** (2016)
20. M. Bazrafshan, N. Gatsis, Decentralized stochastic optimal power flow in radial networks with distributed generation. IEEE Trans. Smart Grid, 1–15 (2016)
21. J. Gong, D. Xie, C. Jiang, Y. Zhang, A new solution for stochastic optimal power flow: combining limit relaxation with iterative learning control. J. Electr. Eng. Technol. **9**, 80–89 (2014)
22. H. Zhang, P. Li, Probabilistic analysis for optimal power flow under uncertainty. IET Gener. Transm. Distrib. **4**, 553–561 (2010)
23. A. Schellenberg, W. Rosehart, J. Aguado, Cumulant-based probabilistic optimal power flow (P-OPF) with gaussian and gamma distributions. IEEE Trans. Power Syst. **20**, 773–781 (2005)
24. G. Verbic, C.A. Canizares, Probabilistic optimal power flow in electricity markets based on a two-point estimate method. IEEE Trans. Power Syst. **21**, 1883–1893 (2006)
25. D. Ke, C.Y. Chung, Y. Sun, A novel probabilistic optimal power flow model with uncertain wind power generation described by customized gaussian mixture model. IEEE Trans. Sustain Energy **7**, 200–212 (2016)
26. M. Oua, Y. Xue, X.P. Zhang, Iterative DC optimal power flow considering transmission network loss. Electr. Power Compon. Syst., 1–11 (2016). http://dx.doi.org/10.1080/15325008.2016.1147104
27. V. Sarkar, S.A. Khaparde, Optimal LMP decomposition for the ACOPF calculation. IEEE Trans. Power Syst. **26**, 1714–1723 (2011)
28. T. Akbari, M. Tavakoli Bina, Linear approximated formulation of AC optimal power flow using binary discretisation. IET Gener. Transm. Distrib. **10**, 1117–1123 (2016)
29. W. Feng, L.A. Tuan, L.B. Tjernberg, A. Mannikoff, A. Bergman, A new approach for benefit evaluation of multiterminal VSC-HVDC using a proposed mixed AC/DC optimal power flow. IEEE Trans. Power Deliv. **29**, 432–443 (2014)

30. S. Bahrami, F. Therrien, V.W.S. Wong, J. Jatskevich, Semidefinite relaxation of optimal power flow for AC-DC grids. IEEE Trans. Power Syst., 1–16 (2016)
31. A.K. Khamees, N.M. Badra, A.Y. Abdelaziz, Optimal power flow methods: a comprehensive survey. Int. Electr. Eng. J. (IEEJ) **7**(4), 2228–2239 (2016)
32. T. Lewis, *A Brief History of Artificial Intelligence, Live Science* (22 June 2015)
33. S.S. Rao, *Engineering Optimization: Theory & Practice*, 4th edn. (Wiley, 2009)
34. E.K.P. Chong, S.H. Zak, *An Introduction to Optimization*, 2nd edn. (Wiley, 2001)
35. W.S. McCulloch, W. Pitts, A logical calculus of the ideas immanent in nervous activity. Bull. Math. Biophys. **5**(4), 115–133 (1943)
36. L. Zade, Fuzzy set as a basis for theory of possibility. Fuzzy Sets Syst. **1**(2), 3–28 (1976)
37. R.B. Devi, E. Barlaskar, O.B. Devi, S.P. Medhi, R.R. Shimray, Survey on evolutionary computation techniques and its application in different fields. Int. J. Inf. Theory **3**(3) (2014)
38. Y. Liu, K.M. Passino, *Swarm Intelligence: Literature Overview* (The Ohio State University Dept. of Electrical Engineering, 2000)
39. J.H. Holland, *Adaptation in Natural and Artificial Systems*, 1st edn. (University of Michigan Press, Cambridge, MA 1975)
40. D. Whitley, A.M. Sutton, in *Genetic Algorithms—A Survey of Models and Methods*. Handbook of Natural Computing, (Springer, 2012), pp. 637–671. ISBN: 978-3-540-92909-3
41. Y. Liu, K.M. Passino, *Swarm Intelligence: Literature Overview* (Ohio State University Dept. of Electrical Engineering, 2000)
42. G. Beni, J. Wang, in *Swarm Intelligence in Cellular Robotic Systems*, NATO Advanced Workshop on Robots and Biological Systems, Tuscany, Italy, 1989
43. A. Chakraborty, A.K. Kar, Swarm intelligence: a review of algorithms, in *Nature-Inspired Computing and Optimization, Modeling and Optimization in Science and Technologies*, vol. 10 (Springer, 2017), pp. 475–494
44. S. Baluja, *Population-Based Incremental Learning: A Method for Integrating Genetic Search Based Function Optimization and Competitive Learning* (Technical Report CMU-CS-94-163), Carnegie Mellon University, USA, 1994
45. S. Yang, X. Yao, Population-based incremental learning with associative memory for dynamic environments. IEEE Trans. Evol. Comput. **12**(5), 542–561 (2008). https://doi.org/10.1109/TEVC.2007.913070
46. F. Glover, Tabu search: part I. ORSA J. Comput. **1**(3), 190–206 (1989)
47. F. Glover, Tabu search: part II. ORSA J. Comput. **2**(1), 4–32 (1990)
48. http://www.springer.com/gp/book/9783662463529
49. A. Ibrahim, Sh. Rahnamayan, M. Vargas Martin, Miguel, Simulated raindrop algorithm for global optimization. Can. Conf. Electr. Comput. Eng., 1–8 (2014). https://doi.org/10.1109/ccece.2014.6901103
50. http://www.cleveralgorithms.com/nature-inspired/probabilistic/boa.html
51. http://www.thewdo.com/
52. http://www.citeuluke.org
53. M. Mitchell, J.H. Holland, S. Forrest, When will a genetic algorithm outperform hill climbing, in *Advances in Neural Information Processing Systems*, (1994), pp. 51–58
54. J.R. Sampson, *Adaptive Information Processing* (Springer, 1976), pp. 131–135
55. T. Kohonen, The self-organizing map. Neurocomputing **21**(1), 1–6 (1998)
56. J. Kregting, R.C. White, *Adaptive Random Search* (Technical Report TH-Report 71-E-24), Eindhoven University of Technology, Eindhoven, Netherlands, 1971
57. http://www.cleveralgorithms.com/natureinspired/stochastic/hill_climbing_search.html
58. M. Sayadi, R. Ramezanian, N. Ghaffari-Nasab, A discrete firefly meta-heuristic with local search for makespan minimization in permutation flow shop scheduling problems. Int. J. Ind. Eng. Comput. **1**(1), 1–10 (2010)
59. http://www.bracil.net
60. J. Duch, A. Arenas, Community detection in complex networks using extremal optimization. Phys. Rev. E **72**(2), 027104 (2005)

61. Z.I. Botev, A. Ridder, L. Rojas Nandayapa, Semiparametric cross entropy for rare-event simulation. J. Appl. Probab. **53**(3), 633–649 (2016)
62. Ch. Ramdane, S. Chikhi, Negative selection algorithm: recent improvements and its application in intrusion detection system. Int. J. Comput. Acad. Res. (IJCAR) **6**(2), 20–30 (2017)
63. J. Kytojoki, T. Nuortio, O. Braysy, M. Gendreau, An efficient variable neighborhood search heuristic for very large scale vehicle routing problems. Comput. Oper. Res. **34**(9), 2743–2757 (2007)
64. B. Khan, P. Singh, Optimal power flow techniques under characterization of conventional and renewable energy sources: a comprehensive analysis. Hindawi J. Eng. (2017)
65. http://futurearchitectureplatform.org/news/28/ai-architecture-intelligence/
66. B. Baydar, H. Gozde, M.C. Taplamacioglu, A research on evolutionary computation techniques in optimal power flow solution. Int. J. Tech. Phys. Probl. Eng. (IJTPE) **9**(33), no. 4, 26–33 (2017)
67. L.L. Lai, J.T. Ma, R. Yokoyama, M. Zhao, Improved genetic algorithms for optimal power flow under both normal and contingent operation states. Electr. Power Energy Syst. **19**(5), 287–292 (1997). (Elsevier)
68. M.A. Abido, Optimal power flow using particle swarm optimization. Electr. Power Energy Syst. **24**, 563–571 (2002). (Elsevier)
69. M.A. Abido, Optimal power flow using Tabu search algorithm. Electr. Power Compon. Syst. **30**(5), 469–483 (2002)
70. T. Bouktir, L. Slimani, Optimal power flow of the algerian electrical network using an ant colony optimization method. Leonardo J. Sci. **6**, 43–57 (2005)
71. S. Sayah, K. Zehar, Modified differential evolution algorithm for optimal power flow with non-smooth cost functions. Energy Convers. Manage. **49**, 3036–3042 (2008). (Elsevier)
72. A.A. Abou El Elaa, M. Abido, S.R. Spea, Optimal power flow using differential evolution algorithm. Electr. Power Syst. Res. **80**, 878–885 (2010). (Elsevier)
73. M. Li, W. Liu, X. Wang, An improved particle swarm optimization algorithm for optimal power flow. IPEMC **25**, 2448–2450 (2009)
74. S.R. Tsai, R.H. Liang, Y.T. Chen, W.T. Tseng, Optimal power flow by a fuzzy based hybrid particle swarm optimization approach. Electr. Power Syst. Res. **81**, 1466–1474 (2011). (Elsevier)
75. I. Oumarou, D. Jiang, C. Yijia, Particle swarm optimization applied to optimal power flow solution. Int. Conf. Nat. Comput. (ICNC) **8**(9), 284–288 (2009)
76. C. Sumpavakup, I. Srikun, S. Chusanapiputt, A solution to the optimal power flow using artificial bee colony algorithm. Int. Conf. Power Syst. Technol. (ICPST) **7**(10) (2010)
77. A. Bhattacharya, P.K. Chattopadhyay, Application of biogeography-based optimization to solve different optimal power flow problems. Inst. Eng. Technol. (IETDL) **5**(1), 70–80 (2011)
78. A.A. Esmin, G.L. Torres, Application of particle swarm optimization to optimal power systems. Int. J. Innovative Comput. Inf. Control (ICIC) **8**(3(A)) (2011)
79. U. Guvenc, S. Duman, Y. Sonmez, N. Yorukeren, Optimal power flow using gravitational search algorithm. Energy Convers. Manage. **59**, 86–95 (2012). (Elsevier)
80. C. Sumpavakup, I. Srikun, S. Chusanapiputt, A solution to multi-objective optimal power flow using hybrid cultural-based bees algorithm. IEEE **2**(12) (2012)
81. O. Herbadji, K. Nadhir, L. Slimani, T. Bouktir, Optimal power flow with emission controlled using firefly algorithm. IEEE **9**(13), 4673–5814 (2013)
82. P.D. Bamane, A.N. Kshirsagar, S. Raj, H.T. Jadhav, Temperature Dependent Optimal Power Flow Using GBest - Guided Artificial Bee, International Conference on Computation of Power, Energy, Information and Communication (ICCPEIC), vol. 1, issue 14, pp. 321–327, 2014
83. S.S. Reddy, S. Rathnam, Optimal power flow using glowworm swarm optimization. Electr. Power Energy Syst. **80**, 128–139 (2016). (Elsevier)
84. A.A. Mohamed, Y.S. Mohamed, A.A. El-Gaafary, Optimal power flow using moth swarm algorithm. Electr. Power Syst. Res. **142**, 190–206 (2017). (Elsevier)

Part III
Planning, Attacks and Recovery in Resilience Systems

Chapter 8
Multi-stage Resilient Distribution System Expansion Planning Considering Non-utility Gas-Fired Distributed Generation

Mehrdad Setayesh Nazar and Alireza Heidari

Abstract This chapter presents an approach for Resilient Distribution System Expansion Planning (RDSEP) considering gas-fired Non-utility DGs (NUDGs) and Demand Side Providers (DRPs). The RDSEP method explores the NUDGs and DRPs impacts on the planning paradigm. The RDSEP problem is decomposed into multi sub-problems that optimize investment, operational and reliability costs. The RDSEP is a complicated problem, and the resilience criteria may encounter different planning schemes that can also be included in the problem modeling. The resilience of a distribution system is the capacity to tolerate the external shocks that may be imposed on the network, and the distribution system must be able to deliver electricity continuously to its consumers. The distribution system may have NUDGs and DRPs that interchange electricity with Distribution System Operator (DSO) and they can dynamically change the distribution system resources. The NUDG and DRP contribution scenarios can significantly change the state space of RDSEP, and they can be utilized for different preventive/corrective measures against internal and external shocks. The RDSEP model is a non-linear programming problem, and a heuristic optimization method is utilized. A nine-bus test system and an urban electric system are used to assess the introduced method.

Keywords External shock · Internal shock · Resiliency · Resilient distribution system · Resilient system expansion planning

M. Setayesh Nazar (✉)
Shahid Beheshti University, A.C., Tehran, Iran
e-mail: m_setayesh@sbu.ac.ir

A. Heidari
School of Electrical Engineering and Telecommunication,
University of New South Wales, Sydney, Australia
e-mail: alireza.heidari@unsw.edu.au

© Springer Nature Switzerland AG 2019

N. Mahdavi Tabatabaei et al. (eds.), *Power Systems Resilience*, Power Systems,
https://doi.org/10.1007/978-3-319-94442-5_8

Nomenclatures

C_{UDG}	UDG costs
C_{DRP}	DRP candidates costs
C_{SW}	Switching device costs
C_{RPS}	Reactive power resource candidates costs
C_{NUDG}	NUDG contribution costs
C_{Sub}	New substation costs
C_{Feed}	New feeder costs
C_{OP_UDG}	UDG operation costs
C_{OP_NUDG}	NUDG operation costs
C_{OP_DRP}	DRP operation costs
E	Energy purchased from upward utility
MCP	Marginal clearing price of the wholesale market
$NESE$	Number of extreme external shock
$NESP$	Number of expected external shock
$NESR$	Number of routine external shock
NIS	Number of internal shocks
N_{SC_UDG}	Number of UDG capacity candidates
N_{SC_NUDG}	Number of NUDG contribution scenarios
$Nyear$	Number of planning years
Np	Number of periods
$Nzone$	Number of distribution system zones
W	Weighting factor
φ^{Inv}	Decision variable for investment
ψ_{Sub}	Decision variable for new substation installation
ψ_{Feed}	Decision variable for feeder installation
ψ_{DRP}	Decision variable for DRP contribution
ψ_{SW}	Decision variable for switching device installation
ψ_{RPS}	Decision variable for RPS installation
ψ_{UDG}	Decision variable for UDG installation
ψ_{NUDG}	Decision variable for NUDG contribution
ϕ_{UDG}	Decision variable of UDG commitment
ϕ_{DRP}	Decision variable of DRP commitment
ϕ_{NUDG}	Decision variable of NUDG commitment

8.1 Introduction

The resilience of a distribution system is the capacity to tolerate the shocks that are external to the electric system, continue to deliver electricity to its customers, can recover from major shocks and resume to its new steady state conditions [1]. The shock sources are external and internal to the system [2]. The external shocks can

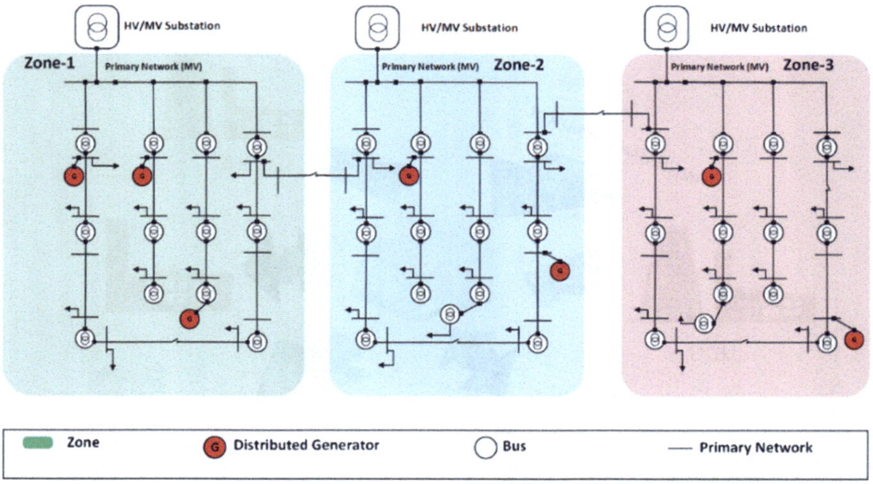

Fig. 8.1 Schematic diagram of an electric distribution network

cease to function the distribution system facilities due to natural events, and the internal shocks are the electric system contingencies [3]. Figure 8.1 depicts the distribution system configuration and its DSO can utilize utility, NUDG facilities and DRPs to mitigate the impacts of external shocks and/or system's contingencies [4].

The RDSEP problem consists of optimizing of the parameters of installation of system devices, depend on load growth conditions, reliability and resiliency criteria, demand response programs and NUDG contribution scenarios [5].

This book chapter is about the RDSEP algorithm that considers the NUDGs/DRPs contribution uncertainties and external shocks scenarios. Further, it considers an optimal reconfiguration procedure to mitigate the impacts of different external shocks on the system performance.

8.2 Problem Modelling and Formulation

The resilient expansion planning criteria can be summarized as:

(1) Minimizing the system costs;
(2) Maximizing the system reliability considering contingencies;
(3) Undertaking resilient planning and operational measures against external shocks.

These described objectives are highly conflicting based on the fact that the high reliability and resiliency levels of the system require a higher investment. The second planning criteria are maximized through minimization of interruption cost [5]. The resilient operational measures can be considered in the operational

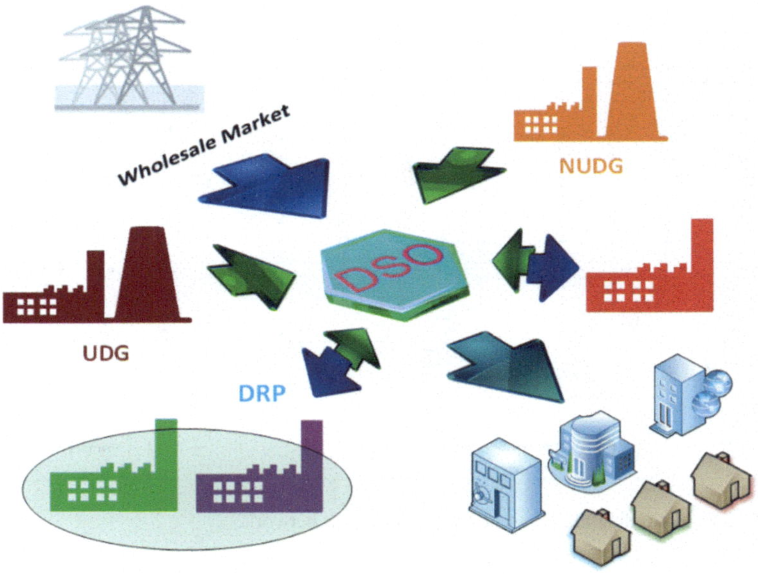

Fig. 8.2 The schematic diagram of NUDGs and loads contributions

preventive/corrective paradigms against external shocks [6]. Thus, the RDSEP may be defined as a problem that the total costs of system are minimized [7]. The resiliency of system is optimized through island formation and system resource coordination in the contingent conditions [8].

As shown in Fig. 8.2, a NUDG can be classified as dispatchable and non-dispatchable DG [9]. The DSO can utilize dispatchable NUDGs/DRPs to mitigate the impacts of internal and external shocks.

The RDSEP problem is subject to the three sources of uncertainty: DRPs and NUDGs contribution scenarios, system electric internal and external shocks [10]. Thus, the uncertainty can be modeled as a scenario-driven model. Hence, the DSO must make optimal decisions throughout planning horizon with incomplete information, and it must determine the optimal values of problem decision variables that consist of the location, the capacity, and the time of installation its system devices [11]. The DSO uses an estimated data of dispatchable NUDGs/DRPs location and contribution scenarios to determine optimal operational paradigms for maximizing its system reliability; meanwhile, it utilizes preventive/corrective actions and coordinate system resources when the external shocks are imposed on the system [12]. The RDSEP takes into account the optimal coordination of control variables such as utility DGs, dispatchable NUDGs/DRPs, and capacitors and lines.

Based on the described planning criteria, the RDSEP decision variables can be summarized as:

(1) Location, capacity, and type of utility-owned facilities and Utility-owned DGs (UDGs),
(2) The volume of energy purchased from NUDGs and DRPs,
(3) The location and capacity of shunt capacitors,
(4) The location of switching devices.

The system costs can be categorized as:

(1) Investment costs of utility-owned resources,
(2) Operation costs of system resources,
(3) Energy purchased from upward utility, NUDGs, and DRPs,
(4) Investment and operation costs of switching devices.

The reliability of the system can be considered as objective functions and constraints. The Energy Not Supplied Cost (ENSC) can be considered as an objective function in RDSEP [13]. The RDSEP is logical in light of demands and system optimal resilient planning and operation. The decision variables are critical due to the utility and NUDGs/DRPs interactions based on the systems constraints. In this chapter, the DSO and NUDGs/DRPs decision state spaces and different parameters uncertainties is adequately modeled to capture the real nature of the problem.

8.2.1 First Stage Problem Formulation

The resiliency evaluation of distribution system can be decomposed into the following steps: (1) the external system shocks are estimated and to be planned for; (2) the intensity and size for each shock is estimated; (3) the recovery plans are prepared, and (4) the performance of RDSEP for each recovery plan is evaluated [14].

The described framework of resilience planning can be done at three levels of magnitude for external shocks that are categorized as extreme, expected and routine. The maximum considered changes of anticipated shock are categorized as an extreme external shock. The expected and routine shock levels are designed and below the expected design levels of shocks, respectively.

Then, the DSO determines the number of system's internal shocks (contingencies), and it estimates the number of contribution scenarios of DRP and NUDG for each shock scenario [15]. Further, the DSO must estimate the volume of energy that will be transacted between its system and the NUDGs and DRPs. The first stage minimizes the present worth of whole investment, operational costs and ENSC for the bi-annually periods of the planning years. The objective function of the first stage problem can be written as (8.1):

$$\min C_1 = \sum_{i=1}^{Nyear}\sum_{j=1}^{Nzone}[W_{invest}.\varphi_{ijkl}^{Inv}.Investment_{ij}]$$
$$+\sum_{i=1}^{Nyear}\sum_{j=1}^{Nzone}\sum_{k=1}^{Np}[W_{NESE_k}.\sum_{l=1}^{NESE}ENSC_{ijkl}+W_{NESP_k}.\sum_{l=1}^{NESP}ENSC_{ijkl}$$
$$+W_{NESR_k}.\sum_{l=1}^{NESR}ENSC_{ijkl}+W_{NIS_k}.\sum_{l=1}^{NIS}ENSC_{ijkl}$$
$$+W_{Purchased\_Energy_k}.MCP_{ijk}.E_{ijk}] \tag{8.1}$$

The investment term of (8.1) will be described in the second stage problem, and the ENSC term will be described in the fifth stage problem. The DC load flow constraints and investment scenario selection constraints are considered as first stage optimization constraints. However, the first stage has a slave problem that optimizes the bi-annually cost allocation of the first stage.

8.2.2 Second Stage Problem Formulation

For each shocks scenario and NUDGs and DRPs contribution scenarios, the optimally distributed generation and network expansion planning problem optimizes cost allocation for different investment alternatives and bi-annual periods. The objective function of second stage problem is as (8.2):

$$\min C_2^a = \sum_{i=1}^{Nyear}\sum_{j=1}^{Nzone}[W_1.(\sum_{k\in Nsub}\sum_{l\in Ntrans}C_{Sub\,ijkl}.\varphi_{Sub\,ijkl}$$
$$+\sum_{k\in Fr}\sum_{l\in feed}C_{Feed\,ijkl}.\varphi_{Feed\,ijkl}+\sum_{k\in RPS}C_{RPS\,ijk}.\varphi_{RPS\,ijk}$$
$$+\sum_{k\in UDG}C_{UDG\,ijk}.\varphi_{UDG\,ijk}+\sum_{k\in SW}C_{SW\,ijk}.\varphi_{SW\,ijk})$$
$$+\sum_{k\in DRP}W_{DRP}.C_{DRP\,ijk}.\varphi_{DRP\,ijk}+W_{Purchased\_Energy_k}.MCP_{ijk}.E_{ijk}$$
$$+\sum_{k\in NUDG}W_{NUDG}.C_{NUDG\,ijk}.\varphi_{NUDG\,ijk}+W_{NESE_k}.\sum_{l=1}^{NESE}ENSC_{ijkl}$$
$$+W_{NESP_k}.\sum_{l=1}^{NESP}ENSC_{ijkl}+W_{NESR_k}.\sum_{l=1}^{NESR}ENSC_{ijkl}$$
$$+W_{NIS_k}.\sum_{l=1}^{NIS}ENSC_{ijkl}]$$
$$a\in \text{First stage problem state space} \tag{8.2}$$

The objective function is decomposed into the following groups: (1) network substation, feeder, shunt capacitor, utility DG operation and investment and switching devices (5 sentences of objective function); (2) the costs of purchased energy from DRPs and NUDGs and upward network, (3) ENSC costs of internal and external shocks.

The second stage problem deals with optimal cost allocation of different investment alternatives. The constraints can be summarized as device loading and DC load flow constraints.

8.2.3 Third Stage Problem Formulation

At the third stage, the DSO estimates the network's electric loads and power exchanges with upward network and NUDGs and DRPs for a quarter of year periods. The third stage problem finds the optimal energy purchasing, operation and investment costs for each quarter of the planning years. The third stage objective function can be presented as (8.3):

$$
\begin{aligned}
\min C_3^b = & \sum_{i=1}^{Nyear} \sum_{j=1}^{Nzone} [W_3.(\sum_{k \in Nsub} \sum_{l \in Ntrans} C_{Sub\,ijkl} \cdot \psi_{Sub\,ijkl} \\
& + \sum_{k \in Fr} \sum_{l \in feed} C_{Feed\,ijkl} \cdot \psi_{Feed\,ijkl} \\
& + \sum_{k \in DRP} C_{DRP\,ijk} \cdot \psi_{DRP\,ijk} + \sum_{k \in SW} C_{SW\,ijk} \cdot \psi_{SW\,ijk} \\
& + \sum_{k \in RPS} C_{RPS\,ijk} \cdot \psi_{RPS\,ijk} + \sum_{m \in UDG} \sum_{n=1}^{Nsc\_UDG} C_{UDG\,ijmn} \cdot \psi_{UDG\,ijmn} \\
& + \sum_{m \in NUDG} \sum_{n=1}^{Nsc\_NUDG} C_{NUDG\,ijmn} \cdot \psi_{NUDG\,ijmn}) \\
& + W_{Purchased\_Energy_k} . MCP_{ijk} . E_{ijk} \\
& + W_{NESE_k} \cdot \sum_{l=1}^{NESE} ENSC_{ijkl} + W_{NESP_k} \cdot \sum_{l=1}^{NESP} ENSC_{ijkl} \\
& + W_{NESR_k} \cdot \sum_{l=1}^{NESR} ENSC_{ijkl} + W_{NIS_k} \cdot \sum_{l=1}^{NIS} ENSC_{ijkl}] \\
& b \in \text{Second stage problem state space}
\end{aligned} \tag{8.3}
$$

It considers network substation optimal capacity determination and allocation, feeder installations, switching and RPS installations, UDG allocation and capacity selection and NUDG and DRP contribution scenarios selection. The third stage objective function is subjected to AC power flow constraints under normal and contingent conditions.

8.2.4 Fourth Stage Problem Formulation

The fourth stage objective function can be stated as (8.4):

$$
\begin{aligned}
\min C_4^c = & \sum_{i=1}^{Nyear} \sum_{j=1}^{Nzone} \sum_{k=1}^{Np} [W_5.(\sum_{m \in UDG} \sum_{n=1}^{Nsc\_UDG} C_{UDG\,ijkmn} \cdot \phi_{UDG\,ijkmn} \\
& + \sum_{m \in NUDG} \sum_{n=1}^{Nsc\_NUDG} C_{NUDG\,ijkmn} \cdot \phi_{NUDG\,ijkmn} \\
& + \sum_{m \in DRP} \sum_{n=1}^{Nsc\_DRP} C_{DRP\,ijkmn} \cdot \phi_{DRP\,ijkmn}) \\
& + W_{Purchased\_Energy_k} . MCP_{ijk} . E_{ijk} \\
& + W_{NESE_k} \cdot \sum_{l=1}^{NESE} ENSC_{ijkl} + W_{NESP_k} \cdot \sum_{l=1}^{NESP} ENSC_{ijkl} \\
& + W_{NESR_k} \cdot \sum_{l=1}^{NESR} ENSC_{ijkl} + W_{NIS_k} \cdot \sum_{l=1}^{NIS} ENSC_{ijkl}] \\
& c \in \text{Third stage problem state space}
\end{aligned}
\tag{8.4}
$$

The DSO uses the scenario-driven information to describe NUDGs and DRPs contribution scenarios. By selection of the best NUDGs and DRPs contribution scenarios, the DSO optimizes the decision variables of (8.4). If the NUDGs and DRPs contribution scenarios are fixed and the DSO contracts with them, the restoration problem is investigated that is considered in the fifth stage problem formulation.

8.2.5 Fifth Stage Problem Formulation

The fifth stage problem objective function is introduced as (8.5):

$$\begin{aligned}\min C_5^e = {} & W_{NESE_k}\cdot\sum_{l=1}^{NESE} ENSC_{ijkl} \\ & + W_{NESP_k}\cdot\sum_{l=1}^{NESP} ENSC_{ijkl} + W_{NESR_k}\cdot\sum_{l=1}^{NESR} ENSC_{ijkl} \\ & + W_{NIS_k}\cdot\sum_{l=1}^{NIS} ENSC_{ijkl} + \sum_{m\in UDG} \Delta C_{OP\_UDG\,ijkm} \\ & + \sum_{m\in NUDG} \Delta C_{OP\_NUDG\,ijkm} + \sum_{m\in DRP} \Delta C_{OP\_DRP\,ijkm} \\ & e \in \text{Fourth stage problem state space}\end{aligned} \tag{8.5}$$

The Weighted Average System Reliability Index (WASRI) as stopping criteria is defined as (8.6):

$$WASRI = w_1'.SAIDI + w_2'.SAIFI + w_3'.MAIFI_E \tag{8.6}$$

8.3 Solution Algorithm

The described RDSEP problem has a large state space that involves thousands of variables in the expansion-planning horizon. The uncertainties of the problem highly increase the state space of the RDSEP problem. Further, the sub-problems are nonlinear and non-convex. Thus, the trade-off between accuracy and computational burden is made to derive the best solution algorithm without oversimplifying the expansion planning process. Hence, the authors try to find the reasonable trade-off between solution quality and acceptable calculation time.

For optimization procedure, an Adaptive Genetic Algorithm (AGA) is used. Figure 8.3 depicts the flowchart of the optimization algorithm. At first, the first stage problem is optimized for wholesale market price scenarios. Then, the second stage problem is solved. At the third stage, the algorithm optimizes device allocation parameters. At the fourth stage, the optimality of contribution scenarios of NUDGs and DRPs is investigated. At the fifth stage, the algorithm optimizes the system restoration procedures.

8.4 Numerical Results

Two test systems were used to assess the proposed RDSEP algorithm. The first was 9-bus test system and the second was an urban electric distribution network. The time horizon was chosen five years into the future.

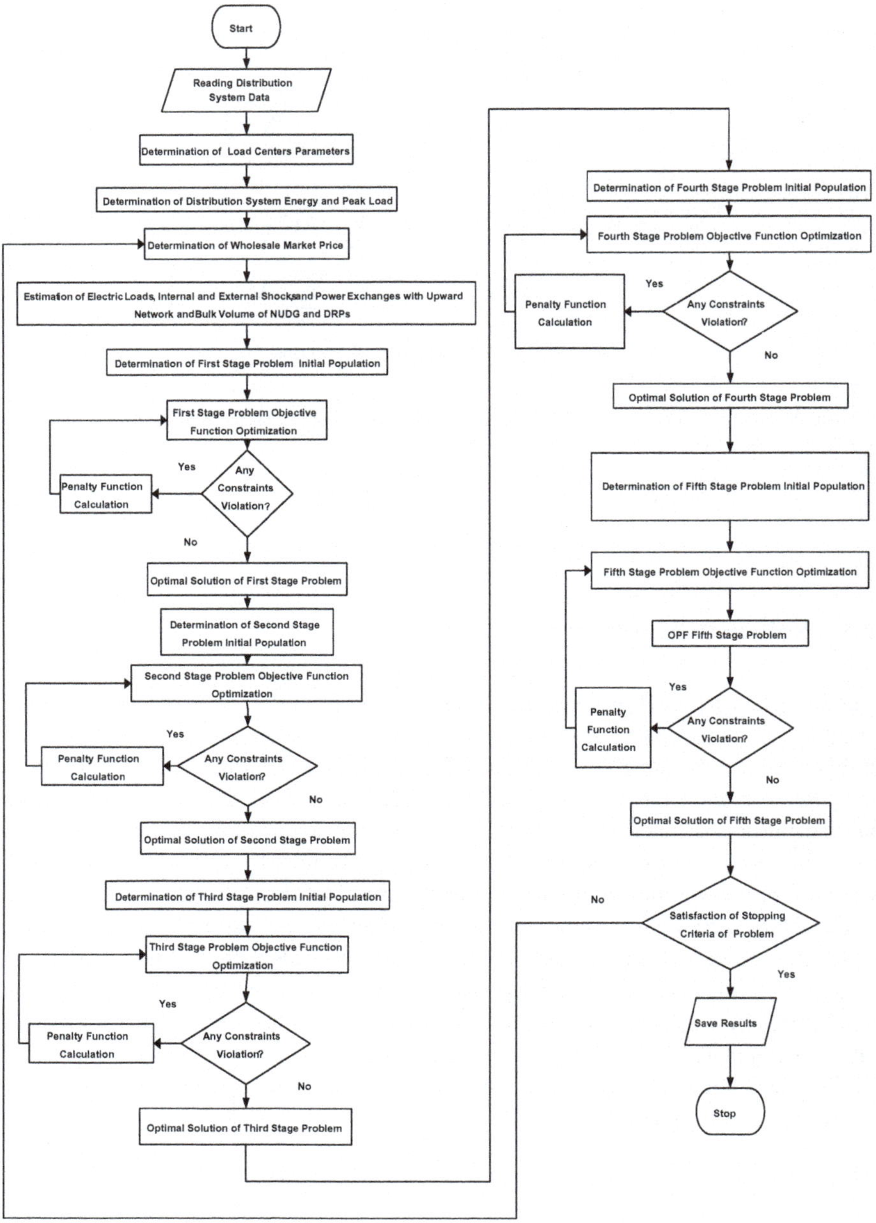

Fig. 8.3 The flowchart of the proposed multi-stage optimization algorithm

A. *The 9-Bus Test System*

Tables 8.1, 8.2, and 8.3 show the system parameters, feeder data, and transformer data, respectively. Figure 8.4 shows load forecasting results for the different year of planning year's horizon. MU stands for monetary unit.

Table 8.1 The 9-bus test system parameters

Parameter	Value
Nominal voltage	20 kV
Maximum voltage drop (ΔV)	3%
Electricity price (*MCP*)	45 MU/MWh
DG operational cost	30 MU/MWh
Power factor	0.9
Load factor	0.8
Feeder outage rate	0.02 year^{-1}
Feeder repair time	2 h

Table 8.2 The 9-bus test system feeder data

Type	Capacity (MVA)	Impedance (Ω/km)	Installation costs (MU/km)	Symbol
1	2.5	1.235	15,000	
2	5	0.8265	32,000	
3	10	0.4126	41,000	
4	15	0.2912	55,000	
5	20	0.2192	62,000	

Table 8.3 The 9-bus test system transformer data

Type	Capacity (MVA)	Installation costs (MU/km)	Symbol
1	2.5	60,000	
2	5	100,000	
3	10	210,000	
4	15	315,000	

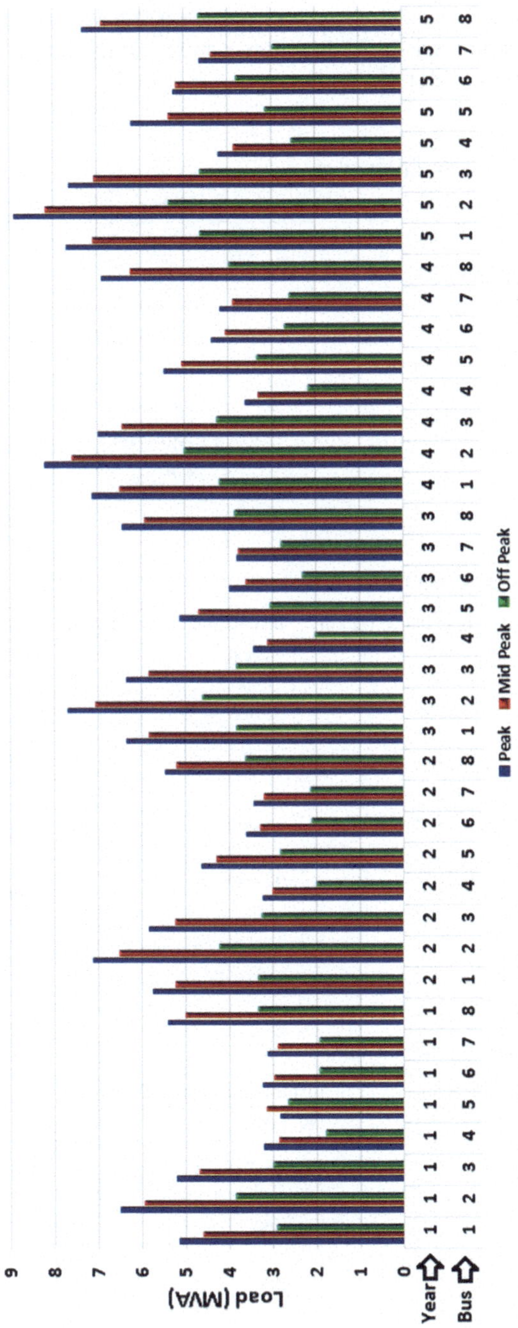

Fig. 8.4 The load forecasting results for the different year of planning year horizon

For the 9-bus test system, three scenarios were considered:

Scenario 1 Optimal RDSEP without UDG and NUDG contributions was performed.
Scenario 2 Optimal RDSEP with UDG contributions was performed.
Scenario 3 Optimal RDSEP with UDG and NUDG contributions was performed.

Double line outages and single line outage were considered as external and internal shocks, respectively. Figures 8.5, 8.6, and 8.7 show the optimal RDSEP results for the 1st, 2nd, and 3rd scenario and the 5th year of planning year's horizon, respectively.

Figure 8.8 depicts the bus voltage of 9-bus test system for the 5th year of planning year's horizon and different scenarios. Figure 8.9 shows the final costs of RDSEP for different scenarios. As shown in Fig. 8.9, the total system's costs were highly reduced by utilizing the NUDG and DRP contribution alternatives in contingent conditions.

B. *Urban Test System*

The urban electric distribution network has about 7000 customers in the 5th year of the planning horizon. The internal shocks (or system contingencies) were categorized, and the external system shocks and their intensity were estimated. Then, the performance of RDSEP for each recovery plan was evaluated. Three levels of magnitude for external shocks were considered that were categorized into extreme, expected and routine external shocks. Table 8.4 shows the external shocks categories and their characteristics.

The DSO estimated the NUDG and DRP locations, contribution scenarios, and their annual electricity transactions with the DSO. Two scenarios of NUDG/DRP contribution scenarios were estimated. Table 8.5 shows the estimated energy consumption of system for two different NUDGs/DRPs contribution scenarios. The estimated costs of energy transactions between NUDGs, DRPs and the DSO are shown in Table 8.6.

The third stage problem determined the optimum allocation of network substations and feeder routing. Table 8.7 shows the final transformer capacity selection results that were determined in the third stage.

Figure 8.10 depicts final monthly expected energy not supplied cost results for different scenarios. Table 8.8 depicts the final optimized RDSEP costs. The investment and replacement costs of the first NUDGs/DRPs contribution scenario, takes on a value 1510 billion MUs, which is decomposed in 0.1980, 28, 612, 723, 146 billion MUs for different planning years. In addition, the RDSEP estimates that the investment and replacement costs of the second wholesale market price scenario will be about 1510.257 billion MUs, which is decomposed in 117, 0.121, 0.136, 624, 769 billion MUs for different planning years.

The fifth stage problem investigates the optimal restoration, and it switches the capacitors and switches. The optimized system topologies of the first scenario are shown in Fig. 8.11a–c for the first, third and fifth year of planning years,

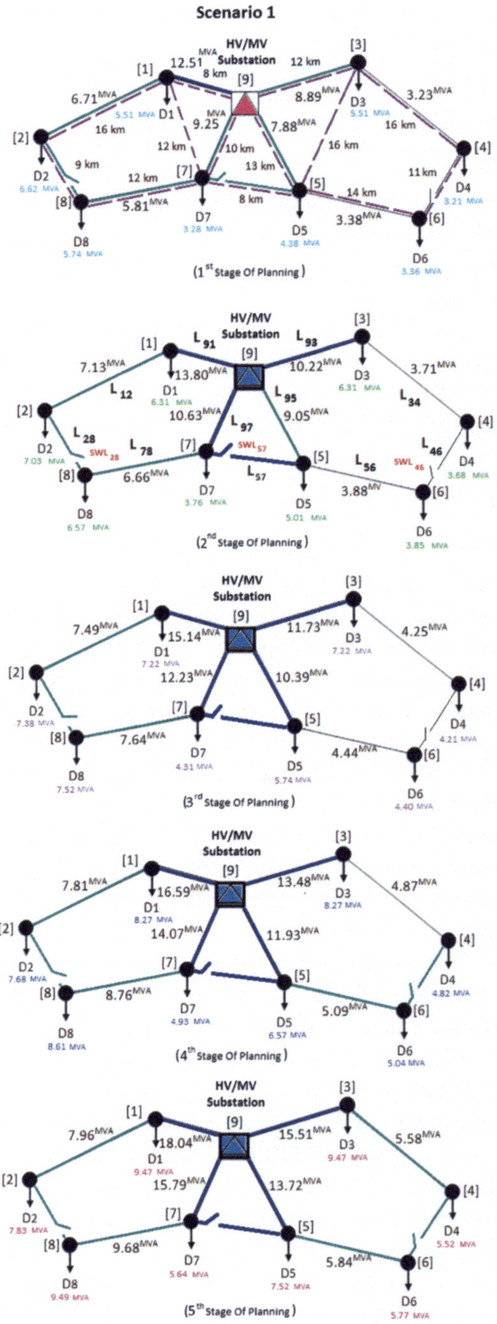

Fig. 8.5 The optimal RDSEP result for the 1st scenario and the 5th year of planning year's horizon

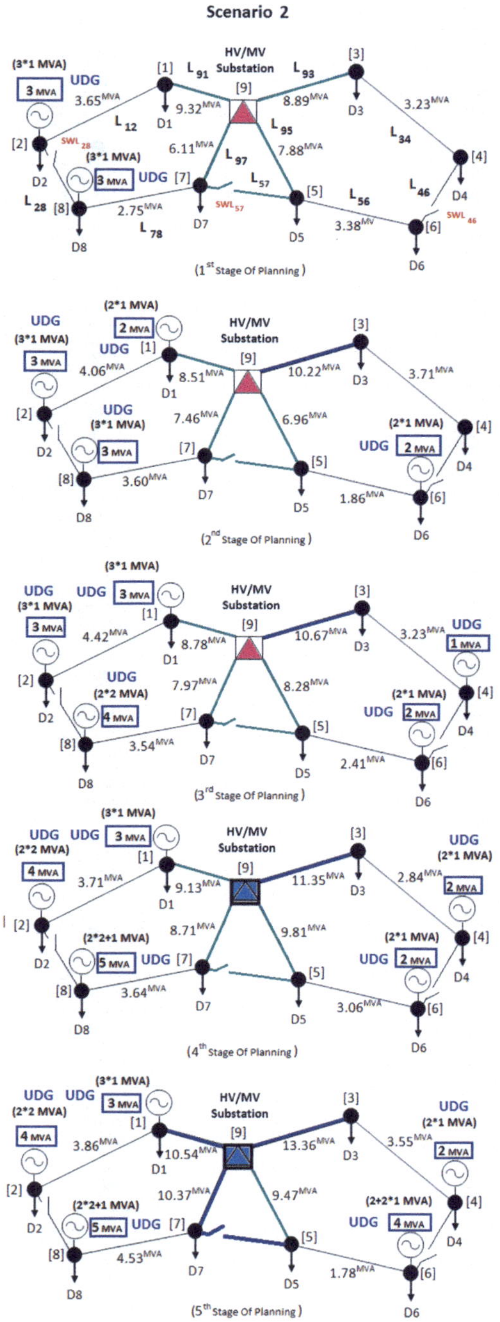

Fig. 8.6 The optimal RDSEP result for the 2nd scenario and the 5th year of planning year's horizon

Fig. 8.7 The optimal RDSEP result for the 3rd scenario and the 5th year of planning year's horizon

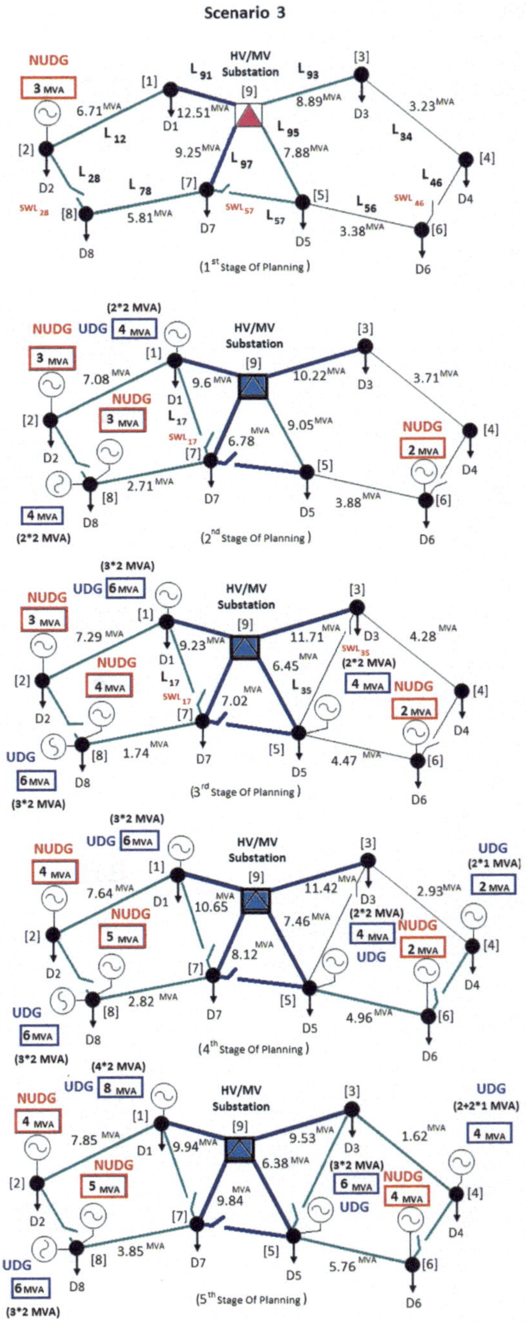

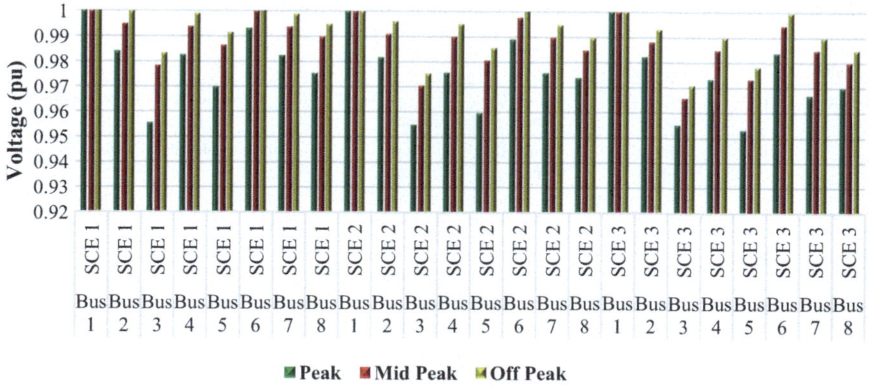

Fig. 8.8 The bus voltage of 9-bus test system for the 5th year of planning year's horizon and different scenarios

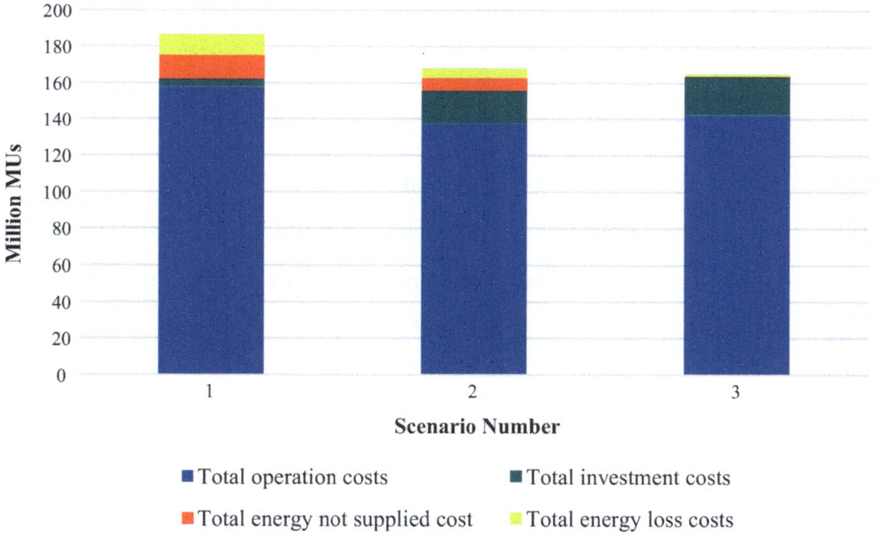

Fig. 8.9 The final costs of RDSEP for different scenarios

respectively. The optimized system topologies of the second scenario are shown in Fig. 8.12a–c for the first, third and fifth year of planning year's horizon, respectively. The NUDGs are shown in red, and the optimal tie switch allocations are illustrated. When the external and internal shocks are imposed on the system, the switches and capacitors are switched to restore the system. The WASRI indices and their components are shown in Fig. 8.13a, b for the first and second scenarios of the NUDGs/DRPs contribution, respectively. As Fig. 8.13a shows, the WASRI index

Table 8.4 The external shocks categories and their characteristics

External shock category		Zone 1	Zone 2	Zone 3	Zone 4	Zone 5
Extreme	Type 1	Triple line outage	Triple line outage	Triple line outage	Triple line outage	Triple line outage
	Type 2	Triple DG outage	Triple DG outage	Triple DG outage	Triple DG outage	Triple DG outage
	Type 3	Combination of Type 1 and 2	Combination of Type 1 and 2	Combination of Type 1 and 2	Combination of Type 1 and 2	Combination of Type 1 and 2
Expected	Type 1	Double line outage	Double line outage	Double line outage	Double line outage	Double line outage
	Type 2	Double DG outage	Double DG outage	Double DG outage	Double DG outage	Double DG outage
	Type 3	Single line and double DG outage	Single line and double DG outage	Single line and double DG outage	Double line and DG outage	Double line and DG outage
	Type 4	Combination of Type 1 and 2	Combination of Type 1 and 2	Combination of Type 1 and 2	Combination of Type 1 and 2	Combination of Type 1 and 2
Routine	Type 1	Single line and DG outage	Single line and DG outage	Single line and DG outage	Single line and DG outage	Single line and DG outage
	Type 2	Double line outage	Double line outage	Double line outage	Double line outage	Double line outage

Table 8.5 The estimated energy consumption of the system

Year	First year	Second year	Third year	Fourth year	Fifth year
Estimated first scenario energy consumption (kWh)	8.85E + 07	9.29E + 07	9.75E + 07	1.02E + 08	1.08E + 08
Estimated second scenario energy consumption (kWh)	8.43E + 07	8.94E + 07	9.39E + 07	9.85E + 07	1.04E + 08

Table 8.6 Total costs of NUDGs/DRPs contribution

Scenario	Type	First year cost (Million MUs)	Second year cost (Million MUs)	Third year cost (Million MUs)	Fourth year cost (Million MUs)	Fifth year cost (Million MUs)
1	NUDGs	2.99E + 05	3.47E + 05	4.12E + 05	5.07E + 05	6.28E + 05
	DRPs	6.04E + 05	7.01E + 05	8.27E + 05	9.63E + 05	1.13E + 06
2	NUDGs	2.40E + 05	2.79E + 05	3.31E + 05	4.07E + 05	5.05E + 05
	DRPs	4.85E + 05	5.63E + 05	6.64E + 05	7.73E + 05	9.04E + 05

Table 8.7 Final transformer allocation results

				First scenario of NUDGs/DRPs contribution		Second scenario of NUDGs/DRPs contribution	
#Bus	Zone	Load in base year (kVA)	Load in fifth year (kVA)	First year capacity (kVA)	Fifth year capacity (kVA)	First year capacity (kVA)	Fifth year capacity (kVA)
1	1	142.0504	165.2246	200	200	200	200
2	1	363.7141	411.0843	500	500	400	500
3	1	148.1401	175.6729	200	250	200	200
4	1	294.6799	331.1711	400	400	315	400
5	1	228.2031	264.2409	315	315	250	315
6	1	365.9655	413.0195	500	500	400	500
7	1	351.0099	402.7583	500	500	400	500
8	1	0	54.20679	25	100	25	100
9	1	0	44.3418	25	100	25	50
10	1	62.353	73.77879	100	100	100	100
11	1	256.3221	300.1912	315	400	315	315
12	1	94.52984	108.4372	200	200	100	200
13	2	327.0744	372.2027	400	500	400	400
14	2	252.9137	287.8661	315	400	315	315
15	2	370.7	435.5547	500	630	400	500
16	2	181.6283	204.0692	250	250	200	250
17	2	403.2697	471.495	500	630	500	500
18	2	0	176.4243	25	250	25	200
19	2	0	66.55204	25	100	25	100
20	2	0	35.4577	25	50	25	50
21	2	938.9845	1077.656	1250	1600	1000	1250
22	2	678.9604	773.2172	800	1000	800	1000
23	2	791.8355	936.542	1000	1250	1000	1000
24	2	68.36737	80.50399	100	100	100	100
25	3	332.6869	383.6478	400	500	400	500
26	3	44.44499	50.07625	100	100	50	100
27	3	0	220.387	25	315	25	250
28	3	0	52.7559	25	100	25	100
29	3	0	50.88049	25	100	25	100
30	3	0	73.58828	25	100	25	100
31	3	426.1217	486.9386	630	630	500	630
32	3	35.96624	42.16398	50	50	50	50
33	3	311.1869	360.067	400	500	400	400
34	3	31.022	36.50194	50	50	50	50
35	3	248.8218	282.3736	315	400	315	315

(continued)

Table 8.7 (continued)

				First scenario of NUDGs/DRPs contribution		Second scenario of NUDGs/DRPs contribution	
#Bus	Zone	Load in base year (kVA)	Load in fifth year (kVA)	First year capacity (kVA)	Fifth year capacity (kVA)	First year capacity (kVA)	Fifth year capacity (kVA)
36	4	82.92556	96.92174	100	200	100	200
37	4	189.8291	213.3273	250	315	200	250
38	4	0	82.9816	25	100	25	100
39	4	0	83.5413	25	100	25	100
40	4	593.0075	704.2321	800	1000	630	800
41	4	364.4825	410.6323	500	500	400	500
42	4	986.6974	1169.166	1250	1600	1250	1250
43	4	1055.936	1221.424	1250	1600	1250	1600
44	4	443.1934	515.9738	630	630	500	630
45	4	243.346	278.1039	315	400	315	315
46	5	153.1459	172.3919	200	250	200	200
47	5	167.9018	198.2665	200	250	200	250
48	5	0	155.2084	25	200	25	200
49	5	0	163.614	25	200	25	200
50	5	0	77.1618	25	100	25	100
51	5	0	88.4071	25	200	25	100
52	5	0	72.7228	25	100	25	100
53	5	0	288.4993	25	400	25	315
54	5	777.8657	883.9553	1000	1250	1000	1000
55	5	538.6144	618.7204	800	800	630	800
56	5	471.2034	551.3081	630	800	500	630
57	5	569.8395	640.8661	800	800	630	800
58	5	726.8883	853.0839	1000	1250	800	1000
59	5	0	174.737	25	250	25	200
60	5	0	98.35	25	200	25	200
61	4	0	410.6323	0	500	0	500
62	4	0	1169.166	0	1600	0	1250
63	4	0	1221.424	0	1600	0	1600
64	3	0	36.50194	0	50	0	50
65	3	0	282.3736	0	400	0	315

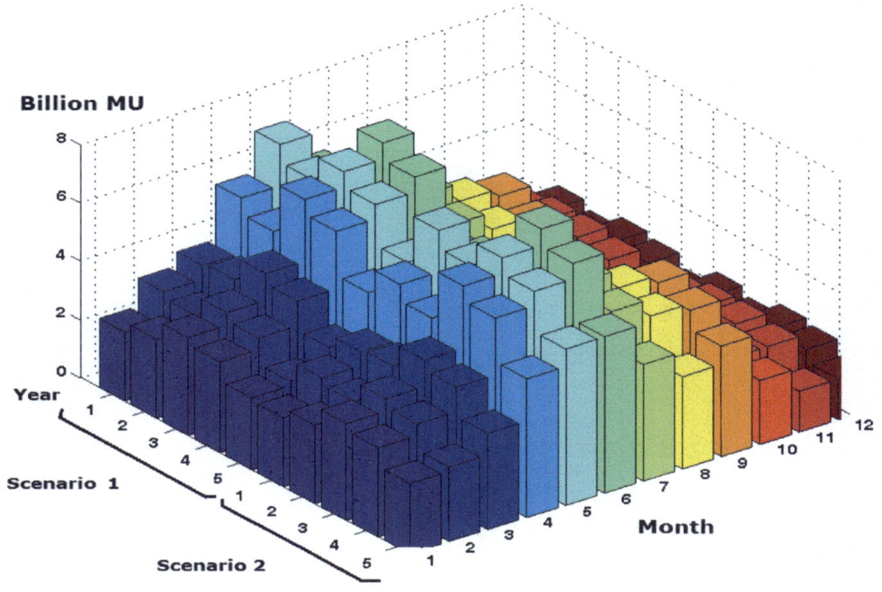

Fig. 8.10 Final expected energy not supplied cost results

takes on a value 6.82 at bus 50 for the first scenario. However, the maximum WASRI indices for the second scenario is 6.53 at bus 6.

8.5 Conclusion

A resilient distribution system expansion planning procedure was reviewed in the present chapter. The introduced method used a model to investigate the NUDGs/DRPs impacts on RDSEP. The RDSEP formulation found the optimum usage of NUDGS/DRPS contribution scenarios and it considered the impact of resilient criteria on the RDSEP. This algorithm decomposed the RDSEP problem into five sub-problems. Based on the iterative fifth-stage optimization procedure, the model of RDSEP was a MINLP problem, and a scenario-driven method with AGA was used. The algorithm was assessed for a nine-bus test system and an urban system with quite acceptable results.

Table 8.8 Final RDSEP results

Costs	First scenario of NUDGs/DRPs contribution					Second scenario of NUDGs/DRPs contribution				
	First year	Second year	Third year	Fourth year	Fifth year	First year	Second year	Third year	Fourth year	Fifth year
A	0.00E + 00	2.38E + 07	4.97E + 08	5.95E + 08	1.18E + 08	9.73E + 07	0.00E + 00	0.00E + 00	5.06E + 08	6.40E + 08
B	0.00E + 00	4.57E + 04	1.13E + 05	1.65E + 04	1.27E + 05	1.16E + 05	7.22E + 04	9.24E + 04	1.31E + 05	2.86E + 04
C	0.00E + 00	4.07E + 06	1.15E + 08	1.28E + 08	2.77E + 07	1.94E + 07	0.00E + 00	0.00E + 00	1.18E + 08	1.29E + 08
D	1.98E + 04	2.97E + 04	4.77E + 04	1.17E + 04	5.24E + 04	7.72E + 04	1.99E + 04	2.92E + 04	1.49E + 05	1.17E + 05
E	9.33E + 04	1.19E + 04	1.14E + 04	1.18E + 04	1.28E + 04	4.49E + 05	2.21E + 05	2.01E + 05	2.21E + 05	2.41E + 05
F	1.55E + 05	3.14E + 04	1.20E + 04	1.39E + 04	2.50E + 04	1.22E + 05	2.50E + 04	9.83E + 03	1.40E + 04	1.69E + 04
G	2.10E + 04	1.51E + 03	3.02E + 03	3.02E + 03	1.51E + 03	2.10E + 04	3.02E + 03	3.02E + 03	3.02E + 03	3.02E + 03
H	1.85E + 03	5.54E + 02	1.11E + 03	1.11E + 03	5.54E + 02	1.85E + 03	1.11E + 03	1.11E + 03	1.11E + 03	5.54E + 02
I	2.16E + 04	7.60E + 04	1.62E + 05	2.93E + 04	1.80E + 05	1.95E + 05	9.32E + 04	1.23E + 05	2.81E + 05	1.46E + 05
J	1.76E + 05	2.79E + 07	6.12E + 08	7.23E + 08	1.45E + 08	1.17E + 08	2.80E + 04	1.29E + 04	6.23E + 08	7.69E + 08
K	1.98E + 05	2.80E + 07	6.12E + 08	7.23E + 08	1.46E + 08	1.17E + 08	1.21E + 05	1.36E + 05	6.24E + 08	7.69E + 08
L	5.78E + 07	6.99E + 07	8.97E + 07	9.99E + 07	1.08E + 08	4.32E + 07	5.46E + 07	6.97E + 07	7.87E + 07	8.70E + 07
M	4.68E + 06	4.21E + 06	3.74E + 06	3.29E + 06	2.83E + 06	6.36E + 06	5.65E + 06	5.06E + 06	4.30E + 06	3.82E + 06
N	3.59E + 07	3.40E + 07	3.25E + 07	3.01E + 07	2.73E + 07	4.35E + 07	4.10E + 07	3.85E + 07	3.57E + 07	3.36E + 07

A = New substation and transformer costs (1000 MUs)
B = Transformer replacement costs (1000 MUs)
C = New line installation costs (1000 MUs)
D = New line replacement costs (1000 MUs)
E = DRPs costs (DRP program starts in the first year) (1000 MUs)
F = Reactive power resources costs (1000 MUs)
G = Total switching device installation costs from DA resources (1000 MUs)
H = Total switching device replacement costs from DA resources (1000 MUs)
I = Total replacement costs (1000 MUs)
J = Total installation costs (1000 MUs)
K = Total installation and replacement costs (1000 MUs)
L = Energy purchased from upward network (kWh/year)
M = Energy purchased from NUDGS/DRPS (kWh/year)
N = Energy purchased from UDG (kWh/year)

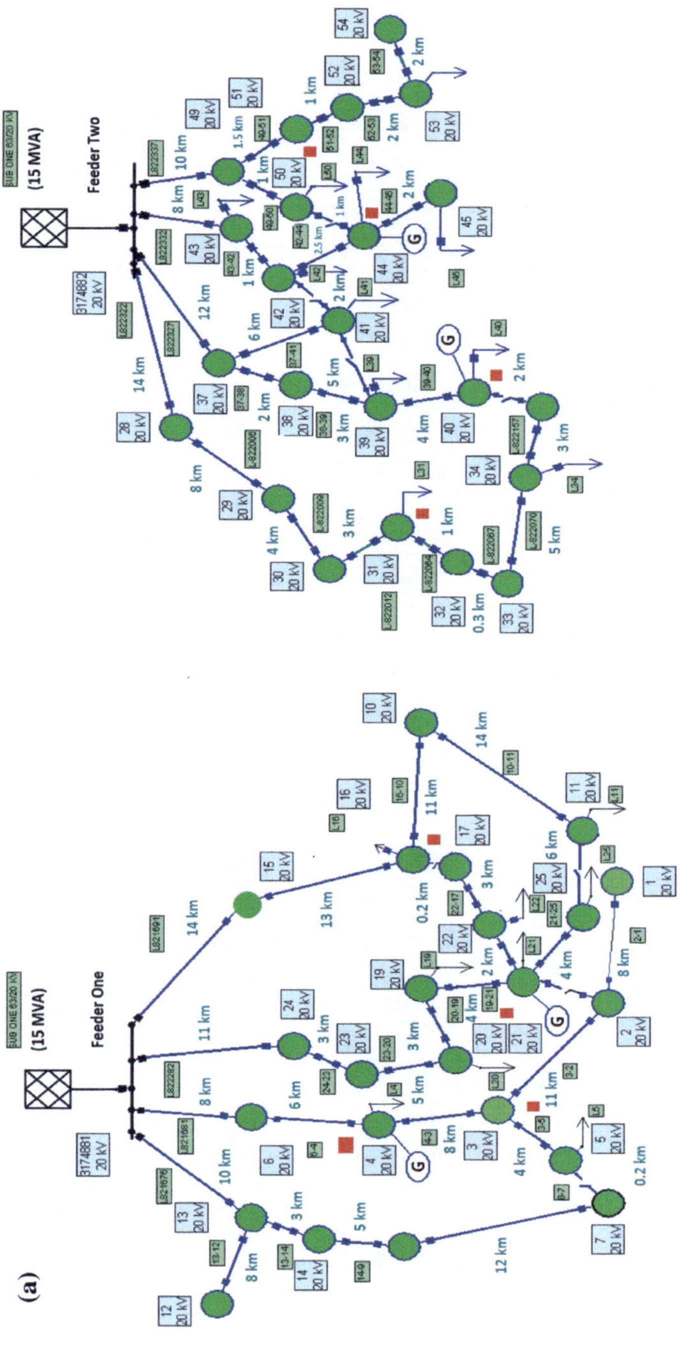

Fig. 8.11 **a** The optimized system topologies of the first scenario for the first year of planning year's horizon, **b** the optimized system topologies of the first scenario for the third year of planning year's horizon, **c** the optimized system topologies of the first scenario for the fifth year of planning year's horizon

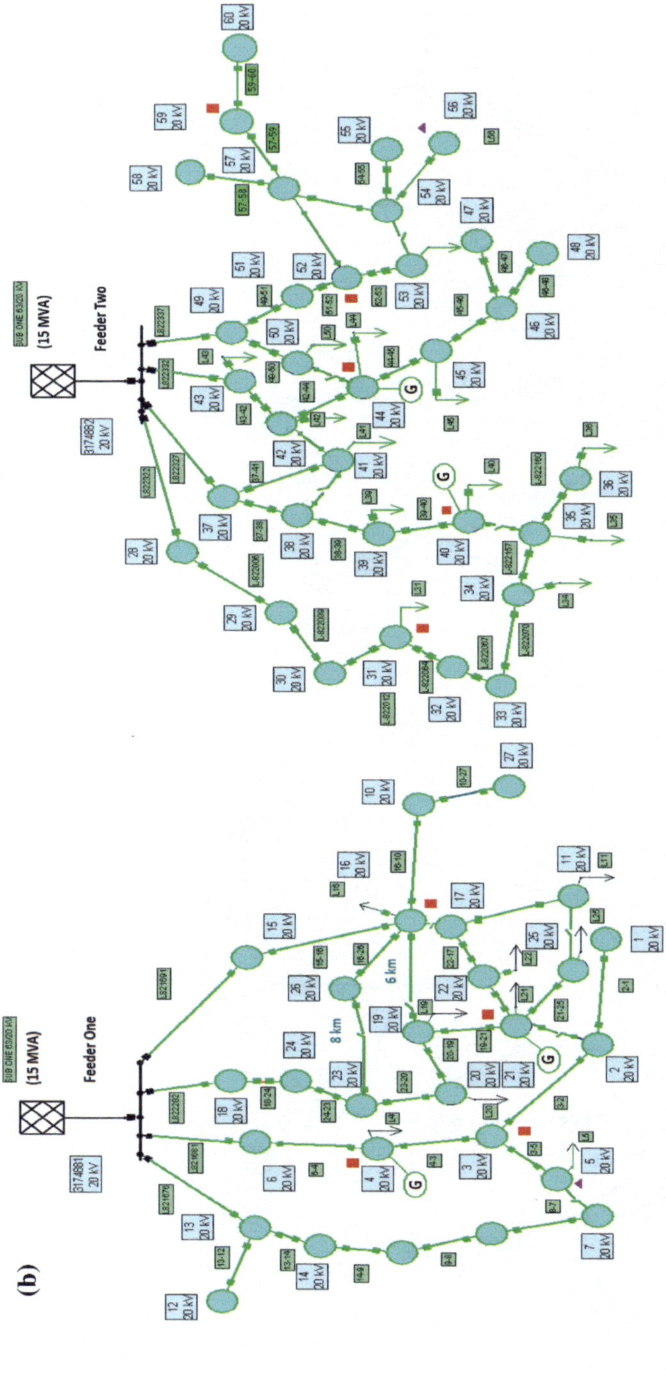

Fig. 8.11 (continued)

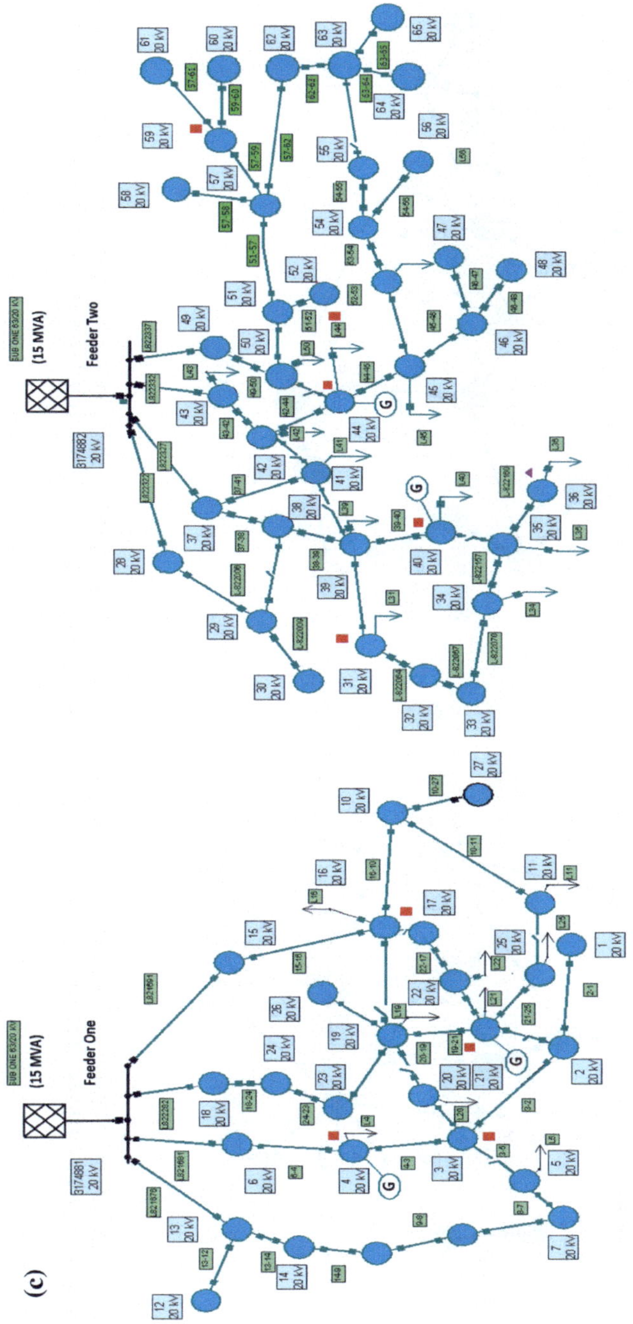

Fig. 8.11 (continued)

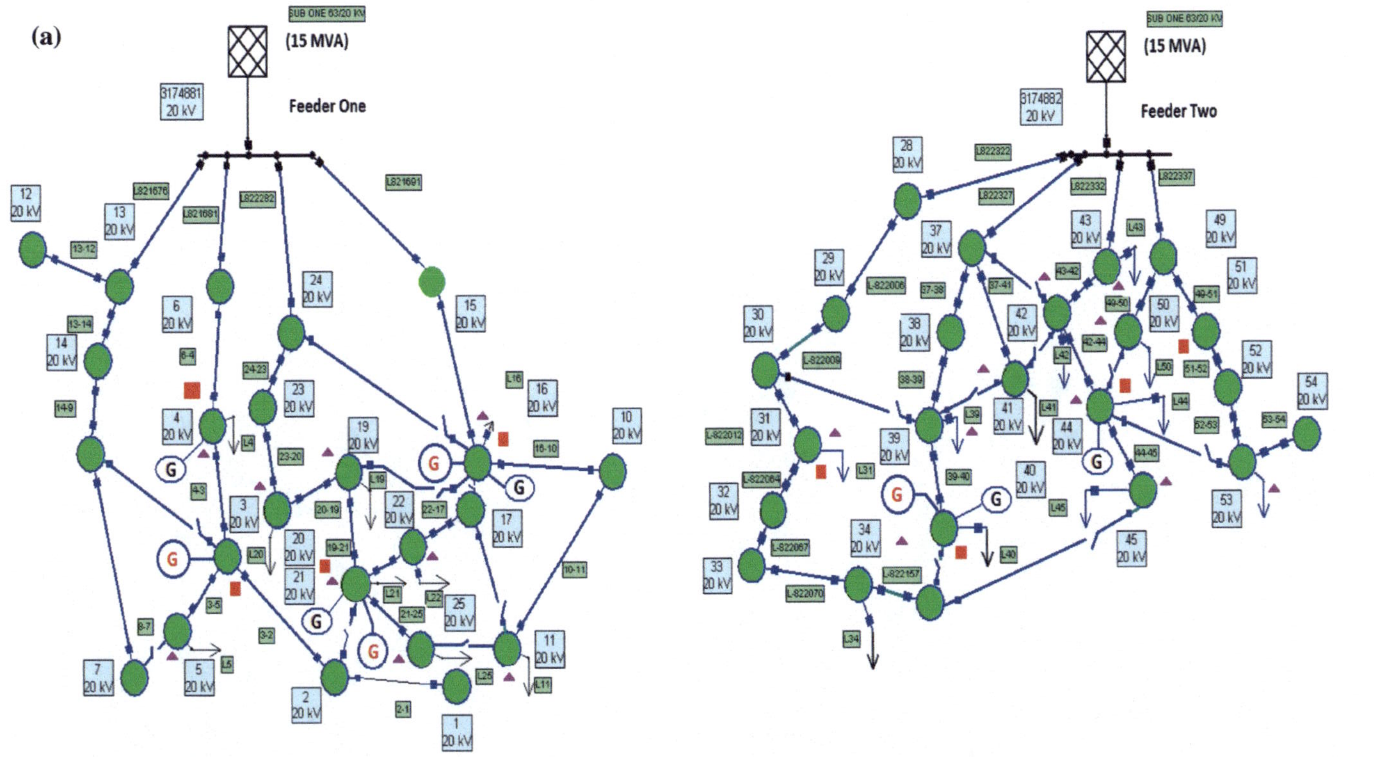

Fig. 8.12 **a** The optimized system topologies of the second scenario for the first year of planning year's horizon, **b** the optimized system topologies of the second scenario for the third year of planning year's horizon, **c** the optimized system topologies of the second scenario for the fifth year of planning year's horizon

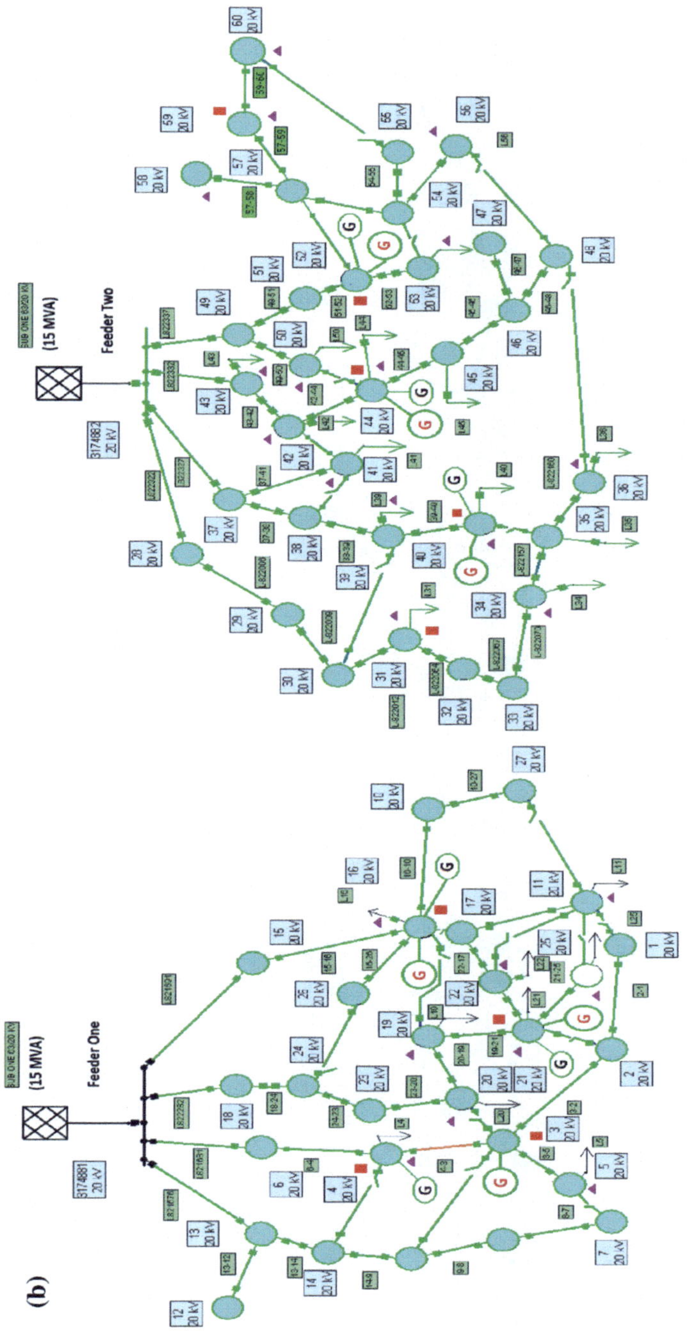

Fig. 8.12 (continued)

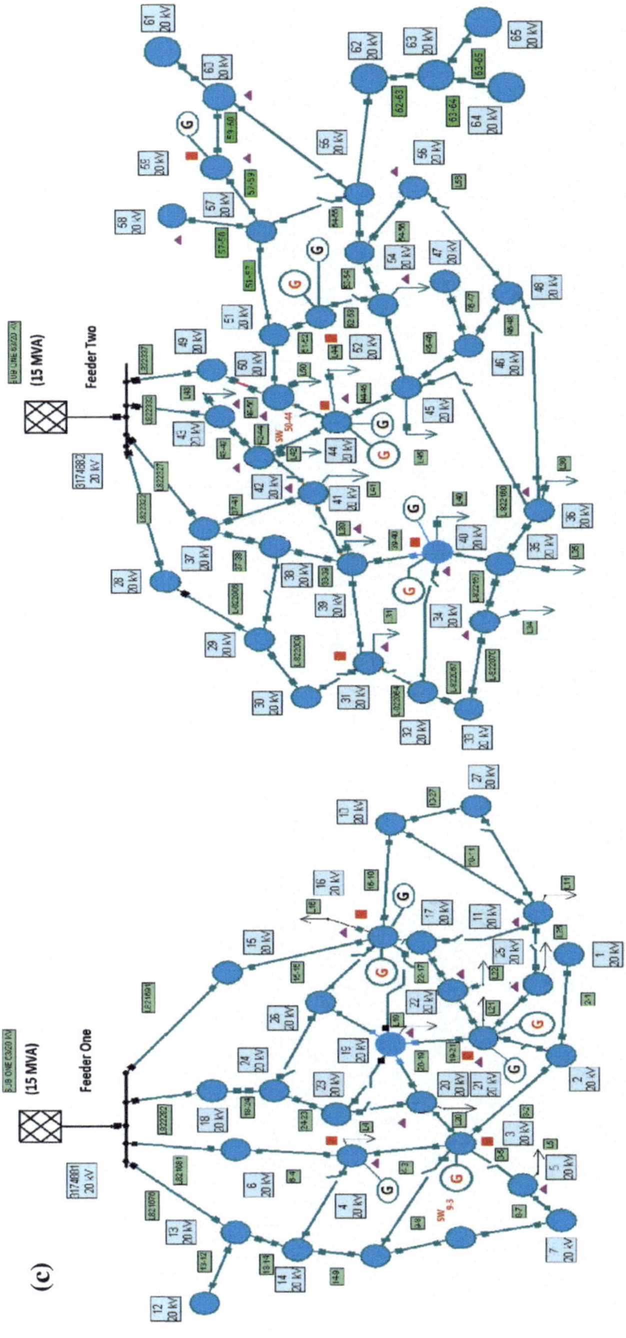

Fig. 8.12 (continued)

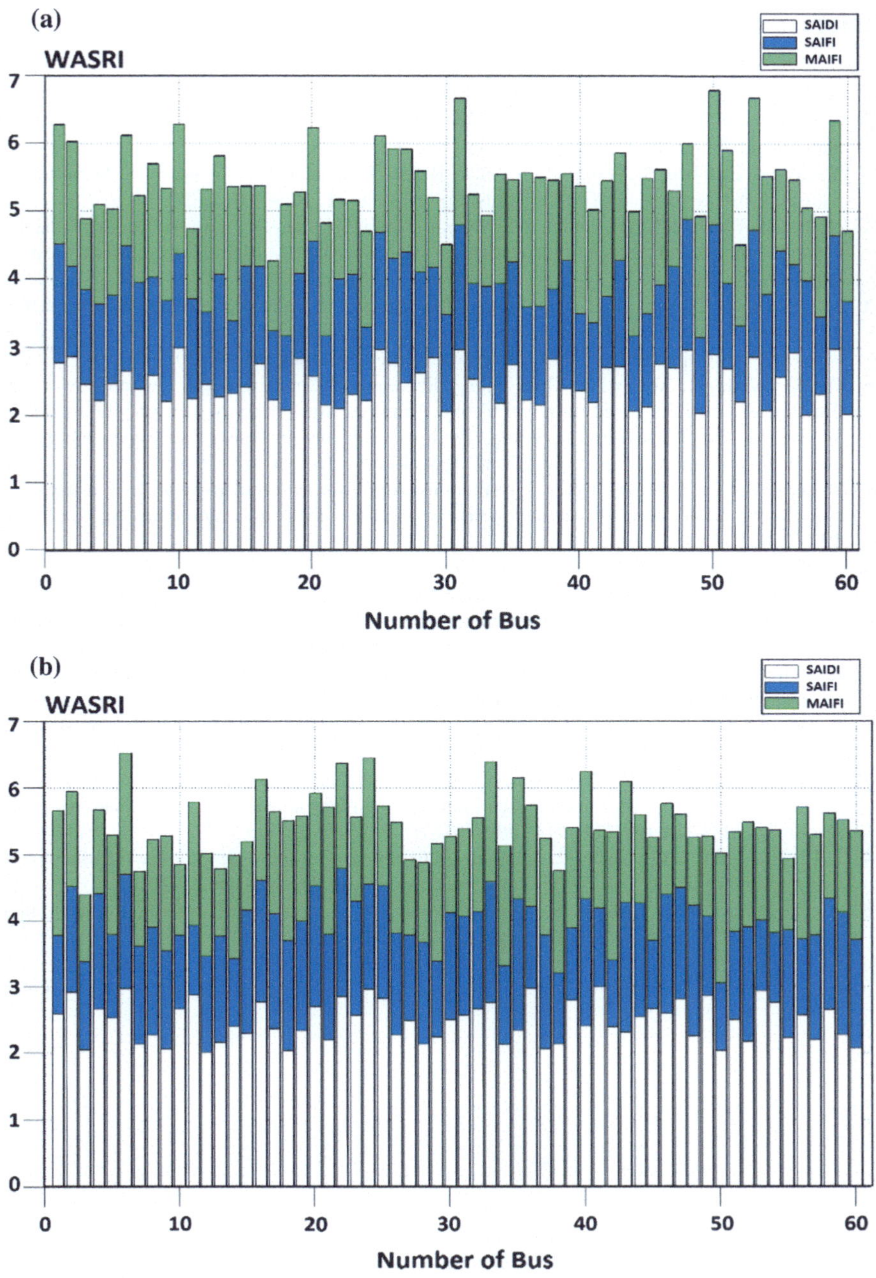

Fig. 8.13 **a** The WASRI index of the first scenario for the 5th of planning year's horizon, **b** the WASRI index of the second scenario for the fifth year

References

1. W. Kroger, E. Zio, *Vulnerable Systems* (Springer, 2011)
2. J.C. Whitson, J.E. Ramirez Marquez, Resiliency as a component importance measure in network reliability. Reliab. Eng. Syst. Saf. **94**, 1685–1693 (2009)
3. Y. Wang, C. Chen, J. Wang, R. Baldick, Research on resilience of power systems under natural disasters—a review. IEEE Trans. Power Syst. **31**, 1604–1613 (2016)
4. H. Farzin Firuzabad, M. Fotuhi, M. Moeini Aghtaie, Enhancing power system resilience through hierarchical outage management in multimicrogrids. IEEE Trans. Smart Grid **7**, 2869–2879 (2016)
5. M. Setayesh Nazar, M.R. Haghifam, M. Nazar, A scenario driven multiobjective primary-secondary distribution system expansion planning algorithm in the presence of wholesale-retail market. Int. J. Electr. Power Energy Syst. **40**, 29–45 (2012)
6. H. Haddadian, R. Noroozian, Multi-microgrids approach for design and operation of future distribution networks based on novel technical indices. Appl. Energy **185**, 650–663 (2017)
7. W. Cao, J. Wu, N. Jenkins, C. Wang, T. Green, Benefits analysis of soft open points for electrical distribution network operation. Appl. Energy **165**, 36–47 (2016)
8. T. Ding, Y. Lin, Z. Bie, C. Chen, A resilient microgrid formation strategy for load restoration considering master-slave distributed generators and topology reconfiguration. Appl. Energy **199**, 205–216 (2017)
9. S. Chowdhury, S.P. Chowdhury, P. Crossley, in *Microgrids and Active Distribution Networks*. IET Renewable Energy Series (2009)
10. Y. Lin, Z. Bie, Tri-level optimal hardening plan for a resilient distribution system considering reconfiguration and DG islanding. Appl. Energy **210**, 1266–1279 (2018)
11. A. Gholami, T. Shekari, F. Aminifar, M. Shahidehpour, Microgrid scheduling with uncertainty: the quest for resilience. IEEE Trans. Smart Grid **7**, 2849–2858 (2016)
12. S. Chanda, A.K. Srivastava, Defining and enabling resiliency of electric distribution systems with multiple microgrids. IEEE Trans. Smart Grid **7**, 2859–2868 (2016)
13. S. Mousavizadeh, M.R. Haghifam, M.H. Shariatkhah, A linear two-stage method for resiliency analysis in distribution systems considering renewable energy and demand response resources. Appl. Energy **211**, 443–460 (2018)
14. J. Li, X.Y. Ma, C.C. Liu, K.P. Schneider, Distribution system restoration with microgrids using spanning tree search. IEEE Trans. on Power Syst. **29**, 3021–3029 (2014)
15. A. Delgadillo, J.M. Arroyo, N. Alguacil, Analysis of electric grid interdiction with line switching. IEEE Trans. on Power Syst. **25**, 631–641 (2010)

Chapter 9
Malicious and Deliberate Attacks and Power System Resiliency

Fernando Georgel Birleanu, Petre Anghelescu and Nicu Bizon

Abstract Modern embedded systems control sensitive data and information depending where these systems are installed to accomplish required tasks. Due to this aspect, cyber criminals or hackers are motivated and determined to rob intellectual property of these systems through more and more sophisticated attacks. A huge problem in defending against these massive and various types of attacks is that in the last years attacks increased their complexity while the knowledge of an attacker decreased significantly because of the tools and devices they can find in the online world and free market. The most important challenges to defend against an attack are represented by these factors: speed of the attack, complexity of the attack and the simplicity of the tools that attackers used. A very often question that most of designers and developers of embedded systems ask is: Why cyber criminals commit attacks and what motivates them? Is it money? Is it celebrity? The answer starts with simple entertainment and extends to material benefits and finding, very often, valuable sensitive information that can cause serious damages to a system and its dependencies or even terrorism acts. Best case scenario is when the attacker is exactly the owner/the developer of the system or when he is demanding various attacks in order to figure out how defense mechanisms resist when facing attacks, how these can be improved and what are the challenges in building new ones. Therefore, this chapter is focused on two main ideas considering modern embedded systems based on Field Programmable Gate Array (FPGA) technology such as communication networks or cryptographic systems. The first idea refers to malicious and deliberate attacks performed against embedded systems starting with risks, threats and vulnerabilities that motivated hackers find and exploit and the second idea is about power system resilience and how attacked systems respond

F. G. Birleanu (✉) · P. Anghelescu · N. Bizon
Faculty of Electronics, Communications and Computers,
University of Pitesti, Pitesti, Romania
e-mail: birleanu.fernando@gmail.com

P. Anghelescu
e-mail: petreanghelescu@yahoo.com

N. Bizon
e-mail: nicu.bizon@upit.ro

© Springer Nature Switzerland AG 2019
N. Mahdavi Tabatabaei et al. (eds.), *Power Systems Resilience*, Power Systems,
https://doi.org/10.1007/978-3-319-94442-5_9

and decide what to do next. This chapter is organized in six parts as follows. The first part of this chapter is an introduction about attacks on embedded systems and a background that provides all the necessary information of how attackers and attacks evolved in the last years. The second part is focused on who performs these attacks and how systems are attacked. The third part refers to the main attacks on embedded systems and how these are classified depending on different criteria such as interlinking features, integration level or programmability level. The fourth part of this chapter is about power system resilience and how actual systems react or how they should react in case of malicious attacks. The fifth part refers, with examples, to the vulnerabilities existing in modern equipment that surrounds us and how these are or can be attacked such as mobile and communication systems and social apps that we use every day. The last part concludes the chapter and draws some goals for future research directions. The main purposes of this chapter are: to review and categorize all types of attacks against embedded systems based on FPGA, to show how attacks evolved from their beginnings until present, to bring to light who are the attackers as well as what motivates these hackers and to picture how "resiliency" feature should operate or operates during life-cycle of embedded systems when someone wants to perform an attack or succeeds one. Another important goal that this chapter aims for is to find and show others vulnerabilities existing in modern systems, especially communications, that most of us can not live without them.

Keywords Hacker · Attack · Deliberate · Malicious · Threat · Risk
Vulnerability · Power · System · Resiliency · Interlink · Communication
Complexity · Reliability · Embedded

9.1 Introduction

Today's modern systems that empower this world to behave in proper conditions, starting with simple communications systems and finishing with complex and innovative systems such as Smart Grid or Internet of Things systems, have vulnerabilities, thus, they are exposed to risks, sometimes to major ones than can cause serious damages.

The first two main goals for developers are to build efficient defense mechanisms that can counter attacks and to decide how the systems will respond, react and recover in case of successful attacks. These two targets represent the fundament in achieving increased resilience of critical systems and critical power infrastructures. Still, the efficiency and complexity of implemented mechanisms depend on the technology used to develop certain systems.

Field Programmable Gate Array (FGPA) technology has become one of the first options in most of modern embedded systems. The solid arguments for this choice are that these chips are right in the center of flexibility, fast time to market, performance, cost efficiency, simple design cycle and real-time guarantees. These are typical specifications of most of the embedded systems that surround us [1].

Main advantages of FPGA chips are the usage of fast clock rates, reconfigurability, Intellectual Property (IP) protection, parallelism and reliability. Malicious and deliberate attacks against FPGA-based embedded systems are designed to modify, interrupt, intercept or destroy the activity of targeted systems [1].

Nowadays, the big problem regarding these attacks is that the complexity level of attacks is very high and increasing while the knowledge and aptitudes of attackers are more and more at a low level and decreasing, in the last two decades. This happens because of the tools that can be found in the open market or on-line and using them usually comes down to selecting the type of the attack desired to perform. If in the 90s an attack was involving guessing passwords, breaking passwords, self-replication or backdoors, today an attack means chip cutting, chemical attacks, optical attacks or fault injection [1–4].

No matter whether an attacker wishes to break FPGA-based embedded systems or a computer network, commonly he is following a pattern to attempt an attack. This starts with the attacker, that can various from simple amateurs to experts or organizations, who dispose of dedicated tools and then he identifies the vulnerabilities that can be found depending on the design, configuration or implementation. Next is the action or the attack (hardware, software or firmware attack) that has a certain target with obtaining unauthorized results (reverse engineering, cloning, corrupt data). All this ends with different objectives from financial gain to stealing sensitive data or even worse, terrorism acts [1, 5].

In simple words an attack is a deliberate act that exploits a vulnerability, which represents a weakness in a system regarding its configuration, implementation or design. Associated with attacks are the threats that represent anything that can cause a potential danger to a system such as people or objects that can disturb the functioning, confidentiality, availability or integrity of that system. This can happen with bad intention, as an incident or as an act of nature [6, 7].

Most common types of threats include human errors, hardware or software failures, deliberate software attacks, deliberate acts of theft, sabotage, vandalism, espionage or information extortion, compromises to Intellectual Property and forces of nature [6, 7].

In 2013 researches by McKinsey (www.mckinsey.com) made a global survey on cyber risk-management maturity, involving 100 institutions from Europe, North and South America, Africa and the Middle East. The maturity level, on a scale of 1 to 4 (where 4 is the strongest), was divided is four sectors: nascent (less than 2.00), developing (between 2.00 and 3.00), mature (between 3.00 and 3.95) and robust (greater than 395). While 34% of the companies were rated nascent and 61% developing, only 5% obtained mature and none of them managed to obtain the highest level [8].

So, fighting just with the idea of cyber-attacks is a big concern nowadays with notable negative inferences in different areas than can slow down today's modern embedded systems such as mobile technologies. A critical concern is what happens right in the middle of an attack and after that when the system should be in the resilient status.

In the following we will see who the attackers or the hackers are, the tools they use and which are the main malicious and deliberate attacks against FPGA-based embedded systems. Also, we will find out how these systems react after major attacks and what are the risks and responsibilities in a hyperconnected world.

9.2 Who Are the Attackers and How Systems Are Attacked?

"Attacker" is a general word to describe those who perform attacks against all kinds of systems, that embraces several more specific terms such as phreak, cracker, cyberpunk, cyber-warrior and different others forms using the hacker term. All these terms appeared gradually along with the increasing development of modern technologies [9, 10].

The "phreak" refers to the person who breaks secured telecommunication system despite all the complex security mechanisms that all providers use. It started in the 1970s where phone phreaks were breaking telephone networks by matching the tones to steal long-distance services using custom made tools.

The "hacker" term, used for the first time back in the 1960s [11] to describe a programmer, was more often practiced in the 1980s and at the beginning they were computer researchers at prestigious research companies such as Massachusetts Institute of Technology. It was when the first home computers appeared and also the old modems. A hacker is defined as a person with computer programming and technology skills who likes to dig in the source code of different programs and see how it works who could increase the capabilities of computer code by removing, at that time "hacking", surplus machine code instructions from those programs.

The 1990s brought the cracker because the hacker membership wished to part the malicious attacks emphasized by the media from heavy hacking research and development performed by underground groups at that time. A cracker or criminal hacker, who possesses good technical aptitudes, wants to gain unauthorized access to a system by cheating or defeating its security mechanisms.

So far, the cyberpunk is the worst and most dangerous category and appeared in the 2000s. It is defined as a combination between the knowledge and technical aptitudes of the phreak, the cracker and the hacker.

IBM (International Business Machines) categorized these hackers in three types: clever outsiders, those with poor equipment that usually exploit vulnerabilities of a system, knowledgeable insiders, those with experience and sophisticated equipment and funded organizations that refer to specialist groups capable of designing extreme complex attacks using state-of-the-art technologies [1, 12].

In 2001, John Chirillo in his book [9], labeled hackers in five categories: the "communal hacker", the "technological hacker", the "political hacker", the "economical hacker" and the "governmental hacker". While the first category is about the most common type of hackers, the second category refers to actions of hacking strengthen by the absence of technology evolution. The "political hacker" is associated with those who want to be heard and who points the mass-media. The "economical hacker" is always looking for money and the "governmental hacker" is correlated with the common terrorist because of the involvements built on its actions.

Modern types of hackers refer to "wannabe" or "lamer", script kiddie, ethical hacker, Quiet, Paranoid, Skilled Hacker (QPS), cyber-warrior or mercenary, industrial spy hacker, government agent hacker and military hacker. The "wannabe" is that person who uses some tools or mechanisms from the free internet without having any idea or curiosity about their functioning. This type of modern hacker is often called "I would love to be a hacker". The "boy from the scripts" or the "script kiddie" is based on UNIX/Linux scripts that are created by others and his interest relies only on the consequences rather than to figure out how he actually got those results. The "ethical hacker" possesses very good technical skills and usually prefers building his own software tools in order to help others, the community or companies, in discovering bugs, no matter whether in the past he was on the "good" side or on the "bad" side.

The "QPS" is related to the ethical hacker using as few as possible tools made by others and at the lowest sense of being caught he will vanish. The "cyber-warrior" or "mercenary" is that type of hacker that is looking after financial gains and he is very skilled especially on focused areas, such as Wi-Fi or Web Defacing. Often, he is associated with extremist groups. The "industrial spy hacker" is about those spies who infiltrated in companies and managed to steal precious data and information with the advantage brought by Information and Communication Technologies (ICT). The "government agent hacker" is an external attacker from governments that performs highly-sophisticated attacks targeting business markets in different nations. Actions of the "military hacker" are very often combined with "state-sponsored attack". The term of "military hacker" appeared in 2004 and it wasn't received very well. But in the last years' history confirmed this kind of hacker [10].

In [11], an article about hacking last updated in August 2017, "hacker" is described as "an individual who uses computer, networking or other skills to overcome a technical problem" and "who uses his or her abilities to gain unauthorized access to systems or networks in order to commit crimes". Also, the article separates hackers in three types: white hat hackers, black hat hackers and gray hat hackers. While white hat hackers are associated with ethical hackers, black hat hackers perform malicious and deliberate attacks as Distributed Denial-of-Service (DDoS) or identity theft and violate laws for many reasons, including money or increased fame. Gray hat hackers are in the middle between white hat hackers and black hat hackers, who ca offer to fix bugs they found rather than exploiting those bugs for different illegal advantages.

A report from Fortune (http://fortune.com/section/tech/), in [13], shows that hackers are regular people between 20s and early 30s who poses technical aptitudes and who own above average computers. If once they were looking for money, today their best targets involve confidential information or intellectual property.

But how much cost these actions performed by today's hackers? In another report from Fortune in [14], in the United States the average total cost of data breach was $5.85 million only in 2014 and this year, in 2017, it is estimated that this cost will reach $7.35 million. Last year, in 2016, malicious cyber-attacks cost the world's economy $450 billion with $4 billion only from one attack—the WannaCry ransomware—that a few months ago damaged computers in over than 150 countries around the world.

All this type of hackers described until here are on the same side, external attackers. So, what about insiders?

An insider can be anyone from students and employees to anyone authorized or authenticated to perform certain actions in a system, who has serious advantages from external attackers through the information and data inside [15, 16]. Usually insiders are identified using three characteristics: they have access, they are well trusted and they possess enough knowledge and capabilities [15]. Sometimes it can be very hard to say who is an insider in an organization, so, it is necessarily to understand the differences between doing something on purpose and doing something that happens unintentionally, between thievishly and clear actions or between malicious and deliberate attacks and threats versus accidental ones [16].

All these types of hackers, insiders or external attackers, have been debated in the last decades in a many books and press. Among dozens of famous hackers, a short list is presented next.

Kevin Mitnick managed to escape from authorities for two and a half years after he was convicted of a number of criminal computer crimes. Mitnick was arrested in 1995 and served five years sentence in a federal prison after hacking networks of forty high-profile corporations in United States. After his release in 2000 he founded a cybersecurity consultancy company and he named his past activities as "social engineering" and not "hacking" [11, 17, 18].

In the 2000s, "MafiaBoy" alias Michael Calce, a "script kiddie", was arrested in Canada after launching a series of DDoS attacks against Dell, Amazon, Fifa.com, Yahoo! or e-Bay [10, 19]. He was only 14 years old.

In 2000, Jonathan James or the "c0mrade" become the first young person (16 years old) who was incarcerated for hacking into websites such as National Aeronautics and Space Administration (NASA) and the United States Department of Defense. When he was 25 years old, back in 2008, he committed suicide as he believed he would be convicted of several malicious network attacks against big companies that he didn't commit [11, 17].

"Anonymous" debuted in 2003 and it is a group of hackers from around the globe with no hierarchy. They focus on defacing and defaming websites, on making public personal information of their and victims and on denial-of-service attacks. Targets of them include the governments of United States, India and Australia, some parts of the dark web and also Amazon, Sony and PayPal [11, 17].

Adrian Lamo or "the homeless hacker" was arrested in 2003 and convicted in 2004 for hacking companies such as Microsoft, Yahoo and New York Times to exploit their security bugs [11, 18].

To see what the implications are after serious attacks, here are some big companies that were hacked in the last five years [20]. In 2012 LinkedIn said that 6.5 million accounts had been hacked and in 2016 this number raised to 117 million accounts. In 2013 Target declared that after the breach 10 million customers' personal and financial information was exposed. In 2015 hackers stole customer credit card data of dozens of Hilton Hotels' chains across the country. In 2016 attackers stole $3.2 million from 9000 accounts in Tesco Bank. In 2017 an Eastern European criminal organization reportedly used phishing methods to steal the credit card information of millions of customers of the Chipotle.

One the one hand, external hackers attack systems and devices with three methods: hack attack, shack attack and lab attack [1, 4]. In the first method, the lowest level method, the attacker can perform only software attacks. The second method refers to the low budget hardware attacks with free market or online equipment that can't perform complex attacks. The third method is the most dangerous and invasive with access to laboratory equipment that can perform very complex attacks such as transistor-level reverse engineering or attaching microscopic logic probes. One the other hand, insiders' job is easier because their work can resume at stealing sensitive data and information with a simple flash drive or through e-mail.

These hacking techniques must leave no traces about who did the attacks, so, usually hackers try to cover their tracks through different ways such as using proxies to hide their IP address or using tunneling techniques [21].

Nowadays we are "covered" in news and events about hackers, their actions, the latest data and information security breaches and increased complexity attacks. All this gained a new name, called the "cyber warfare" which was amply debated in [22]. In this wide world cyber warfare after we have seen what about hackers from the apparition of the first computers until today it is necessarily to see which are the main and most common malicious attacks that motivated attackers perform against FPGA-based embedded systems.

9.3 Malicious and Deliberate Attacks Against FPGA-Based Embedded Systems—Classification

Although cyber-security through software encryption is becoming more and more widespread in these days, hardware is still the way of choice for most commercial and especially military applications. The reasons are the following [23, 24]:

- *Encryption/Decryption Speed*—usually, cryptographic algorithms contain complicated operations applied to the messages and in hardware these operations can be realized in pipeline or in parallel obtaining real-time encryption/decryption.
- *Security Assured through Hardware Devices*—the hardware can be encapsulated and, in this way, can obtain physical protection.

The FPGAs devices give advantages for reducing time to plan the cryptographic algorithm, energy consumption, reliability, high encryption/decryption speed and safety.

Although, malicious and deliberate attacks against FPGA-based embedded systems are various and in increasing complexity because of the easiness in finding tools and software in the free market and on-line. A new classification of these attacks, discussed in [25], separates them in three categories depending on the programmability level, the integration level and on the life cycle phase.

The first criterion, the programmability level, includes software attacks, hardware attacks and firmware attacks. Software attacks are associated with malicious code or malware and refer to viruses, software Trojans, worms, adware, spyware and time bombs [26].

Viruses are programs that can infect other software programs by changing them in order to introduce a copy of themselves into the software programs and usually there are four phases to describe their lifetime: the dormant phase when the virus is expecting an event to occur and then it activates, the propagation phase when the virus creates a copy of itself and inserts it to the specific program, the triggering phase when the virus starts executing its functions and the execution phase when the virus behaves as desired and damages software programs. The most common types of viruses are: file-deleting viruses, file-infecting viruses, stealth viruses, mass mailers, boot sector viruses, macro viruses, polymorphic viruses and memory resident viruses [27].

Software Trojans are designed for system file access in order to download, rename or delete files from the infected device, steal passwords, disable peripherals, disable virus protection, make registry changes or shut down the system that was infected.

Worms refer to malicious software that are built to extend through the network by self-replicating after infecting a computer. Adware is about web bugs or web beacons that collect information about what users visit via Internet and then offers specific advertisements. Those sites that use adware motivates this as a way of improving customers experience, but experts believe is a way of spending more money and that the real problem is what such web sites do with what was collected.

Spyware describes software that can be installed on personal computers to gather information such as passwords, names, web sites visited, data about personal apps or Operating System (OS) and other personal material. Time bombs or logic bombs is a type of software attack that can be activated in the future due to a specific event that occurs. After it was activated the computer fails or it is affected by malfunctions.

Software attacks are also related to the packet switching protocols that consist of replay attacks, man in the middle attacks and eavesdropping attacks. Aimed to steal sensitive and personal data and information, a man in the middle attack happens when someone intercepts a communication between two systems (social media, e-mail, real-time transactions, etc.). Figure 9.1 reveals a generic life-cycle of malware attacks debated in [28].

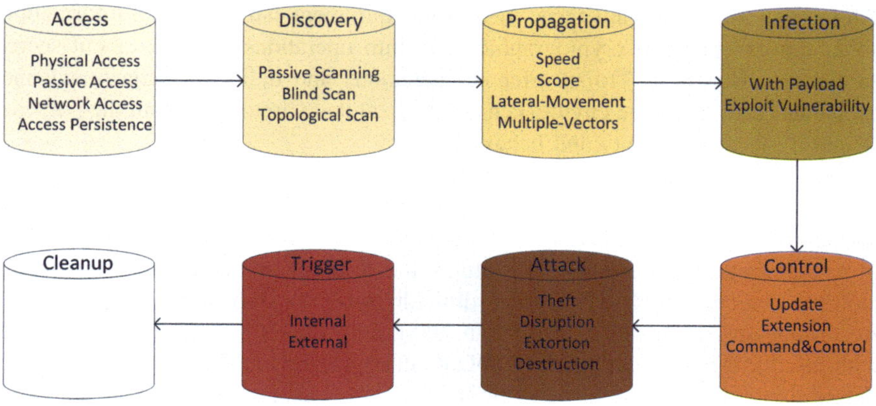

Fig. 9.1 Life-cycle of malware attacks

These eight steps in the picture above show a general life-cycle of malware-based cyber-attacks:

- *Access*—this first step represents the heart of any attack through gaining direct or remote access to critical components of the FPGA-based embedded system;
- *Discovery*—scanning methods and tools are used in the second step to discover new vulnerabilities or new victims if the access is limited to perform a bad intended action;
- *Propagation*—after the second step is accomplished in the third step of the life-cycle the malware uses exploits to propagate to new targets or interest nodes;
- *Infection*—in this next step the target in infected after the malware gained access;
- *Control*—the fifth step refers to modern malware mechanisms that are external controlled;
- *Attack*—the sixth step is the actual attack where its success depends on the amount of access to the compromised resources;
- *Trigger*—step seven shows the fact that modern attacks are very well coordinated and that attack triggers can be remotely turned or they can be hard-coded;
- *Cleanup*—the last step means that attackers usually try to cover and hide their tracks.

Hardware attacks embrace many forms such as reverse engineering, timing attacks, data and traffic monitoring, hardware Trojan, Denial-of-Service (DoS), Distributed Denial-of-Service (DDoS), Electromagnetic Analysis (EMA), Simple Power Analysis (SPA), Differential Power Analysis (DPA) [29]. Targets of hardware attacks are the components of communication infrastructure, industrial and common use control systems, network appliances and surveillance systems [30].

Reverse engineering is when the components of a system or device are separated in order to manage how that system or device functions with the goal to duplicate it.

Timing attacks through advanced measuring equipment can show vital information about the execution of cryptographic algorithm operations and also about secret parameters. Hardware Trojan refers to malicious and deliberate adjustment and attachment to a hardware circuit such as an Integrated Circuit (IC) with the purpose to modify its functionality and behavior.

In a Denial-of-Service attack the targeted system or its resources are unavailable to its users for a limited or indefinite period of time. This is obtained usually by flooding the system or network with superficial traffic in the attempt to overload it. In a Distributed Denial-of-Service attack the flooding traffic comes from many multiple sources which makes very difficult trying to stop this kind of attacks. According to [31] there are more than 2000 daily DDoS attacks world-wide and with just $150 one can buy a week DDoS attack on the black market. Also, Cisco (www.cisco.com) reports that the number of DDoS attacks increased by 172% in 2016 in the entire world and believes that 3.1 million attacks will happen by 2021 (a grow by 2.5 times). In the first quarter of 2017 Nexusguard (www.nexusguard.com) observed a 380% increase in the number of DDoS attacks compared with 2016 [14].

Firmware attacks refer to attacks against the OS kernel. FPGA-based embedded systems store the firmware in flash memories to allow internet updates that can led to unwanted privilege break. Electromagnetic radiation analysis attack is based on measuring electromagnetic signals as a fact of currents flow in a device and physical access to the system is not a key point. After measurements are accomplished attacks are similar to power analysis attacks.

Simple power analysis and differential power analysis attacks demand physical measurements of the current consumption depending on the time. This is made with probes and everything is based on the relation between the current used by a chip and specific performed instructions in a particular time. While in a SPA attack the hacker observes the chart of current consumption over time and tries to correlate this with cryptographic operations, a DPA attack uses statistical tests to isolate an interest signal from noise and other additional power signals in order to find the relation between processed instructions and power traces. SPA and DPA attacks had been used to break implementation of cryptographic algorithms such as Ron Rivest, Adi Shamir and Len Adleman, who invented it in 1977 (RSA) or Data Encryption Standard (DES).

The second criterion, the integration level, includes board level attacks, chip level attacks and Intellectual Property attacks.

Board level attacks are divided in invasive attacks, semi-invasive attacks and non-invasive attacks (passive attacks and active attacks). Invasive board level attacks need physical access to the board and the physical properties of the Printed Circuit Board (PCB) are permanent modified. Techniques used to reproduce the PCB layout include chemical and mechanical imaging processing methods, optical or scanning electron microscopes, probe stations and Focused Ion Beam (FIB). Semi-invasive board level attacks can be performed only against single or doubled layered boards using scanning methods and Computer Aided Design (CAD) tools to reproduce the design. Passive non-invasive board level attacks refer to observing and monitoring data traffic without any interaction with the board. This kind of attacks are difficult to detect and for this reason cryptographic methods including bio-inspired ones are used in order to hide the content of sensitive information [32]. Active non-invasive board level attacks are designed to force the chip to act abnormally through wrong operations. Usually clock signals and supply voltage are modified. Active attacks usually modify the messages, modify system files and masquerading as another entity.

Chip level attacks are related to bitstream reverse engineering, chip cloning, fault attacks (radiation-induced faults) and remote reconfiguration. Intellectual Property level attacks are about modifying the source code of the system so that the attackers can gain full control over it and access this system whenever they want.

The third criterion, the life cycle phase, provides three types of attacks: design phase attacks, fabrication phase attacks and after-production attacks.

On one hand, fabrication phase and after-production attacks are related to attacks performed after the device was released in the market and include design cloning, copies or extracting confidential information. On the other hand, design phase attacks that contain design cloning, spoofing, fault generation and adding unwanted components such as a simple kill switch are related to insiders who can obtain information and data access more easily or who can shut down a system with some little effort.

There are three types of attacks assimilated to insiders [15]: misuse of access, defense bypass and access control failure. Misuse of access insider attack refers to the situations when an insider uses his access privileges to perform illegal actions and it is very hard to detect and to be aware of. Defense bypass attack offers insiders some important advantages over external hackers because they are already inside and they skipped some security measurements.

Access control failure usually happens because of the equipment that sometimes breaks or malfunctions causing technical issues. In a report in [31] latest common hacking techniques include cookie theft, bait and switch, eavesdropping, malware, DoS/DDoS, keylogging, watering hole and Wireless Application Protocol (WAP) attacks, man-in-the-middle attack and phishing.

Cookies are very dangerous because among information of web sites visited they can retain personal data such as passwords or financial data and if someone manages to steal them then can very easy decrypt and read those data that are confidential to others.

Common bait and switch hacking technique is about those advertisings or "irresistible" app downloads acquired by hackers that after accessing it redirects the page in the website to somewhere that runs malicious code and where computers get infected with malware.

Keylogging is when someone manages to record in a log file what keys are being pressed on a keyboard. Thus, one can find out passwords or other sensitive personal information. WAP attacks are about creating a fake wireless Access Point (AP) that enables hackers to gain access over a computer, device or system and then they can observe, monitor, intercept and hijack data traffic.

According to [14], in 2016 among all kind of attacks 62% of total global breaches were organized featuring hacking. From all of these 51% included malware and 81% included breaking weak passwords or stealing passwords. Also, 43% of the total breaches involved social attacks and 14% involved privilege misuse, while only 8% occurred after physical actions against embedded systems.

Against all these variate attacks that stress humanity every day, high-tech systems must respond and react in a proper way and also must achieve high power system resiliency.

9.4 Power System Resilience at Complex Attacks—Models and Case Studies

Power system resilience against complex attacks is a big concern and a precious feature to achieve that defines all mechanisms, tools and countermeasures taken to restore after major disrupting events. The term "resilience" means "to leap back" that translates as the ability to recover [33–35]. The basic idea for any existing system is to decrease the chances of being attacked, to manage to deal with attacks that succeed by absorbing then and to recover as fast as possible after increased complexity attacks.

Resiliency has four properties that are rapidity, redundancy, robustness and resourcefulness. Efficiency of these properties depends on technical, managerial, social and economic aspects that all combined should ensure it. A system status has four stages to pass through when events occur: the pre-disaster stage, the disaster stage, the restoration stage and the long-term recovery stage [36, 37]. Between the first two is where the disruptive event happens due to different kind of attacks. For a good resilient system, the goal is that the restoration time is as short as possible.

For all the attacks presented in the previous section exist different mechanisms to try to detect and to counter them. Viruses are detected and cleaned through antivirus software that consists of four generations [27]. The first one can detect only known viruses and it is based on certain virus signature or virus structure. The second generation uses integrity checking and does not relies on virus signatures. The third antivirus software generation manages to see viruses by the malicious actions performed by them. Last generation of antivirus software refers to scanning methods and access control facilities.

If a high resilient system is based on the ability to prevent, react and response, for DoS and DDoS attacks defense methods are categorized in three: preventive methods, reactive methods and response methods. Preventive methods consist of multiple mechanisms such as firewall systems or security patches that must eliminate or at least try do decrease the possibility of DoS/DDoS attacks.

Reactive methods are used to mitigate the shock of attacks with three main mechanisms: pattern attack detection, anomaly attack detection and hybrid attack detection. For pattern attack detection the structure signatures of known existing attacks are kept in databases and communication are observed, monitored and compared with the data from databases in order to detect if any known attack intends to malfunctions a system.

Anomaly attack detection is based on how an embedded system or a device should function in normal conditions. Current behaviors are compared with normal behaviors to see if known or unknown attacks are trying to infect and disrupt the system. In hybrid attack detection which is used by many intrusion detection tools, the first two mechanisms are bounded to detect new attacks signature and to update those databases.

Response methods are related to the reactive methods and use three mechanisms: agent identification, filtering and reconfiguration. The agent identification mechanism offers identity data about the devices from where attacks are initiated to minimize the shock of malicious and deliberate DoS/DDoS attacks.

When using filtering mechanism, the attack flow is filtered out absolutely complete. The reconfiguration mechanism is very useful for response methods because it can change the network configuration to isolate the devices that perform these specific attacks.

For achieving resilience against hardware attacks, Fraunhofer (www.fraunhofer.de) Institute in [38, 39] proposed a solution through counterfeit-proof chips. The idea is that chips have different physical characteristics (density, length, thickness) that provide a unique "fingerprint" using Physical Unclonable Function (PUFs) techniques implemented. With these feature chips as FPGAs or Application Specific Integrated Circuit (ASICs) and further systems can obtain high physical security.

Also, for FPGA-based embedded systems a very good method to detect and protect against attacks is using tampering mechanisms and encryption methods by developing methods to ensure the availability, integrity and confidentiality of these systems.

While WAP attacks can be avoided with simple methods such as not using apps from untrusted sources and also free hotspots, keylogging can be counter with virtual on-screen keyboard which encrypt or scramble input text as soon as someone clicks.

A big concern is how we manage to deal with insiders because this is and will remain a challenge for many years ahead because they have more access and more capabilities to attack. The two main approaches are to defend against them and to detect them after they committed illegal activities.

We can assume that a system can be protected and have a comforting level of resilience against complex attacks if the below security principles detailed in [40, 41] are respected:

- *Economy of Mechanism*—different complex security mechanisms and methods can lead to wrong configuration of parts or components of the system;
- *Least Privilege*—to perform certain actions, a user or software must use a minimum set of privileges;
- *In-Depth Defense*—it requires at least two security mechanisms that should deal with synchronous attacks;
- *Isolation*—critical parts of the system should be isolated from external access;

- *Minimization of Attack Area*—this principle demands a minimum number of access points into a system while maintaining optimal functionality;
- *Separation of Privilege*—multiple access steps;
- *Psychological Acceptability*—security mechanisms and methods can fail;
- *Open Design*—closed designs are more interesting and attractive than open designs in terms of potential attacks.

In [34], the proposed risk optimization model for enhanced power grid resilience against physical attacks aims for more reliable protection of critical components of the transmission systems at minimum costs and with controlled risks. The model starts with some assumptions as the limited budgets and protection schemes of the hackers to tamper with the specific system, the different reliability levels at different costs for the transmission lines and the fact that risks are computed regarding the total load curtailment and continues with complex calculations involving power constants and parameters and decision variables such as the probability of the attacks success. While experimenting different scenarios with increasing numbers of buses, generators and transmission lines, the results show that conservative protection plans, in which higher risk of unsatisfied demand is not accepted, require protecting a larger number of transmission lines while using also a larger number of the higher-reliability protection methods, that in the end will raise the total investment costs.

Another model [28] for a secure and resilient power system against complex attacks, in this case the smart grid system, shows a list of measures that must be implemented in order to achieve higher system resilience, as follows:

- *Password Policy*—requires strong password policy for long and non-repeating alphanumerical series;
- *Providing Data Confidentiality, Integrity and Availability*—refers to specific techniques and protocols that are implemented to prevent different types of attacks;
- *User Management*—this measure is similar to the least privilege principle;
- *User Education*—this is a basic security measure;
- *Reduced Remote Access*;
- *Security Updates*—this measure intends to prevent vulnerabilities and to decrease the probability that a system can be attacked;
- *Network Segmentation*—the network is divided in subnetworks that have specific purposes;
- *Strict Firewall Rules*;
- *Anti-Virus*;

A recent paper [42] proposed a graphical model (Fig. 9.2) to quantify the security of cyber-physical systems (systems detailed in [41]) when dealing with various attacks that can damage or overstress critical devices inside the system. The model starts with the normal state of a system and continues with different possible states such as warning or penetrated state as a result of different parameter changes

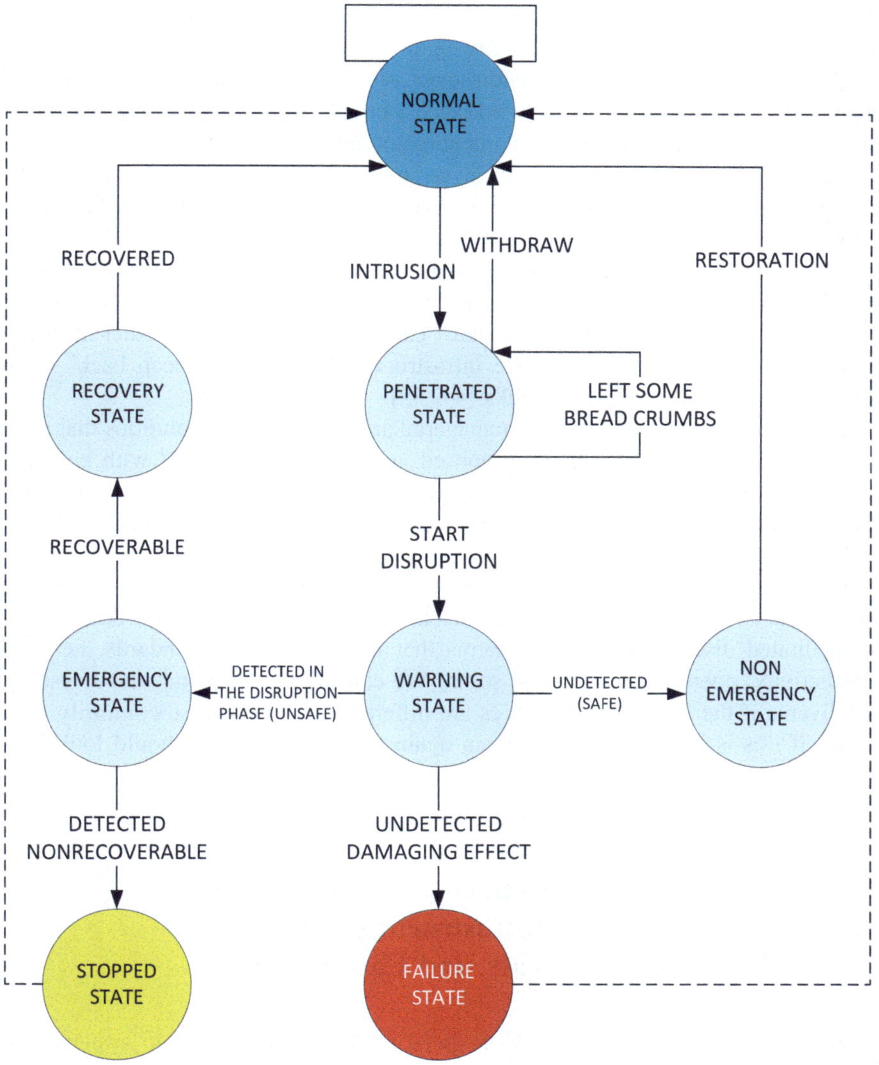

Fig. 9.2 Graphical model for the security and protection of cyber-physical systems against attacks

(detection interval, detection probability, hacker's level of knowledge regarding the targeted system, time interval until disruption).

Insisting on the particular case of cyber-physical systems, in [43] were recommended a list of countermeasures mechanisms (for each of the three layers specific for cyber-physical systems) so that these systems can achieve a higher resilience and a higher level of protection.

For the first layer, the perception layer, recommended countermeasures mechanisms are: access control, data encryption, certification, authentication, trust management, environment monitoring, secure routing protocol, key agreement, sensor data protection and lightweight encryption.

The middle layer, the transmission layer, comes with: attack detection mechanism, robust routing protocol, network access control, hop by hop data encryption and across heterogenous network authentication and key agreement.

The third layer, the application layer, must ensure the following methods: intrusion detection, user authentication and authorization, end to end encryption, trust management and Peer-to-Peer (P2P).

After all, it seems that the two keywords to ensure high power system resilience and robustness for embedded systems and critical infrastructures are innovation and diversity. With higher robustness systems can better prevent and predict intrusion incidents and with higher resilience infrastructures are able to "leap back" or to "bounce back" from complex attacks [44, 45].

In this case innovation refers to managerial and technological solutions that must be combined in order to achieve proposed goals for a system and with ensuring security, protection and power system resilience for FPGA-based embedded systems or other critical infrastructures that surround us and that are already part of our daily activities.

Modern infrastructures incline to be more diverse but this should come in line with greater resilience against cyber-attacks. Diversity brings the introduction of premeditated discrepancies into systems that include pointed standards, network connectivity, operating system and so on. We can say that two embedded systems are diverse if their key characteristics are different and not diverse contrarily.

So, if this is how a resilient system against complex attacks should look like, why most of the apps that we use every day are so vulnerable?

9.5 Opportunities and Challenges for Smart Grids and Other Critical Infrastructure—Risks, Vulnerabilities, Threats, and Case Studies

Nowadays we live in the era of high-speed modern communication systems and social media platforms such as Facebook, Twitter, Instagram, WhatsApp. All these modern embedded systems possess vulnerabilities and have risks and threats that hackers try to exploit.

A top of the biggest and more dangerous social media security threats refers to the lack of a social media policy, to employees, to social networking sites, to social engineering and to mobile apps [46, 47]. It is a fact that social media has become "the gateway for malware" because it is world-wide used and extremely easy to gain access [48, 49]. Basically, these platforms that some say that you don't exist if you don't possess an account, had become the perfect channel for malware spreading and infecting making experienced hackers to love companies who use the facilities of social media [50]. Things that are posted on social media "walls" help attackers to find out sensitive details about the company and other critical data that in the end targets a result consisting in highly sophisticated socially engineered illegal hacking events.

One of the most complex modern system is Internet of Things that is try to interconnect a wide variety of devices via Internet and so far, had partially managed to succeed. IoT risks include code modification, key compromise, password-based vulnerabilities and man-in-the-middle. Researchers found bugs and gaps in the security of IoT that are related to information flow control, access control, authentication, updates and hardware and network layer constraints [51–54]. Also, one of the biggest IoT problems is about devices connected and their huge amount. In [55] risks and threats of IoT devices suggest that 70% of devices will use unencrypted network services and 80% of them will not even use enough complexity for passwords.

In [56] there are presented some security threats specific for IoT sensors that are being integrated into smart cities systems:

- *Data Loss*—sensors must be managed and used properly through efficient policies and procedures. Otherwise data loss appears and the process is compromised;
- *Availability, Integrity and Confidentiality Compromise*—only authorized parties should have access;
- *Remote Exploitation*—the sensor and server channel should be as secured as possible to avoid remote exploitation;
- *Eavesdropping*—transmitted data can be intercepted and manipulated.

Also, another problem is linked with all this data generated in a smart city that it is stored in the Cloud and the main threats and vulnerabilities are:

- *Denial of Service Attacks*;
- *Malware Injection*;
- *Malicious Insider Threats*;
- *Data Locations and Regulation Boundaries*;
- *Application Vulnerabilities*.

The [57] presents a vulnerability assessment overview for a smart street lighting system (Fig. 9.3) and which are the step by step targets of the attackers (Table 9.1), where the main security threats are:

- *Private IP Disclosure*—this allows hackers to find the private address used internally;
- *Clickjacking*—through these trick users click on different web links and after malicious content is activated their computers are controlled by those hackers who tricked them;
- *Cross-Site Request Forgery*—this attack makes the victim to perform undesired actions and requests;
- *Session Hijacking*—through this exploitation, attackers gain unauthorized access to sensitive data in the remote server;
- *Path Disclosure*—this attack makes visible the location of critical data.

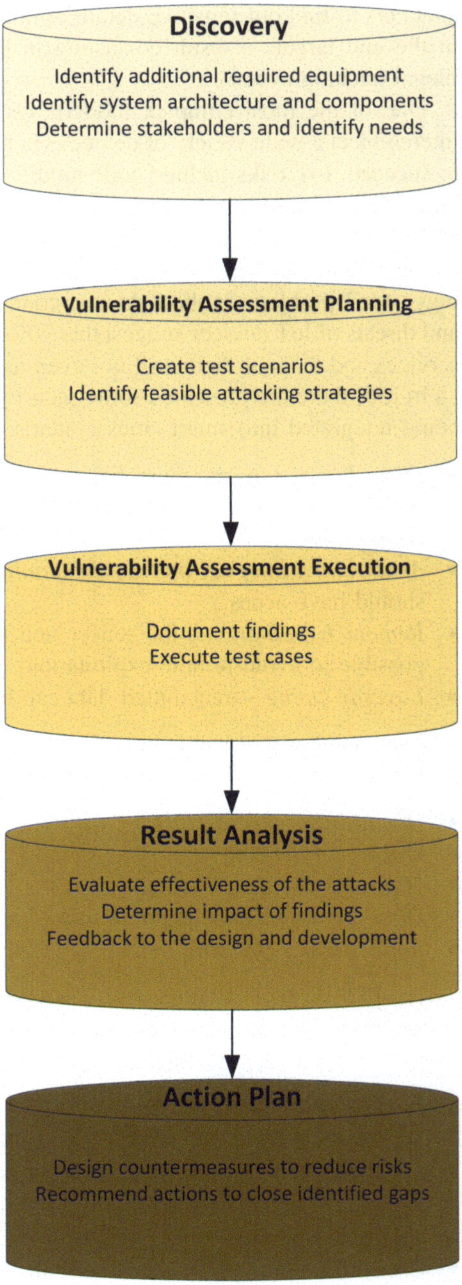

Fig. 9.3 Vulnerability assessment procedure for a smart street lighting system

Table 9.1 Step by step attacks for smart street lighting systems

Forms of access gain to the system	Attack steps	Objective
Network access	Device spoofing	Network compromised
Network access	SYN flood attack	Network compromised
Network access	Eavesdropping, network fingerprinting, network protocol vulnerability, protocol specific exploit	Network compromised
Network access	Session hijacking, exploit neighboring node, packet flood attack, DoS attack	Service interruption
Network access	Session hijacking, routing table overflow attack, DoS attack	Service interruption
External device	Jamming attack, DoS attack	Service interruption
External device	Node replication, exploit neighboring node, packet flood attack, DoS attack	Service interruption
External device	Node replication, routing table overflow attack, DoS attack	Service interruption
External device	Outside eavesdropping, eavesdropping attack	Info gathering and privacy
Physical access	Slander attack, physical tampering	Service interruption
Physical access	Node displacement, physical tampering	Service interruption
Physical access	Node destruction, physical tampering	Service interruption
Physical access	exploit, routing table overflow attack, DoS attack	Service interruption
Physical access	Exploit, exploit neighboring node, packet flood attack, DoS attack	Service interruption
Web server access	Session hijacking	Info gathering and privacy
Web server access	Insider attack	Web server compromised
Web server access	Session sniffing, session hijacking	Web server compromised
Web server access	XSS attack, session hijacking	Web server compromised
Web server access	MITM attack, session hijacking	Web server compromised

After all these issues a recent article calls IoT as “The Internet of Hackable Things” [58] in which it is believed trying to secure a “tsunami” of devices will represent a “disaster” making each one “remotely hackable”.

Because there is no perfectly secured system in this world, Smart Grid has its own vulnerabilities as follows: customer security, physical security, huge number of intelligent devices, the lifetime of power systems, more stakeholders, the use of Internet Protocol (IP) and commercial off-the-shelf hardware and software and the implicit trust traditional power devices [59]. Also, Smart Grid has a lot of issues as authenticating and authorizing users to substation, consumers to smart meters, maintenance personnel to smart meters, users to outdoor field equipment and Home Area Network Devices (HAN) to or from gateways. Other issues include smart meters' key management, securing communications, side channel attacks on field equipment, insecure firmware updates and patch management [60].

Along with some security issues such as data tampering, privacy issue, compromising and malicious code, identity spoofing and so on, [61] comes with several security challenges for IoT-based Smart Grid, that requires to be taken into consideration:

- Deployment;
- Constrained resources;
- Mobility;
- Scalability;
- Legacy systems;
- Time and latency constraints;
- Bootstrapping;
- Trust management;
- Interoperability;
- Heterogeneity.

In paper [62] the security vulnerabilities for IEC 61850, an international standard for communication networks, especially suited for smart electrical grids, are:

- Protocol vulnerability (poor access control, missing encryption of sensitive data, improper authentication);
- Improper security service mapping (inequality of security level, improper protecting data mapping, improper protecting service mapping);
- Improper protocol mapping (improper service mapping, improper data object mapping);
- Network design weakness (nonexistent firewall, weak firewall rules, missing encryption for critical data communication);
- Insecure gateway system (weak password policy, improper user authentication, poor log management, missing integrity check, lack of event audit, poor system access control);
- Insecure configuration tool (software vulnerability, insufficient logging, improper user authentication).

All these risks, vulnerabilities and threats that exist in critical infrastructure should be managed with efficient and reliable methods so that users could enjoy the services of highly secured modern systems. In the end of Malicious and Deliberate Attacks and Power System Resiliency, the next section concludes this chapter.

9.6 Conclusion

We live in a hyper connected world where information is the key that makes everything work in normal conditions. All these communication systems such as telecom or social media that exist today or the future ones that will be invented someday due to continuous technological advances must accomplish a major feature which is security insurance of confidential data flow.

Malicious and deliberate attacks against FPGA-based embedded systems come in many variate forms and their complexity increases with every day that passes due to the easiness in finding tools and software for executing them by bad intending hackers or insiders. From simple malware to advanced monitoring techniques of data traffic between different chips performed by low level hackers to founded organized groups, attacks always are in the pursuit of different gain and advantages with the main goal of accessing unauthorized data or to produce serious damages to a system or to multiple systems.

It is very important how systems respond and action against attacks. Thus, developers and designers must think and built innovative solutions and defense mechanisms to prevent and to deal with these attacks through high power system resiliency. Also, they must practice all kind of intense attacks against their products to see how them resist, how different solutions can be improved, what are the possible risk and threats and where are the bugs that indicates vulnerabilities that hackers can exploit.

While everything points to a worldwide Internet of Things, attackers don't miss any occasion to target it and critical devices with extreme sensitive data are unlocked only when asset owners pay money. Worst case scenario is when devices are still locked and unavailable even after huge amounts of money are delivered which can lead to serious damages through stopping important industrial areas on which the functioning of critical system relies.

Out there in this world ideal secure system does not exist and to make such one relies in finding an optimum solution with specific compromises that can provide the exact characteristics and features that the system was designed for. Big companies are working and trying every day to find this optimum solution and to improve existing defense mechanisms for a better secured world where people can feel protected and not "watched".

A good path to follow starts with encouraging developers to build efficient and securable code and transposing security competences and features we know to use in other technology domains and continues with reliable device and personal identity methods.

Our communications embedded systems must ensure confidentiality and integrity while also ensuring appropriate levels of availability with dynamic patching of vulnerabilities and upgradable security mechanisms [63].

References

1. F. Birleanu, N. Bizon, Reconfigurable computing in hardware security—a brief review and application. J. Electr. Eng. Electron. Control Comput. Sci. (JEEECCS) **2**(3), 1–12 (2016)
2. M. Ciampa, *Security Awareness: Applying Practical Security in Your World*. IEEE Design & Test of Computers (2010)
3. P. Gregory, *CISSP Guide to Security Essentials* (Course Technology, Cengage Learning, Boston, 2010)
4. http://infocenter.arm.com/help/index.jsp?topic=/com.arm.doc.prd29-genc-009492c/ch01s03s04.html
5. M. Chowdhury, A. Apon, K. Dey, *Data Analytics for Intelligent Transportation Systems* (Elsevier, UK, 2017)
6. http://web.cse.ohio-state.edu/~champion.17/4471/4471_lecture_2.pdf
7. http://sourcedaddy.com/networking/threats-vulnerabilities-and-attacks.html
8. http://www.mckinsey.com/business-functions/digital-mckinsey/our-insights/the-rising-strategic-risks-of-cyberattacks
9. J. Chirillo, *Hack Attacks Revealed* (Wiley Computer Publishing, New York, 2001)
10. http://f3magazine.unicri.it/?p=306
11. http://searchsecurity.techtarget.com/definition/hacker
12. D.G. Abraham, G.M. Dolan, G.P. Double, J.V. Stevens, Transaction security system. IBM Syst. J. **30**(2), 206–229 (1991)
13. http://fortune.com/2017/07/26/who-are-hackers/
14. http://fortune.com/2017/06/22/cybersecurity-business-fights-back/
15. S.J. Stolfo, S.M. Bellovin, S. Hershkop, A. Keromytis, S. Sinclair, S.W. Smith, *Insider Attack and Cyber Security: Beyond the Hacker* (Springer, New York, 2008)
16. C.W. Probst, J. Hunker, D. Gollman, M. Bishop, *Insider Threats in Cyber Security* (Springer, New York, 2010)
17. http://www.makeuseof.com/tag/5-of-the-worlds-most-famous-hackers-what-happened-to-them/
18. http://www.telegraph.co.uk/technology/6670127/Top-10-most-famous-hackers.html
19. https://en.wikipedia.org/wiki/MafiaBoy
20. http://fortune.com/2017/06/22/cybersecurity-hacks-history/
21. J. Graham, R. Howard, R. Olson, *Cyber Security Essentials* (CRC Press, Boca Raton, 2011)
22. J. Andress, S. Winterfeld, *Cyber Warfare: Techniques, Tactics and Tools for Security Practitioners* (Amsterdam, Syngress, 2011)
23. P. Anghelescu, S. Ionita, G. Iana, High-speed PCA encryption algorithm using reconfigurable computing. J. Cybern. Syst. (Taylor & Francis) **44**(4), 285–304 (2013)
24. P. Anghelescu, FPGA implementation of programmable cellular automata encryption algorithm for network communications. Int. J. Comput. Syst. Sci. Eng. (CSSE) **31**(5) Sept (2016)
25. H. Elmiligi, F. Gebali, M.W. El-Kharashi, Multi-dimensional analysis of embedded systems security. Microprocess. Microsyst. **41**, 29–36 (2016)
26. M. Erbschloe, *Trojans, Worms, and Spyware* (Elsevier, Amsterdam, 2005)
27. A. Singhal, *Data Warehousing and Data Mining Techniques for Cyber Security* (Springer, New York, 2007)

28. P. Eder-Neuhauser, T. Zseby, J. Fabini, G. Vormayr, Cyber attack model for smart grid environments. Sustain. Energy Grids Netw. **12**, 10–29 (2017)
29. http://www.berkes.ca/archive/berkes_hardware_attacks.pdf
30. http://resources.infosecinstitute.com/hardware-attacks-backdoors-and-electronic-component-qualification/
31. https://blog.finjan.com/9-common-hacking-techniques-and-how-to-deal-with-them/
32. P. Anghelescu, E. Sofron, S. Ionita, L. Ionescu, in *FPGA Implementations of Cellular Automata for Pseudo-Random Number Generation*. The 29th International Semiconductor Conference, CAS 2006, Sinaia, Romania, 27–29 Sept 2006, pp. 371–374
33. E. Zio, Challenges in the vulnerability and risk analysis of critical infrastructures. J. Reliab. Eng. Syst. Saf. **152**, 137–150 (2016)
34. N. Nezamoddini, S. Mousavian, M.E. Kantarci, A risk optimization model for enhanced power grid resilience against physical attacks. J. Electr. Power Syst. Res. **143**, 329–338 (2017)
35. M. Donohoe, B. Jennings, S. Balasubramaniam, Context-awareness and the smart grid: requirements and challenges. J. Comput. Netw. **79**, 263–282 (2015)
36. https://ics-cert.us-cert.gov/sites/default/files/ICSJWG-Archive/QNL_MAR_16/reliability%20and%20resilience%20pdf.pdf
37. N.M. Tabatabaei, N. Bizon, A.J. Aghbolaghi, F. Blaabjerg (eds.), *Fundamentals and Contemporary Issues of Reactive Power Control in AC Power Systems* (Springer Verlag London Limited, London, 2017)
38. https://www.fraunhofer.de/en/press/research-news/2011/february/fingerprint-makes-chips-counterfeit-proof.html
39. Counterfeit-Proof Chips, *Fraunhofer Magazine*, vol. 1.12 (Fraunhofer-Gesellschaft, Munchen, 2012)
40. J.H. Saltzer, M.D. Schroeder, The protection of information in computer systems. Proc. IEEE **63**(9), 1278–1308 (1975)
41. F. Birleanu, N. Bizon, Principles, architectures and challenges for ensuring the integrity, internal control and security of embedded systems. J. Electr. Eng. Electron. Control Comput. Sci. (JEEECCS) **3**(7), 37–45 (2017)
42. H. Orojloo, M.A. Azgomi, A game-theoretic approach to model and quantify the security of cyber-physical systems. J. Comput. Ind. **88**, 44–57 (2017)
43. Y. Ashibani, Q.H. Mahmoud, Cyber physical systems security: analysis, challenges and solutions. J. Comput. Secur. **68**, 81–97 (2017)
44. A.V. Gheorghe, M. Masera, M. Weijnen, L. De Vries, *Critical Infrastructures at Risk—Securing the European Electric Power System* (Springer, Dordrecht, 2006)
45. E.G. Amoroso, *Cyber Attacks—Protecting National Infrastructure* (Elsevier, Amsterdam, 2011)
46. https://www.networkworld.com/article/2177520/collaboration-social/5-top-social-media-security-threats.html
47. W. Wang, Z. Lu, Cyber security in the smart grid: survey and challenges. Comput. Netw. **57**, 1344–1371 (2013)
48. https://www.csoonline.com/article/3106292/social-networking/social-media-the-gateway-for-malware.html
49. K. Pipyros, C. Thraskias, L. Mitrou, D. Gritzalis, T. Apostolopoulus, A new strategy for improving cyber-attacks evaluation in the context of Tallinn manual. J. Comput. Secur. (2017)
50. https://www.forbes.com/sites/sungardas/2015/02/24/why-hackers-love-companies-who-use-social-media/#5348fc6a71a9
51. https://www.networkworld.com/article/3200030/internet-of-things/researchers-find-gaps-in-iot-security.html
52. https://www.networkworld.com/article/3202767/internet-of-things/the-fight-to-defend-the-internet-of-things.html

53. C. Tu, X. He, Z. Shuai, F. Jiang, Big data issues in smart grid—a review. J. Renew. Sustain. Energy Rev. **79**, 1099–1107 (2017)
54. M. Ficco, M. Chora, R. Kozik, Simulation platform for cyber-security and vulnerability analysis of critical infrastructures. J. Comput. Sci. (2017). https://doi.org/10.1016/j.jocs.2017.03.025
55. https://www.networkworld.com/article/3217664/internet-of-things/how-to-improve-iot-security.html
56. Z.A. Baig, P. Szewczyk, C. Valli, P. Rabadia, P. Hannay, M. Chernyshev, M. Johnstone, P. Kerai, A. Ibrahim, K. Sansurooah, N. Syed, M. Peacock, Future challenges for smart cities: cyber-security and digital forensics. J. Digit. Investig. **22**, 3–13 (2017)
57. D. Jin, C. Hannon, Z. Li, P. Cortes, S. Ramaraju, P. Burgess, N. Buch, M. Shahidehpour, Smart street lighting system: a platform for innovative smart city applications and a new frontier for cyber-security. Electr. J. **29**, 28–35 (2016)
58. https://arxiv.org/pdf/1707.08380.pdf
59. https://smartgridawareness.org/privacy-and-data-security/smart-grid-vulnerabilities-a-more-detailed-review/smart-grid-security-threats-vulnerabilities-and-solutions/
60. http://www.energy.ca.gov/2012publications/CEC-500-2012-047/CEC-500-2012-047.pdf
61. C. Bekara, Security issues and challenges for the IoT-based smart grid. J. Procedia Comput. Sci. **34**, 532–837 (2014)
62. H. Yoo, T. Shon, Challenges and research directions for heterogeneous cyber-physical system based on IEC 61850: vulnerabilities, security requirements, and security architecture. J. Future Gener. Comput. Syst. **61**, 128–136 (2016)
63. https://www.cablelabs.com/remote-phy-reality

Chapter 10
Power Systems Recovery and Restoration Encounter with Natural Disaster and Deliberate Attacks

Horia Andrei, Paul Cristian Andrei, Marian Gaiceanu, Marilena Stanculescu, Iulian Nicusor Arama and Ioan Marinescu

Abstract This chapter presents a theoretical analysis of the impact of the extreme weather events and deliberate attacks on the power systems, which is accompanied by several examples taken from existing reports. The power systems resiliency for these cases are presented and the used practices are being assessed. Optimized models to improve the power system reaction time to these new risks are also discussed and proposed.

Keywords Cyber-attack · Deliberate attacks · Resiliency improvement
Natural disaster · Power systems resiliency · Recovery and restoration

H. Andrei (✉) · I. Marinescu
Doctoral School of Engineering Sciences, University Valahia of Targoviste, Targoviste, Romania
e-mail: hr_andrei@yahoo.com

I. Marinescu
e-mail: ioan.marinescu@yahoo.com

P. C. Andrei · M. Stanculescu
Department of Electrical Engineering, University Politehnica Bucharest, Bucharest, Romania
e-mail: paul.andrei@upb.ro

M. Stanculescu
e-mail: marilena.stanculescu@upb.ro

M. Gaiceanu · I. N. Arama
Department of Control Systems and Electrical Engineering, University Dunarea de Jos Galati, Galati, Romania
e-mail: marian.gaiceanu@ugal.ro

I. N. Arama
e-mail: iulian.arama@ugal.ro

© Springer Nature Switzerland AG 2019

N. Mahdavi Tabatabaei et al. (eds.), *Power Systems Resilience*, Power Systems,
https://doi.org/10.1007/978-3-319-94442-5_10

10.1 Chapter Overview

In the first part of this chapter the authors present the importance and the actuality of the addressed major topics. The increase with over 40% of the number of extreme weather events in 2015–2016 over the last 20 years and the doubling of their frequency, but, at the same time, the unprecedented increase, starting with September 11, of deliberate (terrorist) attacks, direct or cybernetic, imposed the necessity of analyzing their impact on the energy infrastructure and on the energetic security of each country. All these risks must be incorporated in finding some technical and design solutions for the development of a safe and resilient power system that should have the ability to anticipate, to absorb, to adapt and to quickly recover after the occurrence of a disturbing event.

Furthermore, in Sects. 10.2 and 10.3 of this chapter, there are defined two types of extreme weather events (severe natural conditions, disasters) respectively deliberate attacks on power systems. Impact analysis of these events on power system operation and the system resiliency under these highly disruptive conditions, are extremely important for achieving a good risk management. This is why examples of extreme weather taken from international reports and complemented by several situations that took place in Romania, over the past 10 years, are being studied. The statistics made in Romania are the basis of the predictive estimations related to the occurrence frequency of extreme weather events. Also, these statistics offer information related to the damages produced to system, the type of defects occurring in the system and the time necessary for the system to return to its normal operation state. The interruption period of the power system in extreme weather conditions can be effectively eliminated by using secondary generator systems.

Such a secondary system which uses fuel cells and renewable energy sources is proposed and analyzed by the authors in this chapter. The operation, the response time and the performance of the extended secondary system are described in details. Concerning the risk of deliberate attacks, Sect. 10.3 describes the potential terrorist threats ranging from cyber-attacks under their multiple aspects, to direct attacks through physical destruction. The authors analyze IT attacks which took place especially in USA and which targeted in a proportion of 40% energy companies, which could suffer losses of about 1.87 billion dollars by 2018. There are also mentioned some terrorist events and their impact on the US power system.

The improvement of the power system resiliency by adopting on one hand, preventive measures and on the other hand re-design of the damaged infrastructure is studied in the last part of this chapter. Continuous monitoring solutions based on the intelligent systems designed in an integrated approach are proposed. The chapter ends with a list of specific bibliographic references.

10.2 Introduction in New Risks: Power Systems Against Natural Disaster and Deliberate Attacks

The Power System (PS) is part of the critical infrastructures because of its vulnerability to risks which target him directly or are headed against the components and processes of which it consists. The risks to PS depend and are favored by three factors:

- the fixed location of all the system's components, starting with every type of electrical energy production plant, going on with the routes of the overhead or underground power lines for transport and distribution and transforming and distribution stations;
- the multitude and in homogeneity of hazards against PS;
- the various ways of manifestation and the unpredictability of the threats against PS.

The risks against PS can be grouped according to their occurrence location, their way of manifestation, and the way they develop malfunctions of the system. Some of these hazards are part of the system's natural structure (e.g. old equipment, used conductors, repeated transient processes, the connection of several types of energy sources at the same transport network, the simultaneous connection of several important consumers) and they can be called system or process threats, being an effect of the PS erosion/development or a product of the dynamic and transient processes evolution from PS. Then, the influence of climatic and geophysical factors against PS is extremely important. For example, hurricanes, tsunamis, earthquakes, geomagnetic storms, long lasting drought which strongly affected the PS functioning had been more and more often in the last decade all over the world. Other hazards are caused by human intervention either by error (for example wrong technical interventions in the PS equipment, wrong manual switching) or deliberately, provoked by direct attacks against PS and its components as well as through indirect, cybernetic attacks, carried out in the virtual space through informatics networks to which PS is connected.

Having in mind all the hazards against PS, the authors consider that these can be grouped as following (Fig. 10.1):

- PS system and process hazards;
- hazards due to climatic and geophysical factors;
- hazards resulting from human activity.

System and process hazards can be predictable and modeled, while climatic and geophysical factors and human intervention, especially the deliberated one, are inhomogeneous, hard to predict hazards, which have different manifestation forms.

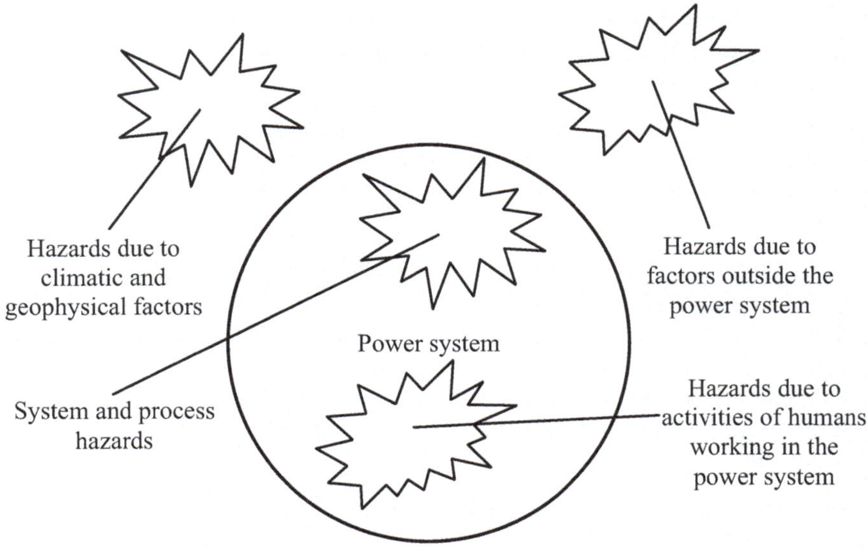

Fig. 10.1 The power system (PS) and associated hazards

The decrease of the risks produced by these hazards against PS and the increase of the security grade of PS, but also the return of PS to the natural functioning state after such hazards take place represent one of the great challenges of specialists.

On a long term, in the PS structure one will substantially find Renewable Energy Sources (RES) which together with their energy storage systems will contribute to the increase of the energetic security grade. But the variability of the main RES types—wind, photovoltaic, biomass—raise fitting problems regarding PS, which must be ensured with conventional flexible capabilities, and of these, the ones based on natural gas are the most efficient.

The hazards of extreme climatic and geophysical conditions require maintaining in operation coal-based groups, and as an alternative, especially at European level, is the integration of PS in the regional energy markets for equalizing the production-consumption balance and ensuring the supply in extreme conditions.

Deliberate terrorist or informatics human attacks or made by one of the states who possess satellites which evolve around the earth by means of electromagnetic waves are extremely dangerous and with potentially catastrophic effects.

The short term energetic security depends on the ability of a state to manage short term crisis of energy supply, caused by natural disasters, physical or cybernetic attacks, or the deliberate action of a state and needs the planning of strategic stocks, back-up systems and regional cooperation arrangements.

10.3 Natural Disaster—New Risk Management, Case Studies: Banqiao Dam Failure (China), Vidraru Dam Valve Malfunction and Belci Dam Failure (Romania)

Hydropower is one of the oldest and most widely-used renewable sources of energy. The way a dam is operated represents a critical factor in the determination of the probability of catastrophic failure. The lower the water level in a dam, the greater capacity for control of severe floods, but this approach also lowers its capacity of generating electricity and providing water for agricultural needs. Also, the water level cannot be kept at the maximum in order to maximize its efficiency because its purpose in controlling floods would become obsolete.

The operator must assess the dam excess capacity based on the trade-off accepted between the benefits of having stored a large quantity of water versus the value of flood control.

The purpose of a dam for flood control is a balance where the consequences of small floods are preferable to the damage following a catastrophic failure of the dam because the water stored behind it will be instantly added to the flood, leading to the failure of other dams located downstream, by a cumulative effect [1].

The largest hydroelectric power plant is Chinese, the Three Gorges Dam is located in Yichang, Hubei province, on Yangtze river, as seen in Fig. 10.2. The construction of the concrete gravity type dam begun in 1994, the dam body (without the locks) and the project was complete and functional in 2012, as seen in Fig. 10.3, at a cost of almost 28 billion dollars.

The dam's electric generating capacity consists of thirty-two main turbines (and two generators of 50 MW each in order to power the plant itself), with a total of 22,500 MW.

More than 1.2 million people from 13 cities, 140 towns and 1350 villages were forced to relocate, as the more than 600 km long reservoir filled. The Three Gorges Reservoir has a total capacity of 39.3 × 10^8 m^3 with a surface of 1084 km^2 [2, 3]. The dam is 2309.5 m long and 185 m in crest elevation, the normal water level is

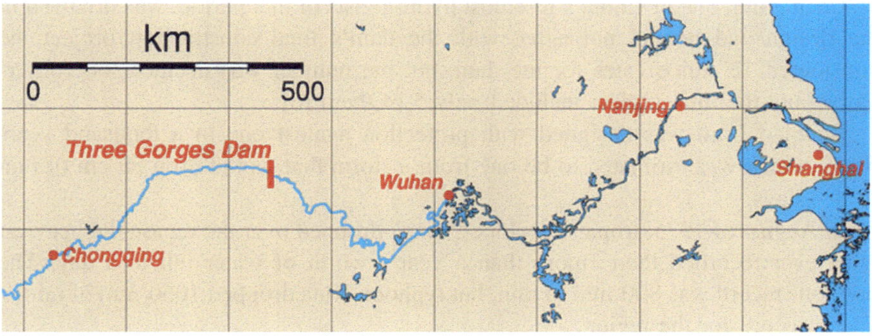

Fig. 10.2 Location of the Three Gorges Dam

Fig. 10.3 Site sketch of Three Gorges Dam

135 m and according to the check-flood calculations (1000 years design flood plus 10%) the maximum reservoir water level should be 180.4 m, the maximum downstream water level should not exceed 83.1 m at a maximum discharge flow of 102,500 m^3/s [4].

A few days after the reservoir filled, fine surface cracks emerged in the structure of the dam. However, 163,000 concrete units of the Three Gorges dam passed quality testes and the deformation was within design limits so a group of experts granted the project with a good quality rating, according to Living on Earth—Three Gorges Dam [5].

The largest natural disaster caused by the failure of a dam happened in 1975 in China. In the fifties, they built several dams (including the Banqiao Dam, completed in 1952) in the Huai river basin of the Henan province, as presented in Figs. 10.4 and 10.5 to ensure electrical power generation and in order to control and contain severe floods which regularly affected the country [1].

The Banqiao Dam was originally designed to pass 1742 m^3/s through sluice gates and a spillway. The storage capacity was set at 492,000,000 m^3 with 375,000,000 m^3 of this capacity reserved for flood storage. The height of the dam was 116 m.

Due to design flaws and construction errors, cracks appeared in the dam after the construction completed. Those were repaired with know-how by Soviet engineers between 1955 and 1956, at that time the dam being considered as unbreakable.

Chen Xing, one of China's foremost hydrologists of that period, was involved in the design and he did not agree with the dam's final construction project; he considered 12 sluice gates for the dam, but the number was deemed too conservative and the final project included only 5 of them [6].

Banqiao Dam was designed with protection against one in a thousand years flood, which was estimated to be one from a storm that would drop 53 cm of rain over 3 days.

In August of 1975, tropical cyclone "Nina" dropped over the region the heaviest rains ever recorded there: more than a year's worth of water fell in a day. The previous record was 800 mm of rain, but typhoon Nina dropped 1060 mm of rain—a new record for the region.

Fig. 10.4 Banqiao Dam location

The Banqiao Dam collapsed in the night of August 8, 1975, caused a huge wave, 10 km wide and 3–10 m high to rush downstream at a speed of approximately 50 km/h, adding another 600,000,000 m^3 of more water to the floods [7].

Altogether, 62 dams broke by the cumulative effect. The dikes and flood diversion projects located downstream could not resist in front of the flood wave, as presented in Fig. 10.5. They broke as well and the flood spread over more than a million hectares of farm land throughout 29 counties and municipalities, severely affecting eleven million people throughout the region.

The flooding caused around 26,000 fatalities and another 145,000 died afterward as a result of epidemics and famine; material damages were also severe, nearly 6,000,000 buildings collapsed. Unofficial estimates of the number of people killed by the disaster have run as high as 230,000 people, but the conservative regime kept the disaster outside public attention [8].

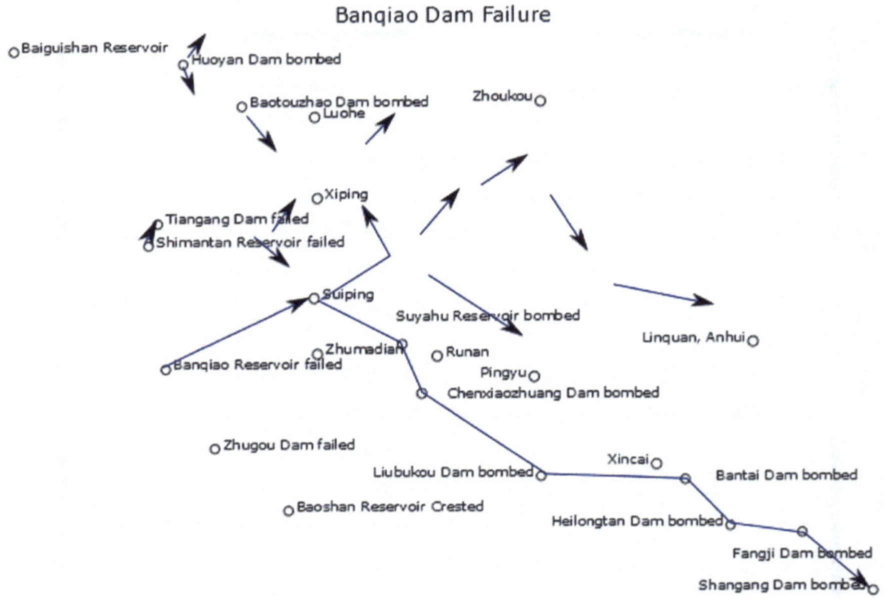

Fig. 10.5 Banqiao Dam failure cumulative effect

The flood broke all telecommunication lines, being difficult for the authorities to know the gravity of the disaster and their response was slow, as they did not expect the possibility of an "iron dam" (indestructible dam built with soviet know-how) to collapse under any circumstances.

In order to satisfy the growing energy demand and to reduce the pollution, China is building massive hydro power plant systems. Even though this kind of energy is clean and cheap, building such ample projects taking into consideration only the economic benefits is dangerous. The largest hydroelectric power plant in the world has proven it's safe since it has gone operational, but outside the risk of dam collapse and catastrophic flooding as a result, the controversial project already caused massive landslides due to erosion in the reservoir, induced by the rising water.

In Romania on July 6, 1974, there were leaks, water jets and gravel falls on the slope upstream of a Johnson valve, above the riverbed, at Vidraru Dam site, an engineering and architectonic masterpiece of Romania's communist era, as seen in Fig. 10.6 [9].

Soon, blocks of rock dislocated from the slope began to fall, followed by a rupture of the gallery upstream of the Johnson valve, the shearing of the gallery, the expulsion of a portion of the mountainous massif of approximatively 10,000 m^3, expelling the metallic armour to the rupture area, breaking the concrete block of the

Fig. 10.6 Vidraru Dam

Johnson valve, producing an artificial flood of about 600 m^3/s, which destroyed the power supply lines for the manoeuvring of the valves, the passage viaduct located at the foot of the dam, and several other destructions downstream, including portions of the road, bridges, dwellings, and several people lost their lives.

The most severe hydro technical accident in Romania's history occurred in 1991. The flood following the dam failure wiped out 250 houses and 25 people lost their lives when Belci Dam, constructed between 1958 and 1962 near Onesti, on Tazlau river, broke [10].

The causes of the accident were as follows: during the design phase, the dam was classified as first importance class for which insurance calculations flow of 0.1% and verification calculation flow of 0.01% were well below the values determined on the basis of 25 years of exploitation and hydrological observations, respectively 1515 and 2450 m^3/s.

Although known, the modification of the maximum flows did not lead to the increase of the evacuation capacity. Under these circumstances, the overflowing and then the failure occurred due to the exceptional flood of the night of the accident, estimated at 2800–3000 m^3/s and the impossibility of manoeuvring the hydro-mechanical equipment caused by the interruption of the power supply.

10.4 Deliberate Attacks—New Risk Management, Cases Study

From its beginning, during the various types of natural causes emergencies, the energy industry has been able to re-establish its essential services very quickly. Later, during the World Wars the industry faced potential sabotage. The sabotage threat continued, with a lower level during the Cold War, but the main concerns for physical security were those related to domestic problems such as vandalism, theft, etc. The latest international events highlighted the high threat of terrorist attack on infrastructure, mostly the threats facing the electrical power and distribution system.

The transmission and distribution of electricity is at high risk of being subjected to an attack which pose minimal risk from the attacker, a matter well-known by possible attackers or saboteurs. The fact that power transmission lines, power generating stations or communications facilities are located in remote locations or, for example, gas pipelines fuelling facilities are in less populated areas, allows a potential attacker to carry out his operations with minimal detection risk. Selecting potential attack points and estimating the resulting consequences are the capabilities of antiterrorist specialists [11].

The electricity system has various components such as: cybernetics, physical systems etc. and people who support these systems. Every day, threats are becoming more and more sophisticated and lasting, so the danger is increasing. Because of their complexity, threats have turn to one of the most effective weapon of our century. Increased use of IT-type products which grow more diverse and complex by the day lead to an increase in threatened areas if some entities decide to search and exploit their vulnerabilities. As utilities switched from electromechanical to digital equipment and, moreover, to interconnected digital equipment, the risks of external penetration increased accordingly.

Threats can be divided considering the sources, as:

- *Potential external threats* from terrorist organizations employing criminal organizations to attack public utility networks (affecting the control and generation networks of these utilities);
- *Potential internal threats* (people inside the firewall of the utility company and network operators that may jeopardize the functioning of parts of the system), including employees who can be co-opted without their knowledge of these threats.

Attacks on an electrical network may be *physical* or *cyber*-attack type, caused by people who do their job neglectful, either through omission or with the intention of doing harm.

Cyber-attacks are more frequent and possess the following criteria:

- Can appear at any given moment;
- Can come from anywhere;
- This type of risk cannot be eliminated, just attenuated;

- Its prevention costs significantly, even though it's difficult to measure the benefits or to estimate the economic costs of energy disruptions.

As a result, counter cyber-attacks strategies must be a priority for any government and for public utility companies [12].

A major modern challenge is to maintain cyber reliability. An attack from a unsatisfied employee or from an opponent (another company or a state agency) can produce negative economic and social consequences, as an electrical power outage may take hours, days or even months until recovery.

Power distribution systems are subject to potential risks that can have a natural or artificial origin. These risks were analysed by means of two large-scale studies conducted by North American Electric Reliability Corporation (NERC):

- High-Impact, Low-Frequency Event Risk to the North American Bulk Power System (2010) [13];
- Severe Impact Resilience: Considerations and Recommendations (2012) [14].

These studies concentrated on:

- risks with the possibility of causing catastrophic effects for the energetic system but are less likely to appear;
- risks which haven't occurred yet but are prone to occur in the future. Here we can add:
    - Coordinated cyber-attacks;
    - Physical attacks upon the power grid;
    - Mixed attacks;
    - Power network overload due to electromagnetic impulses created by an electromagnetic impulse weapon or the detonation of a nuclear weapon;
    - Major natural disasters: *severe weather* (extreme temperatures, floods, drought-induced fires, electric storms, massive snowfalls, strong winds, avalanches) or *interferences in the magnetic field of the Earth* (coronal mass ejections, geomagnetic storms, severe proton events).

Physical vulnerabilities sources are:

- severe weather phenomena's consequences (power lines discontinuances caused by fallen trees or other objects carried by strong winds);
- earthquakes;
- physical attacks (terrorist attacks or sabotage);
- physical power lines theft or electricity theft.

Cyberspace Vulnerabilities

Cyber security represents all the actions and measures taken to protect systems, networks and data against deliberate attack or accidental compromise, covering all aspects from the preparation to recovery. Cyber security is an important element of information technology systems, making possible the safe existence of utilities in the future.

A survey conducted on chief executives of public utilities, conducted by Black & Veatch in 2014 [15], on strategic directions in the energetic sector shows that cyber security is considered the fourth most important threat after reliability, environmental regulations and economic regulations.

Utility companies try to avoid the problems caused by cyber breaches, which carries major damage (high repair costs alongside low reputation) but fail to invest in a strong cyber infrastructure in order to prevent events which don't occur constantly, such as the usual risks a power grid is used to face.

This represents a challenge when experiencing a massive power outage, in which case the probability of occurrence is low, but the cost can be very high.

Cyber-vulnerabilities may have the following locations:

- power plants or any other associated facility for electricity production;
- distribution and supply systems (digital interfaces, SCADA, frequency and voltage control, etc.);
- digital measuring equipment, user interface;
- pricing, billing and bidding systems.

The 2014 Bipartisan Policy report shows that cyber-threats to the North American electricity grid are on the rise, making cyber security a priority at national and international level. At the same time, the Federal Bureau of Investigation notes that cyber-attacks have become a greater risk than terrorism, being now the primary security threat in the United States.

Electric lines that lead to users may be a vector for cyber-attacks on military and industrial targets. Some military organizations worry about power lines or systems used for electricity distribution may offer ways to gain access to military systems and to classified information [16].

The threats can be human, the systems themselves or the environment. Threat agents may attack an infrastructure either deliberately or inadvertently.

Deliberate attacks are caused by threat agents like:

- malicious employees;
- economic espionage agents;
- hackers;
- thieves;
- vandals;
- terrorists.

Neglectful attacks are caused by threat agents like:

- neglectful users;
- inadequate or superficial security controls;
- consequences with unintended effects;
- natural hazards;
- equipment malfunction;
- poorly designed systems.

The most dangerous type of threat agent is represented by an unsatisfied employee who knows exactly the vulnerabilities of the system, which are the security measures that can be bypassed and what actions can be carried out in such a way as to cause the greatest possible damage.

The most common threat agent is the neglectful user which makes data errors, wrongly or unsafely connects to networks or equipment or leaves software programs unprotected by strong passwords, leading to an easy cyber-attack.

Still, the most destructive threat agent is represented by nature (natural hazards).

Case Studies: United States of America

Romania experienced a serious blackout shortly after the devastating earthquake that affected the country on March 4, 1977. The cascading effect led to a generalized power shortage on May 10, 1977 because of a human error, with important financial losses but there are no solid data available regarding the magnitude of the losses.

Until now such an undesirable event hasn't happened, but as nowadays the electro-energetic system is interconnected (a fact with many advantages), if there would occur any significant disturbance in the electro-energetic system, it would not only manifest itself locally but it would affect the whole country, including Romania's energetic export capacities, which would translate into loss of lives and financial losses of billions.

On August 14, 2003, the power grid from the North America experienced its most significant blackout, event which caused more than 50 million people to suffer from this shortage and more than 70 GW of electrical load in USA (states from USA and regions from Canada being affected).

Power was successfully restored to most customers within hours but it took up to two days for some areas in the USA and parts of Ontario experienced rotating blackouts for up to two weeks [17].

The cascade sequence of the power outage is presented in Fig. 10.7.

The common factors responsible for the black-out include [18]:

- poor implementation of vegetation growth management (this further resulted in the contacts between the conductor and the trees)
- system operators incapacity to visualize the events on the system;
- incapacity and failure of operating the given system within known established safe limits;
- inadequate operational coordination and communications;
- ineffective operators training to respond and recognize the emergencies in the system emergencies;
- inappropriate resources of reactive power.

As presented above, the power outage was caused by natural causes and human oversight. Sabotage, terrorism or criminal activity aimed at part of the grid could produce a similarly devastating cascade if appropriate safety measures won't be implemented.

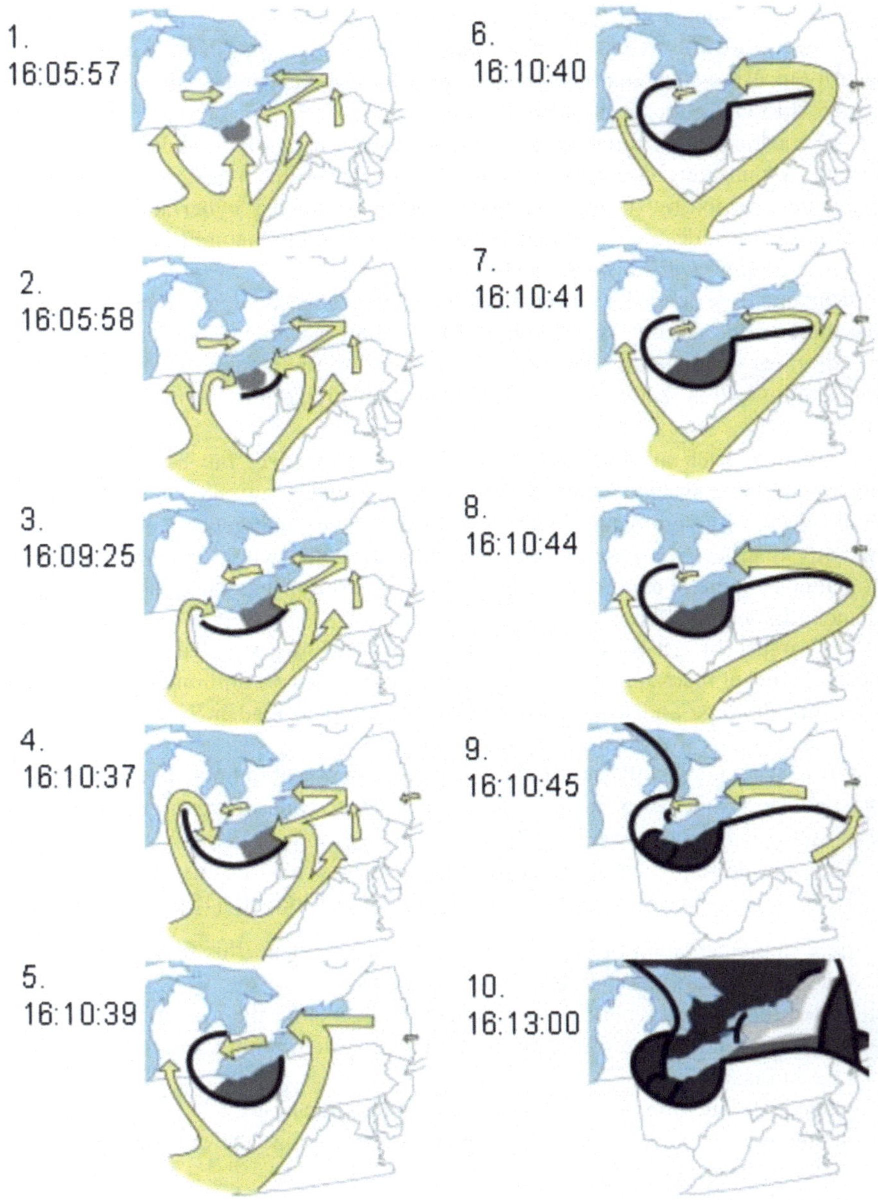

Fig. 10.7 Power outage cascade sequence

The entire grid is impossible to defend against physical threats as many of the assets are in remote areas, but as with other forms of terrorism and organised crime, the most effective solution is pro-active intelligence gathering and law enforcement

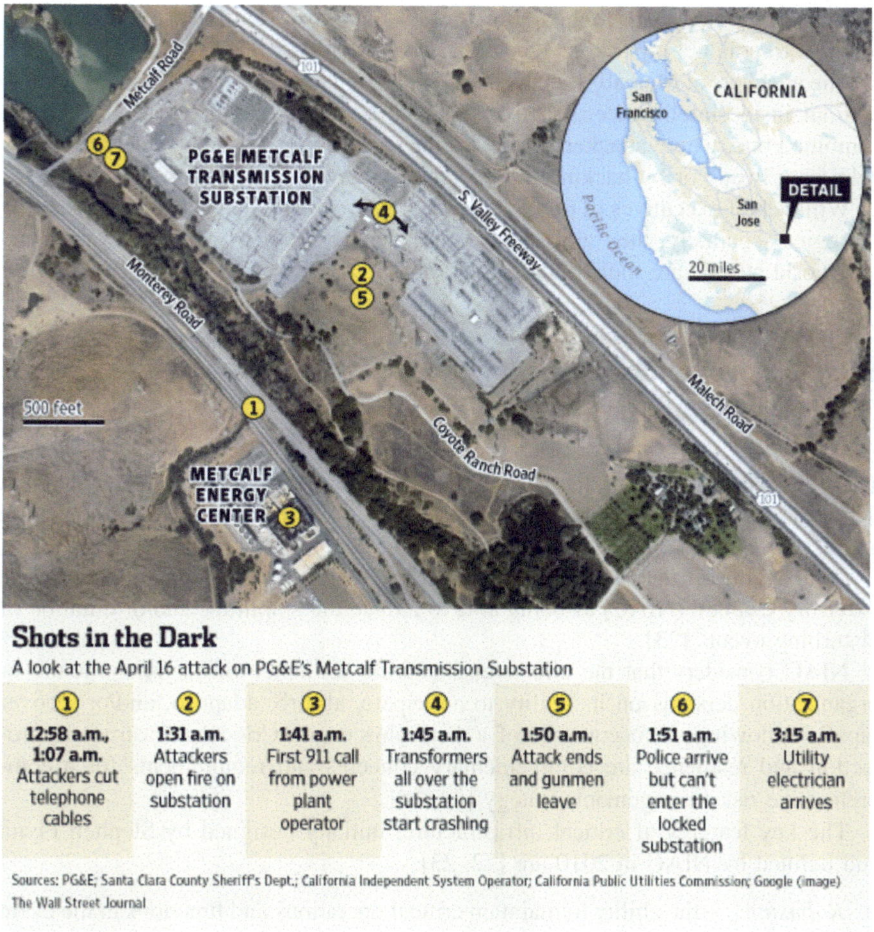

Fig. 10.8 Metcalf sniper attack scene [20]

activity to prevent, disrupt and deter attacks on the electric grid before any hostile actions can take place.

On April 16, 2013, the so called "Metcalf sniper attack" aimed at a power outage by physically attacking Pacific Gas and Electric Company's Metcalf Transmission Substation from Coyote, California, USA as presented in Fig. 10.8. Following the course of events, the attackers cut the fibre optic telecommunication cables before opening gunfire at the 17 electrical transformers inside the station. The sabotage attempt was a professional job, as authorities couldn't find fingerprints or any other clues in order to solve the case [19].

In the attack, no one was injured or lost its life, nobody was arrested and the reasons behind the sabotage remain unknown. While the attack failed its intended purpose of generating a power outage as grid officials' successfully rerouted power generated by other plants to the region, the substation was offline for nearly a month in order to conduct repairs.

The attackers at Metcalf Transmission Substation couldn't have been so effective without inside intelligence, so this cannot be considered a physical attack but a combined one, where attackers exploited the weak points of the power station with help from inside or by hacking the power company cyber systems.

While chief executives of public utilities may look at physical and cyber-security strategies separately, they are most certainly intertwined, as combined cyber and real world attacks are more and more common [21].

10.5 Improve the Resiliency of Power System, Case Studies

The concept of resilience has evolved considerably from Holling's fundamental definition as "a measure of the ability of a system to continue to function by addressing changes in status, management and parameter variables" [22].

Critical infrastructure resilience is defined by the National Infrastructure Advisory Council (NIAC) as being able to reduce the amplitude and/or duration of disturbing events [23].

NIAC considers that the effectiveness of the resilience of an infrastructure or organization depends on its ability to anticipate, absorb, adapt to, and/or recover rapidly following the occurrence of a disturbing event. Critical infrastructure protection and resilience are complementary concepts and requirements for a comprehensive risk management strategy [24, 25].

The key features of critical infrastructure, initially designed by Stephen Flynn and defined by NIAC in 2010 are [22, 23]:

- *Robustness*—the ability to maintain critical operations and functions in the event of a crisis: reflected in physical construction and infrastructure design (buildings, bridges, dams, dikes) or in the redundancy and substitution ability of systems (transport, electricity and communications);
- *Responsiveness*—the ability to adequately prepare, respond, and manage activities in the event of a crisis or disruption: involves identifying how the crisis or disruption develops, planning business continuity, supply chain management, prioritizing actions to control and reduce damages;
- *Rapid Recovery Capacity*—the ability to return to and/or reconstitute normal operations as quickly and efficiently as possible after an interruption. This includes carefully designed emergency plans, appropriate emergency operations, and ways to have the necessary resources at the right place;
- *Adaptability*—the way to absorb new lessons that can be extracted from a catastrophe. It involves reviewing plans, modifying procedures and introducing new tools and technologies to improve robustness, responsiveness and rapid recovery before the next crisis.

In the 110 kV grid in Galati County, the power injection from the Thermoelectric power plant Galaţi and the Movileni hydroelectric power plant are being carried out. On one hand, monitoring and control are the responsibilities of the National Energetic Dispatcher, on the other hand their power injection is made at the hierarchical top of the electricity distribution network, being equivalent to power take-over from the transmission grid from the electrical transformer station 400 kV/110 kV Smirdan and electrical transformer station 220 kV/110 kV Filesti.

From these points of view, the 110 kV grid is a passive one, while the medium voltage grid, benefiting from the power injections from the wind farms, solar power plants and diesel generators, is really active. However, the 110 kV grid has a certain level of complexity both through the very nature of the equipment operating at this level of voltage and by making the interconnection between the electricity transmission grid and the active power distribution network at medium voltage.

Whether active or not, the medium voltage network must be interconnected to the transmission network, which requires that the 110 kV network failure is previously cleared, all the more so as access to voltage of a small number of nodes, allowing them to quickly energize and/or promptly identify the remaining disabled nodes (for a short duration without voltage), which does not interfere with the elimination of the damage in the medium voltage network, and even precipitates this process by knowing promptly the exact situation in the 110 kV network.

Solving damaged 110 kV network can be done:

- alternating priorities through send-by-receive priorities;
- working on the priority of certain classes of nodes, depending on the tension or other considerations such as vulnerability, contractual clauses, etc.

In the first situation, the prior definition of source nodes was preserved, but a kinematic orientation of graphs emerged, altering the priorities with the send-by-receiver mechanism, changing the composition of the node classes in real time. In this way, the operator's intervention for solving the damage is done by discriminating consumers on contractual, vulnerability, opportunity, etc. reasons and by a certain flair in solving up the damage.

In the latter case, the definition and ranking of the source nodes, as well as the discrimination of the other nodes, are made by the nodes themselves, which communicate to the dispatching agent their available injecting power, their own vulnerability, the actual availability of the injection power control, frequency regulation, participation in bandwidth enhancing, security of electricity supply, loss reduction, vulnerability, etc. which is expressed by promptly updating the information base and the self-orientation of the graphs.

For now, the human operator only has the ability to leave the 110 kV grid it's default priority in order to solve the failure or, on the contrary, to prioritize solving the failure in the power distribution network at medium voltage.

As can be seen in Fig. 10.9, the 110 kV the operating mode is similar to the medium voltage network; after the Cudalbi connection of the circuit breaker in the block

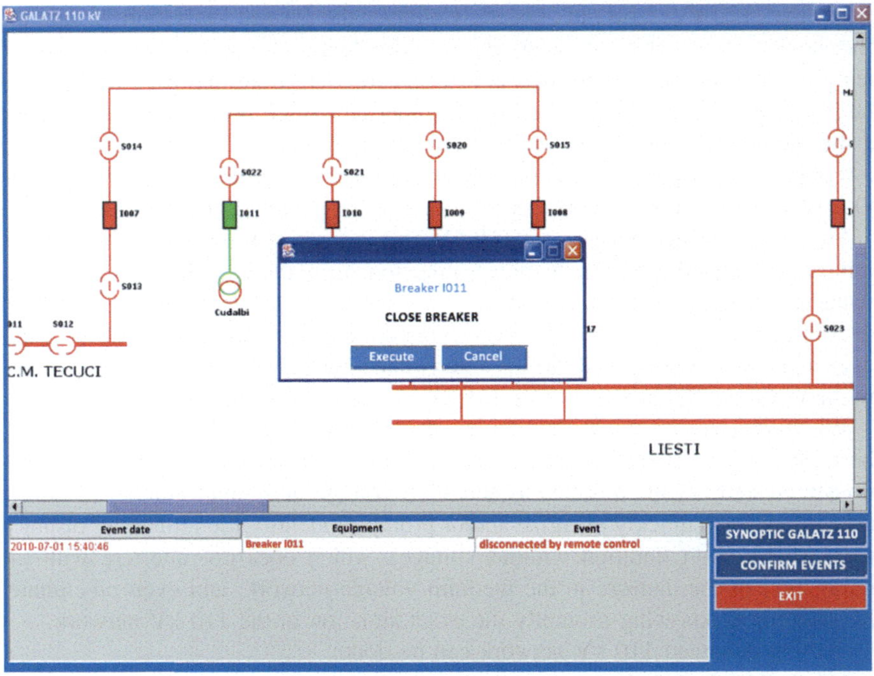

Fig. 10.9 Remote control of a circuit breaker in the 110 kV grid

diagram there were a series of changes on the operator terminal, as shown in Fig. 10.10, the electrical transformer station 110 kV/20 kV Cudalbi being energized.

In the case of the 110 kV Tecuci overhead electrical circuit breaker in the Tecuci 110 kV/6 kV electrical transformer station, due to the operation of some protections, a new situation is displayed on the operator terminal as presented in Fig. 10.11.

In this scenario, the entire 110 kV network between Tecuci and Marasesti has gone blackout, consumers in Tecuci and the rural area of northwest Galati County being deprived of electricity. There are two possibilities, routing the power through the medium voltage distribution network or re-energizing the 110 kV distribution network. Routing the power through the medium voltage distribution network is less cost effective because:

- higher power losses in the medium voltage distribution network in the new consumer re-fuelling configuration following a 110 kV grid failure;
- he difficulty of bandwidth framing in a medium voltage network that charges large electricity losses in an abnormal configuration;
- reduced consumer safety security for the same reasons, and also the two above considerations;
- field intervention teams' operation and their management in real time;
- the costs of electricity are in this case far outweighed by the counter value of the damage to consumers.

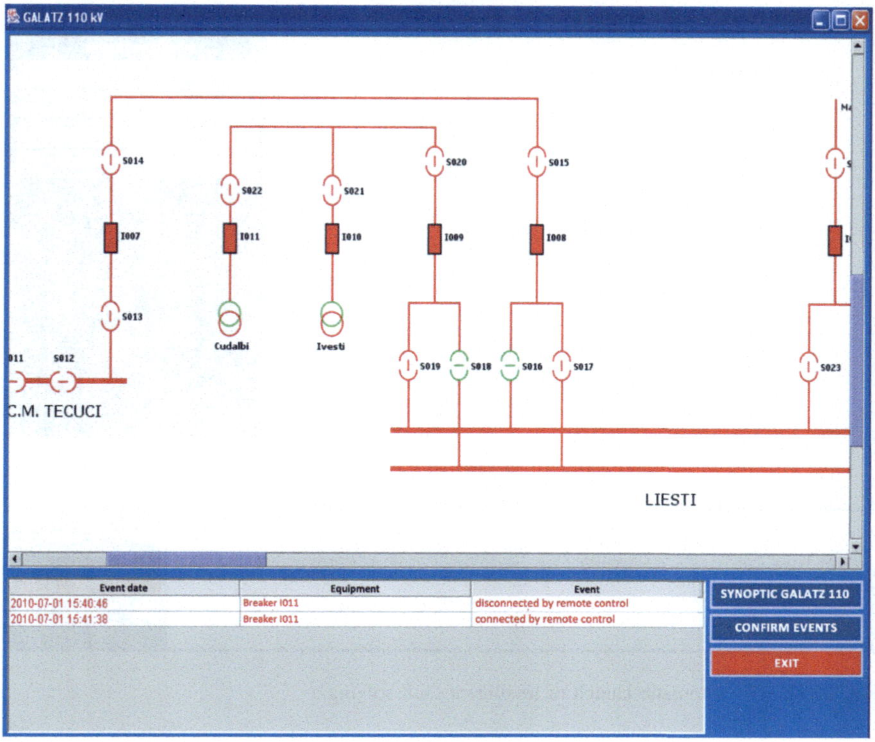

Fig. 10.10 The consequences of connecting a 110 kV switch by remote control

Accordingly, priority is given to resolving the 110 kV failure in order to pass the responsibility for resolving the operational management failure to the medium voltage distribution network as quickly as possible.

The urgency of the "transfer of the operative management competence" function in the context of the existing situation regarding the evolutionary stage of the electricity distribution and the imperative of this function in the context of transition from the static orientation to the kinematic and dynamic orientation of the graphs modelling the electricity distribution network in the future evolution of the electricity distribution inter-condition each other, a situation which must be taken into account in the development of the crash management application.

It is necessary for the nodes to intercommunicate so that an image of the damage is aggregated. The action to be taken recalls the Petri network formula by launching distributed intelligence functions:

- self-orientation of the graph by modelling the electricity distribution network;
- transferring the competence of operative management;
- procedures to clear the damage by seeking access to the source.

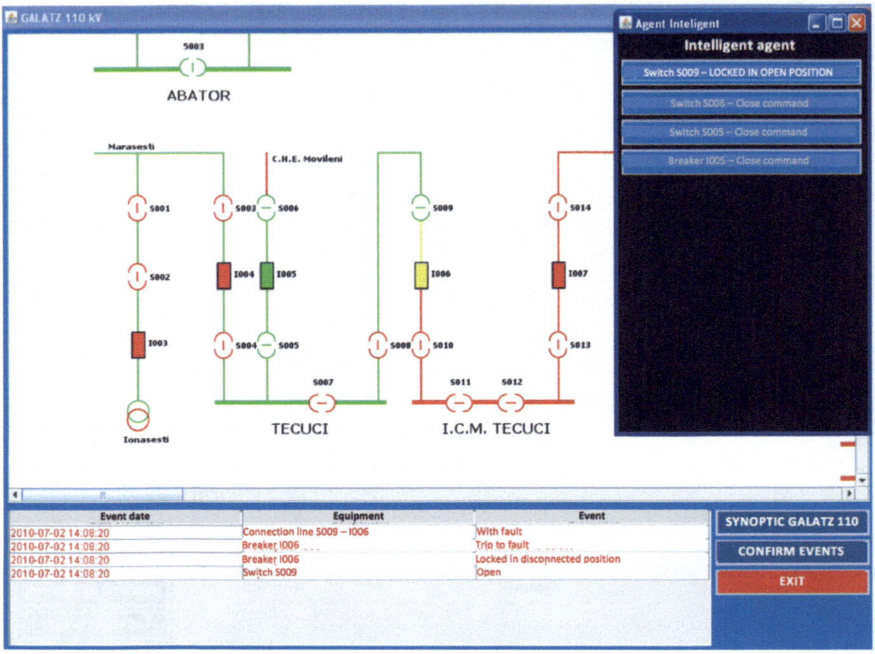

Fig. 10.11 The automatic launch of intelligent fault solving

Self-targeting will be accomplished with send-by-receive mechanisms. Competence transfer for operational management will be similar to the current procedure for damage liquidation by elaborating the decision (on the transfer of competence) by setting the minimum durations of searching for it to sources with the minimization of the resource requirements using the method "for research" aggregating all restrictions of the electricity distribution network on each of the voltage levels [26].

References

1. www.sjsu.edu/faculty/watkins/aug1975.htm. Accessed 14 Sept 2017
2. www.internationalrivers.org/campaigns/three-gorges-dam. Accessed 10 Sept 2017
3. G. Lollino, D. Giordan, G.B. Crosta, J. Corominas, R. Azzam, J. Wasowski, N. Sciarra, *Engineering Geology for Society and Territory, Volume 2: Landslide Processes* (Springer, Switzerland, 2014)
4. www.chincold.org.cn/dams/rootfiles/2010/07/20/1279253974143251-1279253974145510.pdf. Accessed 10 Sept 2017
5. www.loe.org/series/series.html?seriesID=28. Accessed 12 Sept 2017
6. www.engineeringclicks.com/banqiao-dam/. Accessed 13 Sept 2017
7. www.internationalrivers.org/resources/the-forgotten-legacy-of-the-banqiao-dam-collapse-7821. Accessed 12 Sept 2017

8. www.britannica.com/event/Typhoon-Nina-Banqiao-dam-failure. Accessed 14 Sept 2017
9. http://adevarul.ro/locale/pitesti/secretele-barajului-vidraru-gigantul-beton-piept-apei-navalnice-jumatate-secol-1_54db80b2448e03c0fd81ae8a/index.html
10. https://ro.wikipedia.org/wiki/Barajul_Belci
11. Committee on Enhancing the Robustness and Resilience of Future Electrical Transmission and Distribution in the United States to Terrorist Attack, Board on Energy and Environmental Systems, Division on Engineering and Physical Sciences, *Terrorism and the Electric Power Delivery System* (National Academies Press, Washington, DC, 2012) www.nap.edu/read/12050
12. C.W. Draffin, *Cybersecurity White Paper 1, MIT Energy Initiative Utility for the Future*, 15 Dec 2016. energy.mit.edu/research/utility-future-study
13. *High-Impact, Low-Frequency Event Risk to the North American Bulk Power System.* A Jointly-Commissioned Summary Report of the North American Electric Reliability Corporation and the U.S. Department of Energy's November 2009 Workshop, National Academies Press, June 2010. https://energy.gov/sites/prod/files/High-Impact%20Low-Frequency%20Event%20Risk%20to%20the%20North%20American%20Bulk%20Power%20System%20-%202010.pdf
14. North American Electric Reliability Corporation (NERC), *Severe Impact Resilience: Considerations and Recommendations.* Board of Trustees Accepted: May 9, 2012. www.nerc.com/docs/oc/sirtf/SIRTF_Final_May_9_2012-Board_Accepted.pdf
15. *2014 Strategic Directions: U.S. Electric Industry a Black & Veatch Report.* www.bv.com/docs/default-source/reports-studies/14-sdr-electric-report.pdf
16. *White Paper: Cyber Security Issues for the Smart Grid Frances Cleveland* (Xanthus Consulting International). http://xanthus-consulting.com/Publications/documents/White_Paper_Cyber_Security_Issues_for_the_Smart_Grid.pdf
17. North American Electric Reliability Council, *Technical Analysis of the August 14, 2003, Blackout: What Happened, Why, and What Did We Learn?* p. 1
18. North American Electric Reliability Council, *Technical Analysis of the August 14, 2003, Blackout: What Happened, Why, and What Did We Learn?* pp. 93–94
19. https://en.wikipedia.org/wiki/Metcalf_sniper_attack
20. www.theblaze.com/news/2014/04/02/how-has-the-most-significant-incident-of-domestic-terrorism-involving-the-enery-grid-gone-largely-unreported-for-10-months/
21. www.entrepreneur.com/article/240408
22. D.C. Barbu, Improvement of TIC critical infrastructure protection by increase the resiliency. J. Natl. Inst. Res. Dev. Inform. ICI, art. 02 **26**(4), 29–33 (2016) (Romanian)
23. M. Selak, in *Power System Protection—Where are We Today?* PES Student Energy Conference (ZEC 2015), Zagreb, 10 Dec 2015
24. Z.Q. Bo, X.N. Lin, Q.P. Wang, Y.H. Yi, F.Q. Zhou, in *Developments of Power System Protection and Control.* Protection and Control of Modern Power Systems (Springer, Singapore, 2016), pp. 1–7
25. www.dhs.gov/national-infrastructure-advisory-council
26. I.N. Arama, Applications of Multiagent Systems in the Distribution of Electricity, Ph.D. thesis, Dunarea de Jos University, Galati, 2011 (Romanian)

Chapter 11
Resilience Enhancement of Cyber-Physical Systems: A Review

Sanda Florentina Mihalache, Emil Pricop and Jaouhar Fattahi

Abstract Cyber-Physical Systems (CPS) represent a complex class of systems that are implied in electric power generation and delivery, as well as in critical infrastructure operations, traffic flow management or healthcare services. CPS consist of a robust combination of computational and physical components, that implement modern technologies such as wireless sensor networks (WSN), Internet of Things and recently Internet of Everything, machine to machine (M2M) communication, smart devices and even smart everything. Cyber-Physical Systems integrate heterogeneous equipment with computing power, which represent an attractive target for various attackers. Successful attacks can lead not only to data breaches, but also to interruption of CPS functioning, hence reducing their availability. CPS are affected also by technical failures and accidents. In order to enhance the CPS resilience, the proposed methods should mitigate the security risks by implementing powerful and robust security measures, in the same time increasing the fault tolerance and redundancy of the system. In this chapter, the authors analyze the state of the art methods for enhancing resilience of Cyber-Physical Systems. In the first section of the chapter the authors review the threats and vulnerabilities that can affect systems functioning. In the second part of the chapter the existing methods for resilience enhancement (redundancy, fault tolerance, security) are presented based on an extensive literature study.

Keywords Cyber-physical systems · Fault tolerance · Resilience
Redundancy · Security

S. F. Mihalache (✉) · E. Pricop
Automatic Control, Computers and Electronics Department,
Petroleum-Gas University of Ploiesti, Ploiesti, Romania
e-mail: sfrancu@upg-ploiesti.ro

E. Pricop
e-mail: emil.pricop@upg-ploiesti.ro

J. Fattahi
Department of Computer Science and Software Engineering,
Laval University, Quebec, Canada
e-mail: jaouhar.fattahi.1@ulaval.ca

© Springer Nature Switzerland AG 2019

N. Mahdavi Tabatabaei et al. (eds.), *Power Systems Resilience*, Power Systems,
https://doi.org/10.1007/978-3-319-94442-5_11

11.1 Introduction

Modern communities are dependent on efficient, safe and secure operation of critical infrastructures that are key components of building management systems, energy production and distribution, healthcare and life services, water distribution, wastewater management, industrial facilities, transportation, military facilities, banking or public safety and security. The functioning of these infrastructures is relaying not only on IT infrastructure but also on various sensors, actuators and communications equipment. Also, there is a close linking between physical system components and their logical and functional ones.

Apart from the functioning of critical infrastructure, a recently introduced concept, namely Industry 4.0, implies the integration of smart technologies in automation field, generating the need to design Cyber-Physical Systems (CPS). The cyber-physical systems need the collaboration between a variety of science fields, such as systems theory, software and hardware engineering, mechanics, electronics, process engineering and architectural design. In this chapter, the abbreviation CPS will cover both singular and plural form of cyber-physical systems. It is necessary to establish also the main features of CPS and main notations, models and analysis methods.

Also, the resilience term has different interpretations in each science field, so this chapter wants to establish the appropriate significance related to CPS. The CPS heterogeneity makes multiple possible definitions for resilience at every layer. In the following the abilities of detecting anomalies and properly react to these anomalies and to recover the system after an attack will be discussed.

The rest of this chapter is organized as follows. Section 11.2 discusses some important concepts for cyber-physical systems. Section 11.3 describes the cyber-physical systems threats and vulnerabilities. Section 11.4 presents the state of the art methods for cyber-physical system resilience enhancement and trends, while in Sect. 11.5, the main conclusions are drawn.

11.2 Cyber-Physical Systems Overview

The concept of system has emerged and developed over time as a result of emphasizing common features for a series of processes and phenomena from different domains, which allowed them to be treated in a unitary and systemic way. The literature comes with different definitions for the system notion, some of them taking into account its generality, others specific to a field of knowledge. Usually *a system is collection of elements that interacts with each other and its environment, organized in order to achieve a goal* [1]. A system is a connection of elements, each element constituting at its turn a system (subsystem). Interaction between the elements of the system may give it new properties, different from those of the component subsystems. In the case of physical systems, interaction is achieved through mass and energy flows, called stimuli if they come from the environment to the system or responses as system reactions to stimuli.

Cyber-physical systems (CPS) are according to Ref. [2] "*smart systems that include engineered interacting networks of physical and computational components*". These interconnected and integrated systems are usually designed to improve quality of life and to provide technological advances in critical infrastructures from defense and homeland security, health care, emergency response, traffic flow management, autonomous vehicles, robots, intelligent buildings, smart manufacturing, and energy supply and use as mentioned in [3].

The cyber-physical systems (CPS) consists in heterogeneous devices connected via different communication infrastructures. The physical system has interactions with the environment (including human interactions) via its sensing devices and responds to them via its actuating elements. In order to profit from computational capabilities, the physical system migrates into the cyber space within CPS framework.

In addition to CPS, within Industry 4.0 framework there are developed other concepts e.g. Industrial Internet, Internet of Things (IoT), machine-to-machine (M2M), smart cities, and so on that overlap more or less with CPS concept.

The Cyber-Physical Systems Public Working Group [2] was created in mid-2014 in US to establish a CPS Framework that refers to a common vocabulary, structure, and analysis methodology for CPS. There are also European research grants that focus on the development of CPS [4–8].

There are different definitions for CPS, all summarizing the main functional features of CPS [9]:

- interact with physical process via sensors and actuators;
- save and process the signals from sensors, and then act or react to them, interacting with both physical and cyber systems;
- use globally available data and services;
- are connected with one another and in global networks via digital communication facilities (wireless and/or wired, local and/or global);
- have a series of dedicated human–machine interfaces [9].

Due to its architecture, a common property of CPS is heterogeneity, having different communication infrastructures between its compatible elements [3, 10]. CPS have three main parts: the physical layer, networking layer and cyber layer as shown in Fig. 11.1.

The physical layer gathers all the devices that control the real process and provide the control interface to the real process. A typical control network is composed of Program Logic Controllers (PLC), Remote Terminal Units (RTU), Distributed Control Systems (DCS) as well as Wireless Sensor and Actuator Networks (WSANs). The networking layer is used to interconnect the physical layer with the cyber layer. The cyber layer comprises the servers with specialized software for modeling the system, used to generate a forecast; this layer means hardware and software packages used to assist in decision making. The cyber layer

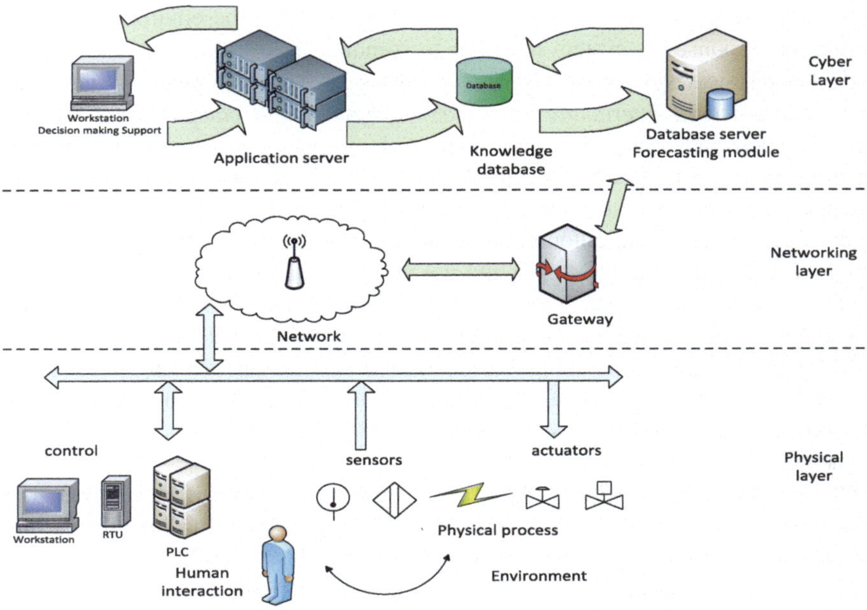

Fig. 11.1 An example of three-layer CPS architecture

is the twin of physical layer. Any decision in the physical layer can be evaluated in cyber layer and a real time forecast on how it might affect the real process is delivered, a very important feature in critical infrastructures. The cyber layer can find the optimal decision for the physical layer, usually the best control strategy. In order to acquire the information, process it and communicate its decision to the physical layer, the cyber layer needs a lot of physical equipment to deal with perception and communication, storage data, computational forecast and transfer the decisions in the real environment. Cyber-physical system means to profit from smart technologies to realize coordination awareness and control of the physical world. Cyber world and physical world are now more integrated [11].

According to Ref. [12], a CPS structure was proposed in 2015. The proposal is known as the 5C architecture. The proposed structure can be used as a guideline in developing of CPS for industrial applications. This CPS structure consists of two main components as mentioned in Ref. [13]: (1) the advanced connectivity that ensures real-time data streamlining between the two layers; and (2) intelligent data analytics that forms the cyber space.

The 5C structure, presented in Fig. 11.2, includes 5 levels for implementation in developing a Cyber-Physical System application: Smart Connection, Data-to-info Conversion, Cyber, Cognition and Configuration levels [12].

Smart Connection. At the first (bottom) level the priority consists in acquiring accurate and reliable data from the real process (e.g. via sensors) and implementing an accurate command to real process (e.g. via actuators). The data obtained at this

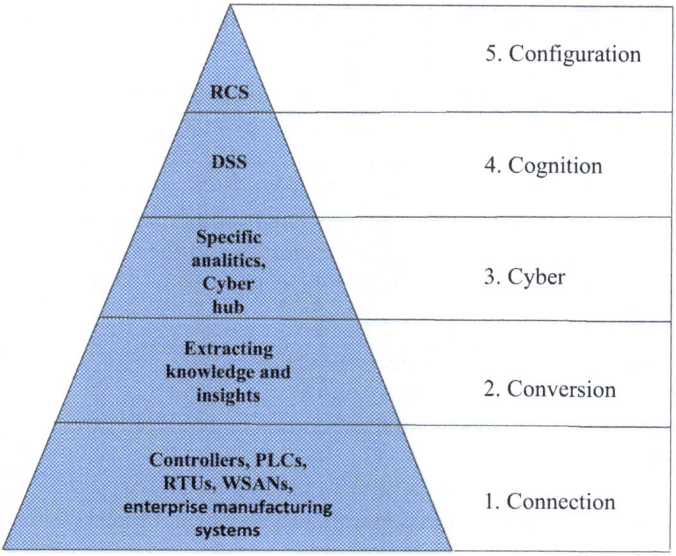

Fig. 11.2 The 5C functional architecture

level can come from sensors, controller or enterprise manufacturing systems. This involves a various type of data and consequently a specific protocol for data acquisition and transfer.

Data-to-info Conversion. In conversion level the focus is on transforming the data into knowledge and insights usually using methods based on artificial intelligence techniques (e.g. data mining techniques). The second level of CPS architecture enhances the process units with self-awareness [12].

Cyber. The cyber level acts as central information hub as presented in Ref. [13]. Information is being sent to it from every connected process unit to form the process network. Specific analytics have to be used to extract additional knowledge about process units' status among the process. These analytics enhances the process unit with self-comparison ability (historical knowledge is used to predict the future behavior of the process unit).

- *Cognition*. In this level, it is generated knowledge and insights of the system. Decision Support Systems (DSS) are developed to take the correct and optimal decisions [14]. The decisions must be prioritized.
- *Configuration*. The configuration level hosts the resilience control system (RCS) that apply the corrective and preventive decisions, which has been computed in cognition level, to the physical system. It works like a supervisory control system that generates the feedback from cyber to physical system.

11.3 Cyber-Physical Systems Threats and Vulnerabilities

It is very important that the CPS is designed in a trustworthy manner. Dependability and security concepts are very important features of CPS. Avizienis et al. [15] define the threats on dependability and security as failures, errors, and faults, long before a CPS framework is developed. A dependable and a secure system is a trustworthy one. The heterogeneity of CPS emphasizes mainly dependability and security issues for the cyber and physical elements and their interactions.

According to Ref. [15] the dependability comprises five "attributes" as mentioned below:

- *Availability*—"readiness for correct service";
- *Reliability*—"continuity of correct service";
- *Safety*—"absence of catastrophic consequences on the user(s) and environment";
- *Integrity*—"absence of improper system alterations";
- *Maintainability*—"ability to undergo modifications and repairs".

The most important threats to dependability and security are: faults, errors, and failures. Their formal definition is presented below:

- *Failure*—meaning the incorrect service of a device;
- *Error*—the deviation of a system's state from correct behavior.
- *Fault*—the known or presumed cause of an error (internal or external) [15].

The other main requirement for a dependable CPS is security [3, 16, 17]. Taking into account its nature, CPS security will inherit both physical and cyber components.

According to Ref. [18] the security aspects of CPS will focus on assuring:

- *Confidentiality*, meaning prevention of releasing undisclosed information;
- *Integrity*, meaning prevention of unauthorized alteration of information (data injection);
- *Availability*, meaning prevention of unauthorized possessing of information or resources and limiting the possibility to make critical resources unavailable.

A security breach within a CPS can cause two types of threats: threats on hardware components availability, or threats on data as indicated in Ref. [19]. Anwar and Ali [17] propose to label trust into internal or external, depending on CPS not having or having an interaction with the environment.

In the following paragraphs, we will analyze the threats on a three-layered architecture of CPS.

11.3.1 Threats in Physical Layer

The physical layer is the source of data from monitored system and the receiver of the control commands. The control network formed by PLC, DCS, RTUs or wireless sensors and actuators has a topography that makes some of the unattended nodes vulnerable to attackers [19–23]. The nodes forming the control network have limited cyber capabilities (memory and computing resources, communications, industrial protocols that must be improved for security issues). Security mechanisms used for traditional IT systems are not always efficient [24]. The main security threats of physical layer are discussed in the following paragraphs.

Two types of attack are identified at physical level:

- threats on device availability;
- threats on device information (control).

In the first category of attacks are: physical capture (physical possession of the nodes), equipment failure (reduced or lost performance due to external forces, wearing or harsh environment), and power line failure [25], physical destruction (physical damage of the nodes) and energy-exhaustion attacks [24]. The second type includes electromagnetic interference (unwanted electromagnetic signals interfere with useful signals, with loss of accuracy), data corruption [25], information disclosure [26], information tracking, tampering (attacker intercepts and modifies the data, then sends modified data to the recipient, [27]), sensing information leakage [28], different types of Denial of Service DoS attack (channel blocked through network bandwidth consumption). Introducing a false node to the network can lead to DoS attack, consuming the energy nodes in the control network as shown in Ref. [29]. This is also the case of Path Based DoS, when the channel is flooded with packets leading to energy consumption and network failure.

The physical layer is the base of a CPS, therefore special measures must be taken to ensure its security. Each node is important as capturing a node creates a breach in the security of the entire system. The security measures must provide a trustworthy CPS, where the signals from sensors are the real ones and the commands executed at this layer are those computed and verified in cyber layer.

In order to increase the CPS security at physical level it must be considered:

- to improve the identification and authentication mechanisms of each node from control network;
- to get better protection of data by applying biometric technology for operator authentication [20];
- to grow the penalties for the illegal activities targeting or involving cyber-physical systems;
- to develop innovative cryptographic technologies that might be applied to cyber-physical systems.

11.3.2 Threats in Network Layer

The network layer in cyber-physical system is formed by heterogeneous networks. The data transmission is wired or wireless and the main threats in this layer appear in the wireless transmission. The massive structure with a lot of nodes and huge amount of information can bring traffic congestion and also increases the chances of a Distributed DoS attack (DDoS). Potential damage to the whole network layer can be done when attacking the synchronization node between the cyber and physical layers. The attacks on this layer can lead to data leakage during the data transmission. An attacker can capture a transmitted message via an interference mechanism such as radio interface, modify and retransmit it, or exchange information between networks [24]. If remote access is allowed in a network with huge number of nodes, the attackers have an increased rate of a successful attack [30]. Usual attacks on this layer include [24]: response and Sybil (multiple identities for malicious nodes), traffic analysis, tampering, exhaustion, collision, black hole (malicious node cheats other nodes to connect with it, and then lose the packet to be forwarded), flooding, trap doors (exploiting exceptions in security policies), node sink, direction misleading sinkhole, wormhole, wrong path selection, tunneling and illegal access.

The main security threats as forms of attacks are presented below:

- *Routing*. Malicious node modifies the data path to create an infinite routing loop, or attack by tampering the routing information with the result of resistant network transmission, increased delays or extended source path as described in Ref. [26, 31].
- *DDoS—Distributed Denial of Service* attack target server with multiple DoS attacks at the same time [27]. This can be the result of a flooding attack that make the networked resource unavailable or not responding.
- *Sink Node Attack*. Malicious nodes interrupt data transmission by attacking the sink node.
- *Wormhole Attack*. Announces false paths through which all the packets are routed as presented in Ref. [32].
- *Jamming*. Block the wireless channel between sensor nodes leading to interruption of communications and subsequently to denial of service or significant performance drop [31, 33].
- *Selective Forwarding*. Makes a compromised node to lose key packets, and forward selected packets [31].
- *Sinkhole*. This attack generates the best routing path to be used for data transmission to normal nodes, attracting a huge amount of data. This attack creates the premises to launch other attacks, such as selective forwarding and spoofing [31].

Security measures must be adopted to the network layer of cyber-physical systems to provide data integrity, confidentiality and consistency. The authentication of nodes must have improved security mechanisms, because a compromised node is a breach in security that can facilitate other type of attacks.

11.3.3 Threats in Cyber Layer

At this layer, used to compute control commands and make decision based on extracted knowledge and insights, there are huge amounts of data (users' data) that need to be protected and have privacy. This level comprises a lot of applications that requires high processing capabilities and high storage facilities. Each application needs a specific set of security measures.

The common threats are:

- Leaking confidential data or credentials due to the vulnerabilities in the data transmission protocols, storage facilities [25].
- Unauthorized access to the network and system data [25].
- Code in the system with no effect (malicious, worms and viruses), with possible security risks [25, 34, 35].
- Attackers may use the system or damage the system by forging control commands as stated in Refs. [25, 34, 35].

Main security measures for cyber layer include the following:

- implement a clear access control policy of the system based on the least privilege rule;
- improve the authentication mechanisms and data security by using powerful encryption algorithms;
- establish a centralized, efficient security management platform.

CPS vulnerability represents the openness to specific attack or damage. There are a lot of vulnerabilities of the CPS and proper security policies can reduce their number. The vulnerabilities can appear in CPS management, platform and network.

11.3.4 Vulnerabilities of Management

The CPS must have a safety and security plan. It is possible that such a plan does not exist or lacks important components. Policy and implementation guideline can be incomplete, inappropriate, or does not exist. The vulnerabilities appear when there are no adequate Industrial Control System (ICS) security policies, no safety training and attack recognition programs, faulty security design, no written safety procedures, no ICS security review (that must be checked periodically), no Disaster Recovery Plan (DRP) or Resilience Control System (RCS).

11.3.5 Vulnerabilities of Platform

In CPS, the vulnerabilities might appear from defects in the platform which can be misconfigured or poorly maintained, or from platform hardware and software. The platform configuration vulnerabilities come from delaying or not updating the operating system or applications with the latest security patches, not saving or backing up the most important settings and data, no protecting data with a strong password policy [27]. The platform hardware vulnerabilities derive from the lack of performing security control, or not physical protecting the critical systems, not having a redundancy scheme for important resources, not preventing physical access to critical facilities, having remote access to physical devices from the process (usually via wireless transmissions). Among platform software vulnerabilities there are running unnecessary services, improper authentication and access control, not maintaining logs for alarms and events. Also, the platform malware vulnerabilities include not installing malware protection software, not updating the definitions for malware protection software and not testing for known attacks.

The security plan should have security controls such as updating and patching the operating system and applications, biometric access control [22] or antivirus software [27].

11.3.6 Vulnerabilities of Network

These CPS vulnerabilities are related to network platform configuration (not using data flow control, not configuring the best security network settings, passwords are not encrypted), network hardware (unsafe physical ports, not having redundancy architecture to important nodes in the network), network perimeter (not having a firewall or having an improper configured one, not having defined a security perimeter), network monitoring and logging (the control network uses industrial protocols with no security components, lack of firewalls and security logs), network communication (no authentication of user or device) and network connection (not enough certification between the client and the server, not enough data protection between the client and server [27]). The security plan should include enhancing the industrial data protocols with security components [23], intrusion detection systems, network traffic encryption, network link limit [27].

11.4 State of the Art Methods for Cyber-Physical System Resilience Enhancement and Trends

The heterogeneity of CPS makes them prone to a variety of attacks. The number of attacks to cyber-physical systems is increasing [2, 36]. Traditional security and protection mechanisms cannot cope with such complex attacks, making necessary new approaches for these heterogeneous cyber-physical systems.

According to Ref. [10] the concept of resilience was originally associated with the fields of ecology and psychology. The main parts of cyber-physical systems determine two meanings for the resilience term: a movement among entities from CPS to improve their ability to quickly recover from catastrophic events, such as natural disasters or terrorist attacks—for the physical part and for the cyber part—the stability and quality of service in face of threats on the computing and networking infrastructure [10, 37].

There is no unanimously accepted definition for resilience. There are researchers that consider resilience as the availability of the underlying system. Others add both the ability to handle threats of an unexpected and malicious nature, and the defense and recovery after cyber-attacks [37, 38]. In literature, there are definitions for CPS resilience with respect to its service sectors: buildings, energy, consumer and home, healthcare and life service, industrial, transportation retail, security, public safety, IT and networks [2]. One definition for resilience includes the ability to accommodate faults or events that otherwise might compromise the stability of the system and the underlying goals [10]. Arghandeh et al. [38] propose a definition for CPS resilience that takes into account the threats to physical, cyber and cyber-physical parts. According to them "the resilience of a system presented with an unexpected set of disturbances is the system's ability to reduce the magnitude and duration of the disruption. A resilient system downgrades its functionality and alters its structure in an agile way". A resilience framework, presented in Fig. 11.3, is proposed in comparison with the Pressure and Release (PAR) risk analysis framework [38].

The specific resilience properties of durability, survivability and self-healing are time dependent. Resilience assessment adds the temporal dimension to risk assessment framework.

Resilient control systems (RCS) are a part of CPS, as shown in Fig. 11.2, and their design must take into account all possible threats. Rieger et al. [39] defined resilient control system as "one that maintains state awareness and an accepted level of operational normalcy in response to disturbances, including threats of an unexpected and malicious nature". The concepts of state awareness and context awareness are introduced, both used to resilience enhancement. Taking into account the difficulty of detecting an attack, monitoring system parameters is the first step in mitigating the attack. The system parameters can be monitored if they can be measured or estimated.

State awareness is a resiliency related concept that has many definitions. In CPS, state awareness is the ability to know or estimate the necessary control system states

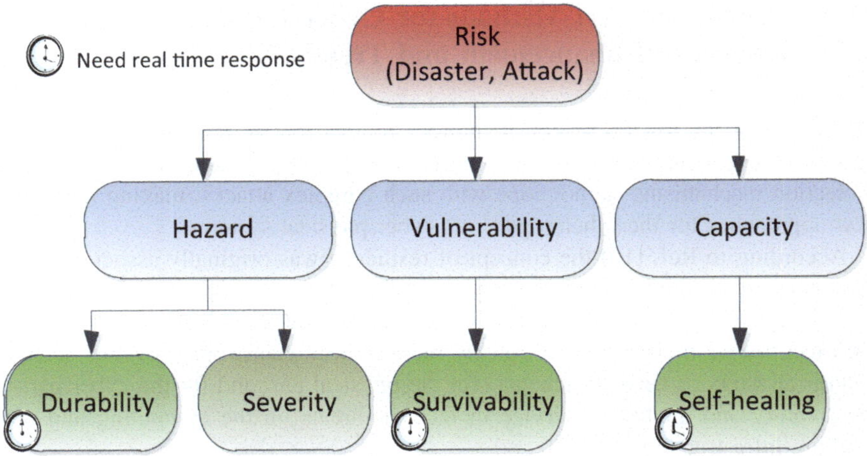

Fig. 11.3 The resilience framework proposed by Arghandeh [38]

to maintain a stable closed loop operation, and the generation of sufficient knowledge and insights of operation to make reliable informed decisions [39, 40].

The observability concept from control theory [1] is connected to state awareness concept. The observability is an internal system property (observability is a measure of how well current internal states of a system can be inferred from knowledge of its external outputs), and state awareness refers to actual measurement or estimation of the internal system states. Melin et al. [40] consider the state awareness as the availability of the internal system's states, either from direct measurement (sensors) or from inferring available outputs (via estimator).

The effects of an attack can be mitigated if the system is capable of maintaining state awareness. This means that if the attack provides false sensorial data, the system is no longer capable of maintaining state awareness, resulting in physical damage and faulty behaviour. The resilient control system is in the top of functional pyramid architecture of 5C which means that its decisions are based on sufficient knowledge and insights of system parameters. Rieger et al. [41] states that in a resilient control system it is necessary to consider everything that might affect the system's normalcy, to be able to maintain state awareness. They propose two cyber control strategies to maintain state awareness: a closed loop and an open loop. The performance of resilient control system is measured in closed loop strategy assessing cyber and physical security, process efficiency and stability, and process compliancy. It is crucial that the knowledge or insights extracted from the available data to permit maintaining the normal operation of the system. Consequently, state awareness can be considered as the availability of the necessary knowledge to maintain an acceptable level of normal operation of the system. The necessary knowledge consists in information about internal states of the physical process and information from cyber level as recommendations to maintain the normal operation of the system.

Context awareness is an important property especially in the field of critical when services must be delivered at the proper time. A cyber-physical system with context awareness is able to sense the changes in the physical environment and adapt/react rapidly to deliver the solutions in real time. Basically, it means that a CPS can get information from its environment and can use it in an intelligent manner to anticipate needs and situations and react with solutions (intelligent adapting to environment [42]). Another definition for context awareness is the ability of systems to link their operation with changes in the environment [43, 44].

The context is "any information that can be used to characterize the situation of an entity. An entity is a person, place or object that is considered relevant to the interaction between a user and an application, including the user and applications themselves" [45]. More definitions for context in cyber-physical systems can be found in Ref. [46–49].

The notion of context is time dependent and characterizes the current situation of the highly dynamic and heterogeneous system. The interactions between two entities of the system creates context, and the lack of interactions between them is also context. The information from the environment is inferred in context inference block and proper preventive and corrective decisions are made to deal with acquired information.

Context awareness refers to detect, interpret and react to aspects of a user's environment. A system has context awareness if it is able to support real time changes in its behavior, as a reaction to known and measured changes in its context [45]. Truong et al. [50] define a context awareness system as that one which is able to extract, infer and use context knowledge in order to adapt its functionality to the current context.

Context awareness systems usually process and reason uncertain, ambiguous and missing information [46]. There is a need of designing trustworthy cyber-physical systems [3] that include more than resilience and robustness in their definition, and context aware systems are now in trend of their design.

The literature presents a variety of context awareness cyber-physical systems corresponding to the area of CPS implementation (buildings, energy, consumer and home, healthcare and life service, industrial, transportation retail, security, public safety, IT and networks). The proposed solutions may vary, but the information flow usually has the same path [10] as presented in Fig. 11.4.

There are context awareness CPS that only react after changing conditions from its normal behaviour that are usually measured with sensors (the context awareness closed loop). Other context awareness CPS prevent those changes taking into account some important disturbances (the context awareness open loop). The most complete context awareness CPS works in a combined manner both feedback and feedforward and have the structure presented in Fig. 11.4.

The information flow goes in a feedback or feedforward manner and must be first collected via sensors or estimators (for those variables that cannot be measured). Next step includes information categorization, data fusion and aggregation, managing missing data, machine learning algorithms and inference [46]. The pre-processed collected information is modeled and stored usually using databases and ontologies. The context model that describes the underlying entities and their relationships can be static or dynamic [51].

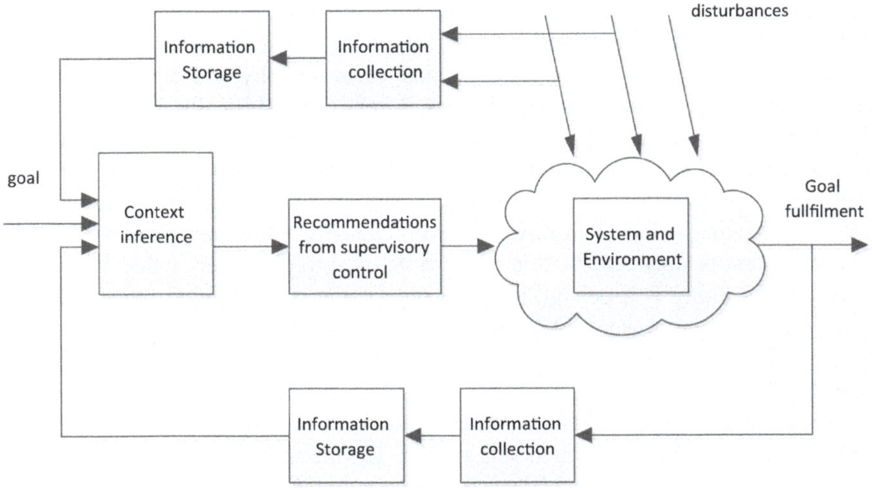

Fig. 11.4 Information flow in context awareness CPS

The context inference step must provide new information based on available information, as recommendations of the supervisory control system. Computational intelligence techniques (fuzzy inference system, artificial neural networks, ANFIS, etc.) are used as context inference techniques in the literature [10, 45]. The context-aware CPS process the available information to identify changes in context and provide solutions to adapt its behaviour, in response to changes. The recommendations to adapt its behavior sent to the real process are taken after formulating and solving of an optimization problem.

There is a need for the resilience enhancement of CPS. Most enhancements are related to resilience planning or resilience operation (response and restoration). The artificial intelligence techniques are in trend to solve different problems from resilience planning or operation.

The resilience enhancement of CPS is dependent on the field in which CPS is developed. A thorough review study on these types of systems applied in smart grids is presented in Ref. [52]. Denker et al. [53] argued the necessity of an "observe-analyze-adapt" methodology for structure adaptation in resilient systems. This methodology can be implemented processing the existing information in a what–if analysis.

In Ref. [10] there is presented an enhanced resilient system response based on a multi-agent based framework. The main advantage of the proposed system is that it is designed for a general CPS, with multiple interactions. However, the framework is tested for a relatively small CPS involved in a continuous stirred tank reactor control. The proposed multi-agent framework had improved responses at execution time, autonomy, services continuity and superior levels of scalability with interesting possibilities of applying it to a large-scale CPS.

Computational intelligence techniques are also proposed in [54] in order to improve state awareness of resilient CPS. The proposed architecture is based on fuzzy-neural data fusion engine and must provide real-time performance monitoring and analysis of complex critical control systems (detection of anomalies). The enhancement is an earlier identifying of intrusive behavior than conventional threshold-based alarm systems.

In [55] it is proposed STPA-SafeSec, a novel analysis methodology for both safety and security. This methodology can be used to build a reactive CPS with the most effective mitigation strategies to ensure safety and security of the system. This methodology is applied in the power grid domain and proves the ability to highlight the physical effects of security vulnerabilities or system flaws quantifying the possible losses.

In [56], an enhanced resilient control algorithm for a wireless networked control system is proposed. The algorithm must ensure operational normalcy in case of wireless link failure, using a Kalman filter approach to predict and estimate the correct wireless sensor output.

In [57], an enhanced resilient system is proposed applied to monitoring complex engineering facilities. The proposed system claims to have the ability to correctly assess facility health within real time decision period despite cyber-physical coordinated attacks. The three-layer system is able to dynamically adapt and reconfigure depending on current conditions. The proposed system testing is made using small scale CPS. The main disadvantage is the need of knowing the precise model of the process in order to make the appropriate decision (a difficult task for complex systems).

Rieger and Villez [58] proposed a generalized design methodology for analyzing and acting upon anomalies in cyber-physical systems (enhancing state awareness), with focus on industrial control systems. The computational intelligence techniques are used to design a method to integrate cyber and physical data, and to appropriately react to both benign and malicious actions. The proposed methods have difficulties in taking the decisions to the lower level in CPS structure in real time manner for a complex heterogeneous system.

11.5 Conclusions

Resilience enhancement in cyber-physical systems is a real necessity these days due to the increased number of attacks. The literature study showed a relatively small number of general solutions for resilience enhancement in CPS. Most of the solutions are particular to the field of CPS application (buildings, energy, consumer and home, healthcare and life service, industrial, transportation retail, security, public safety, IT and networks). The resilience enhancement strategies refer to planning or operation (response and restoration). Another classification of resilience enhancement depends on the resilience property (adaptation and recovery).

Nowadays, the computational intelligence techniques proved to be a useful tool for the planning and adaptation resilience. These techniques are appropriate to contribute to decision making process or early detecting anomalies in CPS behaviour. Nevertheless, the CPS field faces new threats and challenges mostly due to its heterogeneity, complexity and lack of legislation.

References

1. V. Cirtoaje, *Systems Theory—The Elementary Analysis in Time Domain* (Petroleum Gas University of Ploiesti, Romania, 2015) (Romanian)
2. Cyber-Physical Systems Public Working Group NIST, Framework for Cyber-Physical Systems. Technical Report Release 1.0, National Institute of Standards and Technology, May 2016
3. A. Romanovsky, F. Ishikawa, *Trustworthy Cyber-Physical Systems Engineering* (CRC Press, Boca Raton, 2016), p. 462
4. Methodological Guidelines 3. Technical Report, DESTECS, D2.3, 2012. www.destecs.org/images/stories/Project/Deliverables/D23MethodologicalGuidelines3.pdf
5. Convergence Report 3. Technical Report, Compass Deliverable D11.3, 2014. www.compass-research.eu/deliverables.html
6. Structuring of CPS Domain: Characteristics, Trends, Challenges and Opportunities Associated with CPS. Technical Report, CyPhERS (Cyber-Physical European Roadmap and Strategy), D2.2, 2014. http://cyphers.eu/project/deliverables
7. Definitional Framework. Technical Report, TAMS4CPS, D1.1, 2015. www.tams4cps.eu/wp-content/uploads/2015/04/TAMS4CPS_D1-1_Definitional-Framework.pdf
8. Systems Engineering Handbook: A Guide for System Life Cycle Processes and Activities. Version 3.2.2, Technical Report, INCOSE-TP-2003-002-0.3.2.2, International Council on Systems Engineering (INCOSE), Oct 2011
9. Driving Force for Innovation in Mobility, Health, Energy and Production. Technical Report, Acatech, Dec 2011. www.acatech.de/fileadmin/user_upload/Baumstruktur_nach_Website/Acatech/root/de/Publikationen/Stellungnahmen/acatech_POSITION_CPS_Englisch_WEB.pdf
10. F.E. Pais Januario, J. Leitao, A. Cardoso, P. Gil, Resilience enhancement in cyber-physical systems: a multiagent-based framework, in *Multi-agent Systems*, ed. by J. Rocha (InTech, 2017). https://www.intechopen.com/books/multi-agent-systems/resilience-enhancement-in-cyber-physical-systems-a-multiagent-based-framework
11. K. Zhejun et al., Research on human sensory architecture for cyber-physical systems. J. Netw. **8**(11), 2692+, Academic OneFile, 6 Nov (2017)
12. J. Lee, B. Bagheri, H.A. Kao, A cyber-physical systems architecture for industry 4.0-based manufacturing systems Manuf. Lett. **3**, 18–23 (2015)
13. J. Lee, H. Davari Ardakani, Sh Yang, B. Bagheri, Industrial big data analytics and cyber-physical systems for future maintenance & service innovation. Procedia CIRP **38**, 3–7 (2015)
14. M. Oprea, S.F. Mihalache, M. Popescu, Computational intelligence-based PM2.5 air pollution forecasting. Int. J. Comput. Commun. Control **12**(3), 365–380, June (2017)
15. A. Avizienis, J.C. Laprie, B. Randell, C. Landwehr, Basic concepts and taxonomy of dependable and secure computing. IEEE Trans. Dependable Secure Comput. **1**(1), 11–33 (2004)
16. M. Krotofil, A. Cardenas, J. Larsen, D. Gollmann, Vulnerabilities of cyber-physical systems to stale data—determining the optimal time to launch attacks. Int. J. Crit. Infrastruct. Prot. **7** (4), 213–232 (2014)

17. R.W. Anwar, S. Ali, Trust Based Secure Cyber-Physical Systems. Workshop Proceedings: Trustworthy Cyber-Physical Systems, Technical Report Series: Newcastle University, Computing Science, No: CS-TR-1347, pp. 1–10 (2012)
18. C.P. Pfleeger, S.L. Pfleeger, J. Margulies, *Security in Computing*, 5th edn. (Prentice Hall, Upper Saddle River, 2015)
19. E. Pricop, S.F. Mihalache, *Assessing the Security Risks of a Wireless Sensor Network from a Gas Compressor Station*. 2nd International Workshop on Systems Safety & Security (IWSSS 2014) Bucuresti, Romania, Proceedings of the 6th International Conference on Electronics, Computers & Artificial Intelligence (ECAI 2014), vol. 5 (2014), pp. 45–50
20. E. Pricop, S.F. Mihalache, J. Fattahi, *Innovative Fuzzy Approach on Analyzing Industrial Control Systems Security, Capitol Publication Recent Advances in Systems Safety and Security* (Springer International Publishing AG, Cham, Switzerland, 2016)
21. E. Pricop, S.F. Mihalache, *Fuzzy Approach on Modelling Cyber Attacks Patterns on Data Transfer in Industrial Control Systems*. 3rd International Workshop on Systems Safety & Security (IWSSS 2015), Proceedings of the 7th International Conference on Electronics, Computers & Artificial Intelligence (ECAI 2015), vol. 7, no. 2 (2015)
22. E. Pricop, Security of industrial control systems—an emerging issue in Romania national defense. Sci. Bull. "Mircea Cel Batran" Naval Academy, Constanta, Romania **18**(2), 142–147 (2015)
23. E. Pricop, S.F. Mihalache, N. Paraschiv, J. Fattahi, F. Zamfir, *Considerations Regarding Security Issues Impact on Systems Availability*. 4th International Workshop on Systems Safety & Security, 7th International Conference on Electronics, Computers & Artificial Intelligence (ECAI 2016), Ploiesti, Romania, vol. 8, no. 4 (2016)
24. Y. Ashibani, Q.H. Mahmoud, Cyber-physical systems security: analysis, challenges and solutions. J. Comput. Secur. **68**, 81–97 (2017)
25. Y. Peng, T. Lu, J. Liu, Y. Gao, X. Guo, F. Xie, *Cyber-Physical System Risk Assessment*. 9th International Conference on Intelligent Information Hiding and Multimedia Signal Processing, Beijing (2013), pp. 442–447
26. K. Zhao, L. Ge, *A Survey on the Internet of Things Security*. 9th International Conference on Computational Intelligence and Security, Leshan (2013), pp. 663–667
27. Y. Gao et al., *Analysis of Security Threats and Vulnerability for Cyber-Physical Systems*. 3rd International Conference on Computer Science and Network Technology, Dalian, pp. 50–55 (2013)
28. Q. Gou, L. Yan, Y. Liu, Y. Li, *Construction and Strategies in IoT Security System*. IEEE International Conference on Green Computing and Communications and IEEE Internet of Things and IEEE Cyber, Physical and Social Computing, Beijing (2013), pp. 1129–1132
29. R. Mahmoud, T. Yousuf, F. Aloul, I. Zualkernan, *Internet of Things (IoT) Security: Current Status, Challenges and Prospective Measures*. 10th International Conference for Internet Technology and Secured Transactions (ICITST), London (2015), pp. 336–341
30. Q. Shafi, *Cyber-Physical Systems Security: A Brief Survey*. 12th International Conference on Computational Science and Its Applications, Salvador (2012), pp. 146–150
31. S. Raza, Lightweight Security Solutions for the Internet of Things, Malardalen University Press Dissertations, Vasteras, Sweden, 2013
32. N. Gaddam, G.S.A. Kumar, A.K. Somani, *Securing Physical Processes Against Cyber Attacks in Cyber-Physical Systems*. National Workshop for Research on High-Confidence Transportation CyberPhysical System: Automotive, Aviation and Rail, Washington, DC (2008), pp. 2–4
33. Y. Li, L. Shi, P. Cheng, J. Chen, D.E. Quevedo, Jamming attacks on remote state estimation in cyber-physical systems: a game-theoretic approach. IEEE Trans. Autom. Control **60**(10), 2831–2836 (2015)
34. T. Lu, J. Lin, L. Zhao, Y. Li, Y. Peng, A security architecture in cyber-physical systems: security theories, analysis, simulation and application fields. Int. J. Secur. Appl. **9**(7), 1–16 (2015)

35. H. Suo, J. Wan, C. Zou, J. Liu, *Security in the Internet of Things: A Review*. International Conference on Computer Science and Electronics Engineering, Hangzhou (2012), pp. 648–651
36. Y. Yuan, Q. Zhu, F. Sun, Q. Wang, T. Basar, *Resilient Control of Cyber-Physical Systems Against Denial-of-Service Attacks*. 6th International Symposium on Resilient Control Systems (ISRCS), San Francisco, CA (2013), pp. 54–59
37. E. Hollnagel, C.P. Nemeth, *Resilience Engineering Perspectives, Vol. 2, Preparation and Restoration* (CRC Press, Boca Raton, 2016), p. 310
38. R. Arghandeh, A. Von Meier, L. Mehrmanesh, L. Mili, On the definition of cyber-physical resilience in power systems. Renew. Sustain. Energy Rev. **58**, 1060–1069 (2016)
39. C. Rieger, D. Gertman, M. McQueen, *Resilient Control Systems: Next Generation Design Research*. 2nd IEEE Conference on Human System Interactions (2009), pp. 632–636
40. A. Melin, E. Ferragut, J. Laska, D. Fugate, R. Kisner, *A Mathematical Framework for the Analysis of Cyber-Resilient Control Systems*. 6th IEEE International Symposium on Resilient Control Systems (ISRCS) (2013), pp. 13–18
41. C. Rieger, Q. Zhu, T. Basar, *Agent-Based Cyber Control Strategy Design for Resilient Control Systems: Concepts, Architecture and Methodologies*. 5th IEEE International Symposium on Resilient Control Systems (2012), pp. 40–47
42. R. Malekian, K. Wu, G. Reali, N. Ye, K. Curran, Cyber-physical systems and context-aware sensing and computing. Comput. Netw. **117**(1), 1–4 (2017)
43. M. Rosemann, J.C. Recker, *Context-Aware Process Design: Exploring the Extrinsic Drivers for Process Flexibility*. 18th International Conference on Advanced Information Systems Engineering, Namur University Press, Belgium (2006), pp. 149–158
44. B. Bordel, R. Alcarria, T. Robles, D. Martin, Cyber-physical systems: extending pervasive sensing from control theory to the internet of things. Pervasive Mob. Comput. (2017)
45. C. Perera, A. Zaslavsky, P. Christen, D. Georgakopoulos, Context aware computing for the internet of things: a survey. IEEE Commun. Surv. Tutor. **16**(1), 414–454 (2014)
46. O. Yurur, C.H. Liu, Z. Sheng, V.C.M. Leung, W. Moreno, K.K. Leung, Context-awareness for mobile sensing: a survey and future directions. IEEE Commun. Surv. Tutor. **18**(1), 68–93 (2016)
47. K. Wan, V. Alagar, Context-aware security solutions for cyber-physical systems, in *Context-Aware Systems and Applications, ICCASA 2012*, vol. 109, Lecture Notes of the Institute for Computer Sciences, Social Informatics and Telecommunications Engineering, ed. by P.C. Vinh, N.M. Hung, N.T. Tung, J. Suzuki (Springer, Berlin, Heidelberg, 2013)
48. J. Wan, D. Zhang, S. Zhao, L. Yang, J. Lloret, Context-aware vehicular cyber-physical systems with cloud support: architecture, challenges, and solutions. IEEE Commun. Mag. **52** (8), 106–113 (2014)
49. T. Li, J. Cao, J. Liang, J. Zheng, Towards context aware medical cyber-physical systems: design methodology and a case study. J. Cyber-Phys. Syst. **1**(1), 5–23 (2015)
50. H. Truong, S. Dustdar, A survey on context-aware web service systems. Int. J. Web Inform. Syst. **5**(1), 5–31 (2009)
51. G. Adomavicius, A. Tuzhilin, Context-aware recommender systems, in *Recommender Systems Handbook*, ed. by F. Ricci, L. Rokach, B. Shapira, P. Kantor (Springer US, Boston, MA, 2011), pp. 217–253
52. G. Huang, J. Wang, C. Chen et al., System resilience enhancement: smart grid and beyond. J. Front. Eng. **4**(3), 271–282 (2017)
53. G. Denker, N. Dutt, S. Mehrotra et al., Resilient dependable cyber-physical systems: a middleware perspective. J. Internet Serv. Appl. **3**(41) (2012)

54. D. Wijayasekara, O. Linda, M. Manic, C. Rieger, FN-DFE: fuzzy-neural data fusion engine for enhanced resilient state-awareness of hybrid energy systems. IEEE Trans. Cybern. **44**(11), 2065–2075 (2014)
55. I. Friedberg, K. McLaughlin, P. Smith, D. Laverty, S. Sezer, STPA-SafeSec: safety and security analysis for cyber-physical systems. J. Inform. Secur. Appl. **34**, Part 2, 183–196 (2017)
56. K. Ji, D. Wei, Resilient control for wireless networked control systems. Int. J. Control Autom. Syst. **9**(2), 285–293 (2011)
57. H. Garcia, W. Lin, S. Meerkov, *A Resilient Condition Assessment Monitoring System*. 5th IEEE International Symposium on Resilient Control Systems (2012), pp. 98–105
58. C. Rieger, K. Villez, *Resilient Control System Execution Agent (ReCoSEA)*. 5th IEEE International Symposium on Resilient Control Systems (2012), pp. 143–148

Chapter 12
Issues in Securing Critical Infrastructure Networks for Smart Grid Based on SCADA, Other Industrial Control and Communication Systems

Florentina Magda Enescu, Nicu Bizon and Carmen Maria Moraru

Abstract Computer facilities and microprocessor-based technology have been successfully used in the energy industry. For protection and equipment control, this technology has been used in SCADA, remote control and monitoring applications. Particular attention is paid to the cyber security sector for automation and control systems. Protective application of the equipment and control, SCADA, monitoring and remote control, uses the technology with microprocessor. Due to the great importance of the power supply process, there is no question of being left in a state of vulnerability and neglect. Security is not perfect and will never be. For this reason, there will always be security breaches and incidents. Also, for this reason, not only protection mechanisms are in place, but also mechanisms for rapid detection of incidents and which are able to react effectively to the isolation of problems and to ensuring security. Processes security for systems will continue to evolve in the future. By definition, there are no communication systems that are 100% safe. Attacks against critical industrial infrastructures marked an increase not only in terms of number but also of the level of complexity. The destruction of the industrial control system (ICS) and critical processes were interrupted. For many organizations, the security improvement in ICS systems is great. The extreme sensitivity to ensure the availability and performance of industrial processes has led to a more conservative and rigorous approach to how security measures are implemented. Cyber-attacks that could compromise the availability, integrity, and confidentiality of ICS systems may come from within systems or from outside ICS

F. M. Enescu (✉) · N. Bizon
Faculty of Electronics, Communications and Computers, University of Pitesti, Pitesti, Romania
e-mail: florentina.enescu@upit.ro

N. Bizon
e-mail: nicu.bizon@upit.ro

C. M. Moraru
National Institute of Research-Development for Cryogenic and Isotope Technologies, Ramnicu Valcea, Romania
e-mail: carmen.moraru@icsi.ro

© Springer Nature Switzerland AG 2019
N. Mahdavi Tabatabaei et al. (eds.), *Power Systems Resilience*, Power Systems,
https://doi.org/10.1007/978-3-319-94442-5_12

systems. Among the ICS system infection vectors from the perspective of the SANS Institute (2014) include: external threats (state attacks, hacking etc.), malware, exploiting tools, phishing, internal attacks, cyber security protocols, and industrial espionage. This chapter addresses the cyber security issues required for the protection, automation, control and communications systems of transformation stations as well as methods that could be used to prevent computer attacks that can have a significant impact on the availability of the system Electro-energetic effect with serious consequences on extended area interruptions.

The chapter consists of five parts as follows:

- In the first part the concept of critical infrastructure is approached. It starts from the definition of the term, the protection of critical infrastructure and, last but not least, dangers and threats from the virtual space;
- In the second part, the industrial control systems in the power stations are presented in terms of vulnerabilities, how to implement the security measures, and so on;
- The third part is intended for monitoring, control and data acquisition (SCADA). Specifically, the SCADA system is presented at the level of an electrical station, with all the problems that may arise during its operation;
- The intelligent power grids (Smart Grids) are presented in Part Four. A Smart Grid includes software and hardware designed to significantly improve the functionality of the system. Smart Grid faces a number of security situations related to regulatory systems, Smart Meters, status estimation, and communications networks.

Keywords Control and communication · Critical infrastructure
Critical processes · Cyber attacks · Electro-Energetic · Industrial control systems
SCADA · Security situations · Smart grids

12.1 Critical Infrastructure of the Smart Grid

As a rule, critical infrastructures are those infrastructures that depend on the stability, safety and security of systems and processes. Critical infrastructures are those facilities with an important role in ensuring security in the operation and implementation processes of economic, social, political, and military information.

Infrastructures are considered critical due to:

- unique conditions in the infrastructure of a system or process;
- the vital importance they have as support material or virtual (network) in the operation and implementation processes of economic, social, political, informational, military, etc.;
- important role, that they meet the stability, reliability, security, functionality, especially in security systems;

- increased vulnerability to direct threats, as well as the targeting systems to which they belong;
- the particular sensitivity to changing conditions, especially sudden changes of situation [1].

12.1.1 Protection of Critical Infrastructure

The protection of a critical infrastructure consists of all the measures set to reduce the risks of blocking the operation or destroying a critical infrastructure.

Organized cyber-attacks affect:

- national infrastructures;
- economy;
- national security.

Although attacking such a system involves a higher technical complexity, the organized attackers exploited some vulnerability, demonstrating their destructive possibilities [1, 2].

12.1.2 Dangers and Threats from Virtual Space

In virtual space are targeted:

- physical equipment and systems, which include:
    - computers;
    - providers;
    - connections;
    - network nodes;
    - etc.
- other infrastructure hosting such facilities, such as:
    - buildings;
    - electricity networks;
    - cables;
    - optical fiber;
    - etc.
- data base;
- programs;
- storage and distribution systems;
- material support of databases;
- etc.

Types of threats against critical infrastructure of cyberspace:

- competition between big companies for IT supremacy resources and markets;
- dangers and asymmetric threats;
- development of subversive and unconventional IT networks;
- increasing activity of hackers;
- cyber-terrorism [1, 3].

12.2 Industrial Control Systems Within Power Stations

12.2.1 Vulnerabilities Identified

ICS systems have achieved improvements. If initially proprietary systems were interconnected by serial communication technology (RS232), today the connection is made using geographically distributed systems, proprietary commercial protocols, or classic Internet Protocol (IP) protocol [2].

In the SCADA (Supervisory Control and Data Acquisition) control systems, and those in industrial critical infrastructure, an unprecedented level of agility, speed of development and integration of technologies that support such complex networks has been found [3–6].

12.2.2 Solutions

Lately, there has been recorded an increase not only in the number but also in the level of complexity of targeted attacks against critical industrial infrastructures.

In this respect, security measures are implemented with greater responsibility as follows:

- Protecting systems online by introducing Intrusion Prevention System (IPS) or antivirus (AV) solutions is not used due to the possibility of accidental degradation or system lock. These solutions can affect system performance;
- In the case of automated scans, there may be malfunctions or interruptions of critical services from industrial controllers that were not designed to handle such events. But if it works in this way, the level of cyber-attack prevention tends to zero.

Scenario

During a technical meeting on "Conducting Cyber Threat Assessments at Nuclear Facilities", which was held at the International Atomic Energy Agency in the period 9–12 February 2016 in Vienna, the researcher Mark Fabro, Chief Security Scientist at Lofty Peach Canada, conducted a technical demonstration to raise awareness and highlight a number of capabilities by which an attacker can gain access to a critical infrastructure.

The demonstration includes the following steps:

1. Compromising video monitoring systems of the building—CCTV
   Attackers affect video monitoring systems of the building to get to the access control system (CA). Their interest is to affect the CCTV system network to display static images on monitors instead of a real-time video recording of the building.
2. Compromising the access control system
   To search and retrieve information attacker needs physical access to the building critical infrastructure. In this way the agreement for physical penetration in the building is obtained. From the access control unit are taken necessary information to continue the attack.
3. Identify vulnerabilities and developing exploits
   Attackers made models of the system followed using information obtained. By means of tools and operations fuzzing (vulnerabilities discovery technique), the target system can be discovered vulnerabilities of the system.
   Thus they are identified:
   - 0-day vulnerabilities (vulnerability of a software for which there is no solution yet);
   - ways to exploit them.
4. Execution of the attack
   The attacker, to attack and sabotage a facility control facility, has developed a code sequence or set of commands used to exploit a vulnerability to determine an unwanted behavior of the application. At this point, a remote connection can be established through which attackers collect additional information and install custom malware for the control system.
   An attacker can perform: intercepts data between the ICS's Human Machine Interface (ICS) and control components; manipulates the valves and pumps in a manner that can cause damage; modifies the operator display and alarms; modifies the configuration of the control device; finally sabotage the system's operation.
5. Mitigation
   Defense in depth is a key principle of nuclear safety and overall security of computer networks. At every step of the attack scenario, there are opportunities for preventing, detecting and responding quickly against attackers.
   As the scenario illustrates, cyber-attacks can also include physical damage and human damage, therefore computer security should be an integrated part of a comprehensive nuclear safety program which:
   - integrates endpoint security networks and equipment with an analysis system to detect cyber threats;
   - classify application and user traffic, not just based on ports and IPs;
   - supports the granular segmentation of the network, including role-based access;

- block known signature threats;
- detects and prevents unknown malware attacks;
- stop 0-day attacks at the endpoints of the network;
- provide central management and reporting of critical events;
- secure the use of mobile and virtualization technologies;
- manufactures standard industrial management interfaces through a complex API.

With such functional capabilities in a complex ICS system, companies may be able to discourage advanced threats and adapt to the evolution of cyber threats [7–10].

12.3 Systems for Monitoring, Control and Acquisition of Data (SCADA)

12.3.1 SCADA—Smart Grid—Intelligent Network

At the conceptual level, the term Smart Grid, translated as Intelligent Network, referring to the electrical network, represents a symbiosis between the elements of an electrical network in the classical meaning of the word and the elements of the information and communication technology that complement the functionality of that network.

Regardless of the attempt to define the term Smart Grid, we often meet the three aspects:

- the actual network;
- devices with intelligence, whether distributed or centralized, representing the numerical computing systems or parts thereof, specific to information technology and controlling the network elements according to various algorithms implemented by numerical programs (firmware or software);
- communication infrastructure that mediates bidirectional information exchange between component elements (also leading to specific cyber security problems).
  In this broad sense of the smart grid concept, its applicability is at every level: from the production of energy from conventional and renewable distributed to transport electricity to its distribution and not least in the use of it by the consumer—whether it's from industrial or household use.
  Among the most common technologies that refer to Smart Grid, we can remember:
- The power supply level, intelligent meters can be listed, capable to be programmed to make decisions based on time;
- At the power distribution level, distribution automation systems that are correlated with SCADA type-specific technologies at the power station level, based on RTU equipment or other distributed intelligence IEDs (e.g. numerical protections or control and protection modules of re-chillers) which command

primary equipment mounted at stations, transformer stations or power points, respectively re-chillers or remote-controlled separators, and which performs its functions by running specialized software applications for distribution automation functions;
- At the level of transport or even electricity generation, the synchronous phasor measurement systems and the systems that can be developed from them, which are usable both for tracking new renewable sources and for monitoring the state of the energy system at least regional or even for the implementation of protection and automation at the energy system level;
- At the level of electricity generation, a matter of high relevance is the integration of the distributed sources of renewable energy—wind or even solar, which can have a significant impact on the system [11–14].

  Specialists have deployed systems that use several technologies listed above, such as:

  - Telematics systems;
  - Systems for monitoring the consumption of electricity;
  - SCADA systems;
  - Integrated command-control-protection systems;
  - Measuring Systems (on the transmission network).

12.3.2 SCADA System Structure

SCADA system includes hardware and a software system. SCADA offers real-time control allowing optimization of the technological system. In the Fig. 12.1, typical hardware architecture is presented.

Generally, a SCADA server does not connect directly to PLCs connected to the system. Typically, a Remote Terminal Unit (RTU) is being introduced that collects and centralizes data from and from PLCs. Most of the monitoring and control operations are performed by RTUs or PLCs as we can see in the Fig. 12.2 A RTU device is installed in a remote location and collects data from PLCs. A RTU therefore functions as a data concentrator.

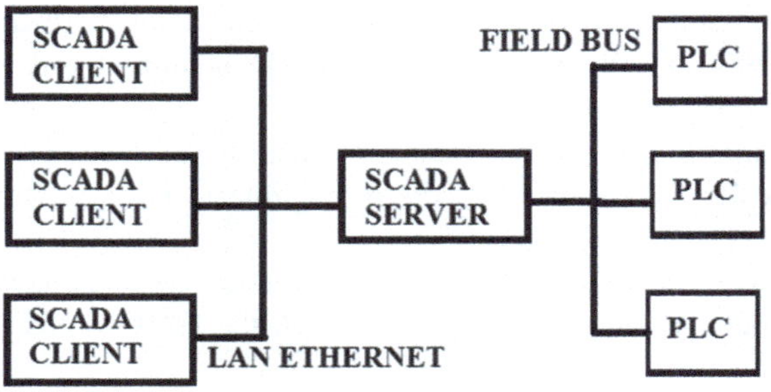

Fig. 12.1 General SCADA network system

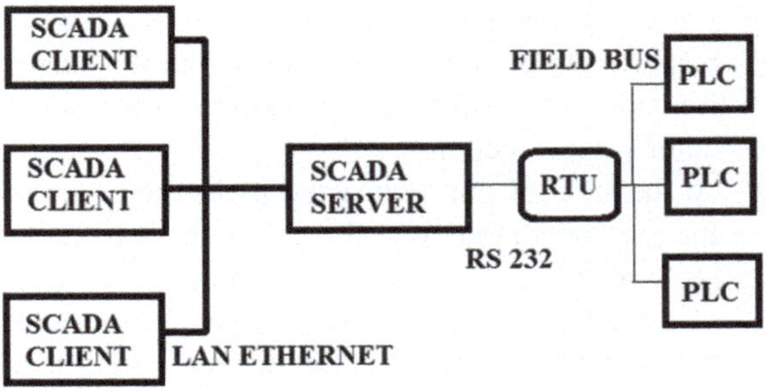

Fig. 12.2 SCADA diagram

The SCADA Server requests data from the RTU, it encodes the data in a format that is transmittable, and then the RTU transmits the data to the SCADA server. RTU also receives commands from the SCADA server, commands it sends to the process.

An RS 485 serial line enables multiple devices to be connected to the same data bus. The SCADA server has only the RS-232 interface, so you need an RS-232/ RS-484 converter. For increasing the system reliability, a lot of servers can be placed, ensuring redundancy for SCADA Servers.

Also, to further increase system reliability, redundancy can be achieved for RTU devices by placing multiple such devices in a master-slave configuration. You can also place multiple field buses ensuring redundancy at this level as well.

In the case of critical technological processes or processes where maintenance costs are high, high redundancy must be ensured to eliminate the incidents caused by equipment failure.

Basically, SCADA software architecture has at least two components: the SCADA server application and the SCADA client application.

The SCADA Server application is typically multi-tasking, being responsible for both data acquisition and storage in a database. In this case, the SCADA server reads data from the RS232 serial port using the MBUS RTU protocol.

Data is stored in multiple tables and the client application uses the database by SCADA server to make graphical user interfaces, so-called HMI—Human Machine Interface (Fig. 12.3).

An HMI mimics a technological process, creates event lists, reports, alarm and warning lists, trending. The hardware part for SCADA systems is robust to withstand temperature, vibration, and voltage extremes, but in cryogenic installations reliability is enhanced by having redundant hardware and communications channels. A failing part can be quickly identified and its functionality automatically taken over by backup hardware. A failed part can often be replaced without interrupting the process. The reliability of such systems can be calculated statistically and is stated as the mean time to failure, which is a variant of mean time between failures.

An HMI also displays processed data to a human operator. At the same time, the human operator can send through the HMI commands to the monitored process.

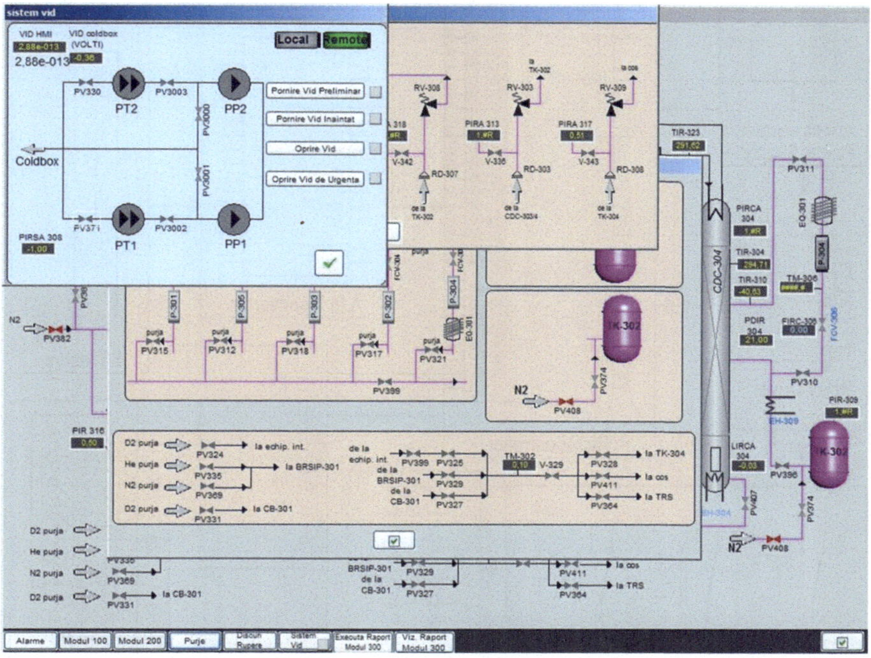

Fig. 12.3 SCADA experimental diagram—HMI

The SCADA client application also offers various facilities in multiple screens that can contain synoptic diagrams and texts to display events, reports, alarm lists, trending [15–17].

SCADA for Remote Industrial Plant

In industrial units, many processes appear simultaneously and each needs to be monitored, which is in fact a complex task. The main purpose of this project is to process data in real time and to control the industrial scale on a large scale on an industrial scale. In real time, a temperature recording system for a remote operation of the plant is made (Fig. 12.4).

For example, in Fig. 12.5 is shown the monitoring scheme of a zone of a medium voltage distribution network which supplies a low voltage network are connected to both users and passive distributed sources.

Intelligent networks not only transfer energy but also a great deal of information. The implementation of smart grids requires measures adopted at the level of each operator in the power system to ensure the required objectives [18–21].

At the user's level:

- efficient energy;
- energy production;
- smart buildings;
- automation of user equipment.

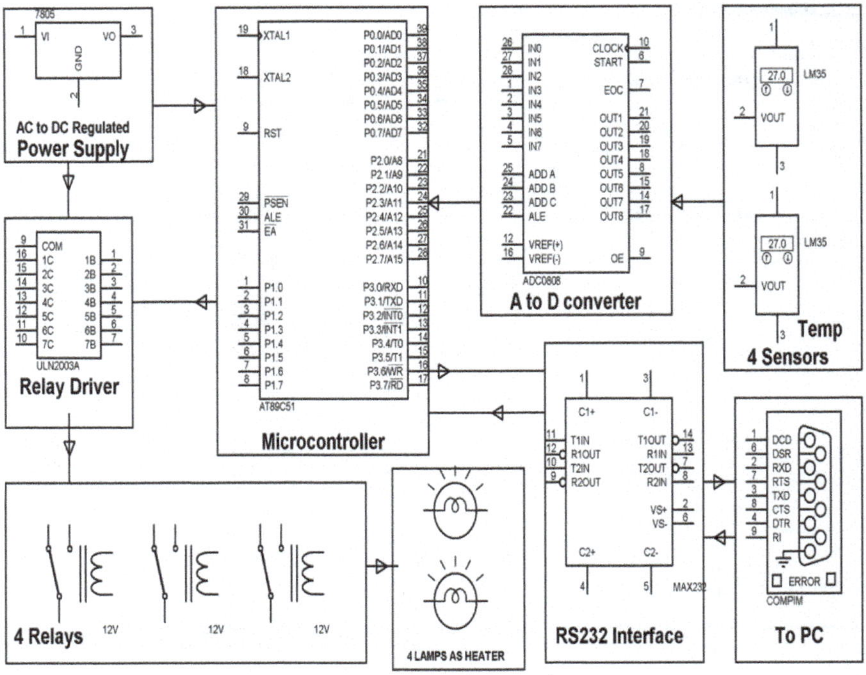

Fig. 12.4 Temperature control for industrial applications

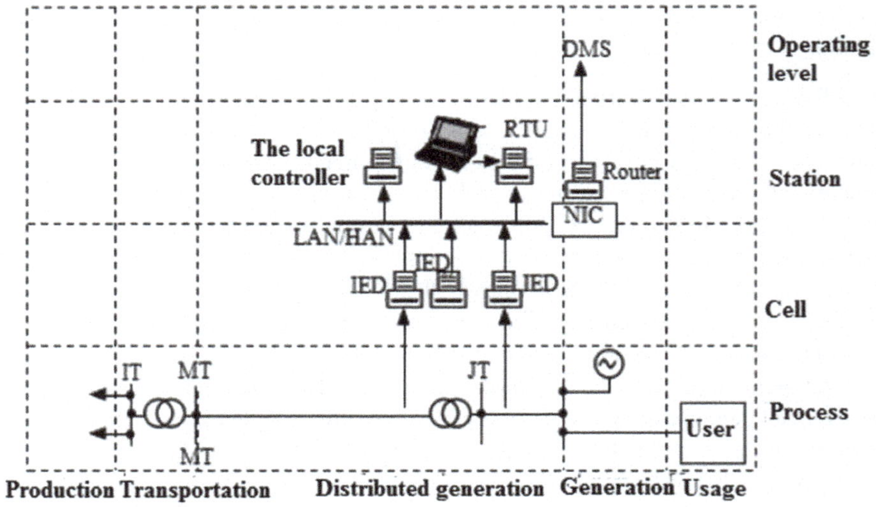

Fig. 12.5 Monitoring system

At generation level:

- adaptive production;
- control of environmental pollution.

At network level:

- automation stations;
- automation of distribution.

12.3.3 Issues in Securing Critical Infrastructure Networks

Developing Smart Grids will allow:

- using new technologies to increase the efficiency, security and reliability of all components of power systems;
- new services, new network users' options, ensuring the proper conditions regarding the quality of the electricity;
- developing the communications system to obtain more accurate, faster information to allow real-time evaluation of events in the system and the adoption of mitigation measures.

The development of smart grids leads to more efficient, flexible, reliable, stable and more interactive energy systems. A basic feature of a firewall is to block unauthorized traffic from entering the protected network. The firewall prevents the establishment of a direct connection from the external Internet to the SCADA local network. The security solution for the Internet is to offer its employees the opportunity access Internet resources, while preventing unauthorized information traffic. The most common way of protecting the internal network is to use a firewall (protective wall) between the Intranet and the Internet (Fig. 12.6). It is possible to transfer data from SCADA systems to the Internet where a simple requirement is that of a fair security policy, including the following security levels:

1. the first level of security is a firewall to ensure a secure Internet connection;
2. the transmission of encrypted data through an Internet security tunnel.

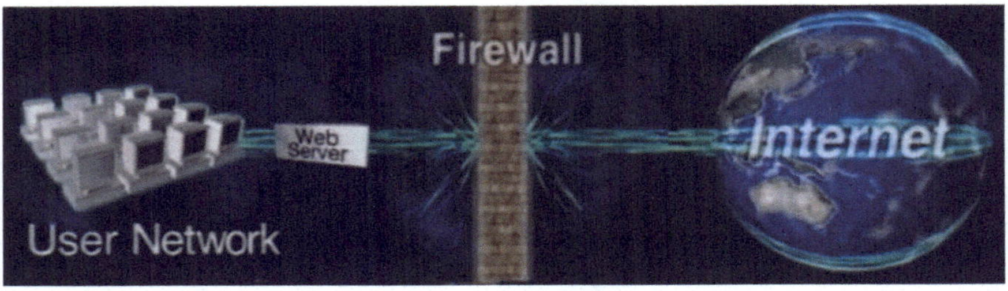

Fig. 12.6 Protecting an intranet network from internet connection through a firewall

Moreover, so-called digital certificates can be used to ensure secure communication with the desired partner.
3. the third level of security is the security at the application level.

Generally, cryptography (information encoding science) uses the so-called keys for coding and decoding information (it is obvious that to decode a message it is necessary to know the key with which it was encoded). Typically, two keys are used to encode a message, one called a public key (publicly known), and the other private key (known only by its user or the person who decodes the message) [22, 23].

12.3.4 SCADA Systems at the Power Station Level

12.3.4.1 Generalities

The main operative functions of the management system are as follows:

- the data acquisition, processing and exchange of data;
- Instant data recording;
- transmission by the command center (to be implemented later);
- remote control (from the control room) and local control of switches, separators, etc.;
- indication of the position of the switching equipment;
- measurement of analogue sizes;
- counting;
- sequential data recording;
- processing and managing alarms;
- adjusting the voltage and controlling the plotter switch;
- remote control and regulation in installations;
- marking;
- archiving of long-term information;
- recording the damage;
- interlocking of primary equipment;
- checking synchronism with the switching sequence of the closing release;
- supervising the state of the information system.

12.3.4.2 Configuration of the Management System

From considerations of reliability and safety the management system will be divided into a number of levels of responsibility, while maintaining complete transparency in terms of exploitation.

The three leadership levels considered are:

- level 1—cell level;
- level 2—station level;
- level 3—network level (management center).

The three levels of management must be operationally independent, so that the lower level can at any time perform the necessary control functions.

Main parts constituting management system are:

- subsystem centrally located in the control room, with general management functions and communication functions;
- local subsystems with 110 and 20 kV cell conduction functions for:

level 110 kV;
cell 1: LEA 1 110 kV;
cell 2: LEA 2 110 kV;
cell 3: Trafo 1-110/20 kV;
cell 4: Trafo—110/20 kV;
level 20 kV;
cell 1: LEA (LES) 20 kV;
11 cells ⇨
cell 11: LEA (LES) 20 kV;
cell 12: 110/20 kV Transformer T1;
cell 13: Transformer T2 110/20 kV;
cells 14–15: section coupling;
cell 16: 1—internal service transformer 20/0.4 kV;
cell 17: Domestic Transformer 2–20/0.4 kV;
cell 18: Measure 1 + 20 kV dischargers;
cell 19: Measure 2 + 20 kV dischargers.

- local subsystem located in the control room (current) for voltage levels 110, 20 kV and limited monitoring functions (position switching devices, measures and some alarms);
- local subsystem located in the 110 kV station's own service building with command functions—supervising its own services;
- operator equipment (monitors, keyboards, printers, mouse).

The following parts of the system will be located in the station control body:

- the central management unit;
- acquisition subsystem at all voltage levels;
- operator equipment;
- telecommunication equipment.

The "cell" control equipment will be located in the relay cabinets. If the equipment at the upper management level is defective, the local control units will be able to take control of the subordinate process.

For parameterization, testing, data readings, evaluations, etc. it is necessary for the system to be provided with a communication software interface at the central level and at the level of each cell. The amount of information required for supervision and operative control at the control rooms of the power stations can be differentiated according to the station's voltage level and how to use it (staffed or not). The volume of information must ensure that the functions assigned to the staff are achieved in normal, incidental, and return to normal station status.

The main information required for supervision and operational management at the control rooms of the power stations are:

- *Parameters*
  - active and reactive powers on lines (for the future situation), transformers (on the primary and secondary);
  - voltages on station bars, lines, service bars, etc. and so on;
  - Indicator of defect locator (for future situation);
  - currents on the 0.4 kV side of their own service transformers on batteries accumulators (220 V);
  - the frequency;
  - active and reactive energy on lines (for the future) and transformers.
- *Indicators*
  Status required for station configuration:
  - position of switching equipment (circuit breakers, separators);
  - transformer switch position;
  - RAR, DRRI and other automation positions;
  - the position of the key to select the commands.

  Alarm, necessary to take preventive measures on the operating mode of the station, such as overcoming limits.
- *Preventive, Plant Status Indication*
  - fault breaker;
  - faulty secondary circuit;
  - defect in its own services/c.a.
  - earthquake in its own services;
  - gas signaling, temperature, overload on transformers;
  - firing of fuses;
  - defect in the data collection or transmission system.

  Incidentally, necessary to take quick remedial action. It is necessary to meet the protection and automation (RAR, AAR and other automation) and configuration changes due to these electric drives.
- *Commands*
  - switching the circuit breakers;
  - switching transformers and coil switches in both directions;
  - switching the automatic circuit breakers;

- removing/putting into operation of local automation (RAR, DRRI, RAT and other automation);
- tripping/shutdown of capacitor batteries;
- triggering of the load;
- canceling self-sustained signals.

The final goal is to supervise and remotely manage, safely and reliably, the specificity of the activity, of geographically dispersed power plants and operating under specific environmental conditions. Providers of equipment for the development of information systems for operative management of the energy installations for the operative management of the power plants must to know and observe a series of specific technical conditions.

12.3.4.3 System Interfaces

- *Interface between Management and Process Equipment*
  - binary input signals (DI)-digital input
    The DI signals are passive type according to IEC 61850. The binary input signal values will be at 220 Vcc. The 48 Vcc voltage can only be accepted if it is generated by the DC/DC converters included in the equipment.
  - (DO)-digital output
    The DO signals are passive type according to IEC 61850. The binary output contacts will have the following parameters:
    - nominal voltage: 220 Vcc;
    - rated current: 2 A;
    - signal duration: = 10 ms;
    - recovery time: = 10 ms;
    - transition time (down/high and high/low): = 8 ms;
    - analog input signals (AI);
    - rated current of the power supply: 1 A, 50 Hz;
    - rated voltage of the voltage source 100 V, 50 Hz;
      Voltages and frequencies will be indicated with increased accuracy for the range of 80–120% (broken feature) [18].
- *The Interface between the Management Device and the Operator Equipment*
  It has the role of providing the necessary means for the operator-process interaction and self-service operation and self-diagnosis of the management system. Information circulates through by means of parallel or serial digital transmissions.
  The information is viewed by the operator using the following equipment:
  - the local control equipment in the cabin of the relay with small diagram for each circuit: shows the primary circuit topology and indicate its status. Can

be used for viewing primary equipment status and launching orders. It consists of elements active and/or passive as multi-position indicators, signaling lamps, switches or switches signaling keys included, buttons/control keys; it is preferable that this equipment must be included in the cabinet of the cell drive unit.

It is recommended that transmission speed, path assignment and transmission parameters be consistent with the appropriate CCITT recommendations. Electrical characteristics (levels signals, input and output impedances, etc.) must be in accordance with:

- Appropriate CCITT recommendations and/or national regulations on data transmission on leased lines and fiber optics;
- IEC 60353 and 60495 for data transmissions on PLC links.

- *The Time Synchronization System*
  A time synchronization system via satellite (GPS) will be provided at the central control point, the signal being distributed to all computing systems and digital equipment (protections, etc.) in the station.
- *Power Supply Interface*
  Two permanent power supplies from two independent sources are included in the control system equipment. Switching from one power supply to another will be done internally for a short enough time so that the system will not be disturbed.
  All system equipment will be powered in alternating current with the following parameters:

  - rated voltage 220 VAC;
  - voltage tolerance −15… +20%;
  - earthing: with insulated poles;
  - degree of curl tension: 10%;
  - power outages: = 50 ms;
  - shock voltage (peak value): 5 kV;
  - auxiliary voltage: 220 VDC +15%… +20%, with insulated poles (for data acquisition).

  The sources are galvanically isolated from earth.

12.3.4.4 Conditions for Interlocking

The design and operation of the interlocking system will lead to reliability and safety in operation. The acquisition and processing of data related to the state of the station equipment must be ensured at all times and uncertain information (intermediate positions, faulty data transfer, etc.) should not allow switching operations.

The command, tuning and synchronization functions must be based on a perfect collection and processing of all station information; the information must be correct and current. Functions such as: action of a load separator, switching without synchronization, intermediate positions of the plotter changer, etc. must be avoided.

Also, defective switching operations must be prevented, both for the protection of personnel and equipment, and to avoid interruption of power supply consumer electricity. In the event of an internal failure of the control equipment, these must block the execution of the switching commands.

The central control unit of the cell will ensure the interlocking between the equipment of a cell and will prevent the simultaneous actuation of primary equipment. In the event of a fall of the central system they will work independently of each other.

The availability of interlocking conditions must be continuously checked by self-testing facilities. The central control unit will provide interlocking conditions for the whole station, especially those related to bar systems.

12.3.4.5 System Operation

The system will satisfy the condition that the level closest to the ordered equipment takes precedence over the upper control levels and the level selected for the command will be indicated at all other levels. Information must be present at the level at which an order is initiated.

The following command options will be provided:

- Local—equipment: equipment selection (circuit-breaker, separator, etc.), manual command without interlocking; command at this level is for evidence, maintenance or repair; the control possibilities at this level are provided by the actuators of the primary equipment;
- Local—cell: selection at cell level; the command is manual with interlocks (or without general interlocks when the computer that manages them is defective), the command at this level is for maintenance/repairs or in situations where the command cannot be given from the central level; the control possibilities at this level will be provided by a synoptic schema panel, control keys, command priority selection and signaling lamps.
- Central—station: selection at central level (station control room) in keyboard; is the normal mode of operation of the station with all the functions, automations and the interlocks that the system supports [46].

12.3.4.6 Process Software

The communication between the RTU in the station and the SCADA will be done by IEC 60870-5-101 protocol and the communication with the central (dispatched) level will be done with a proprietary protocol.

12.3.4.7 Amount of Information

A. Equipment Control in the 110 kV Station

A.1 Requirements Imposed on the Primary Apparatus for its Integration into the Control System at 110 kV Station Level

1. Switches
   The high-voltage switch will be equipped with local power-on and disconnect buttons as well as a control mode selection key. This key will allow the choosing of one of the "local" or "remote" modes, selecting a mode must invalidate the command in the other mode.
   The high-voltage circuit breaker shall be equipped with potential-free auxiliary contacts providing at least the following status information:

   - double signaling on switch position:
     - connected switch;
     - disconnected circuit breaker;
   - double signaling of circuit breaker mode:
     - local command (next to the equipment);
     - remote command (from the control room or from the dispatcher);
   - double sign on the position of the retractable assembly (if a solution of this type is adopted):
     - open circuit breaker;
     - closed circuit breaker;
   - simple indications of the state of the circuit breaker actuator.

   The high-voltage circuit breaker will use the 220 VAC voltage as the operating voltage of the switching and disconnection control and its mechanism will use 220/380 VAC as the supply voltage.
2. Separators
   Separators will be equipped with potential free auxiliary contacts to provide the following status information:

   - double signaling for the position of the separator:
     - closed separator;
     - open separator.
3. CLP (classification, labelling and packaging)
   Earthing knives will be equipped with potential-free auxiliary contacts to provide the following status information:

   - double signaling on the CLP position:
     - closed CLP;
     - open CLP.

4. Measuring transformers
   Current and voltage measuring transformers will have to meet the requirements imposed on measuring equipment according to Romanian and international standards, especially those related to their interfacing with the settlement measuring instruments.
   Transformers measuring current and voltage will have to provide secondary sizes corresponding to the following functions:

   - the measuring function;
   - protection function.
     With regard to the measurement function, in order to be able to interface with the usual types of electricity meters, the current and voltage measuring transformers (together with the related interfaces, if any) will have to provide the following secondary sizes with the accuracy class 0.5 or better on the windings:
   - nominal secondary currents of 5 A c.a.;
   - $100/\sqrt{3}$ V nominal secondary voltages.

5. Transformers 110 kV/MT
   Transformers themselves (together with related measuring devices and accessories) will be equipped with potential free auxiliary contacts that will provide the following status information:

   - Preventive signaling:

     - gases from the Buchholtz gas relay, 1st stage;
     - maximum oil level, from the oil level indicator in the conservatory;
     - minimum oil level, from the oil level indicator in the conservatory;
     - high-temperature step I, from thermometers;
     - temperature stage II, from thermometers;

   - Incident signaling:

     - supported gas protection, from the Buchholtz gas relay, 2nd stage.
       The control device of the load-regulating switch will be equipped with auxiliary contacts that will provide the following status information.

   - Double command:

     - climbs the socket;
     - lower the socket.

     To adjust the voltage on the medium voltage winding, the transformers will be equipped with RATT automation which will be integrated into the control system at the station level.
     Each of the ventilation groups will be equipped with potential auxiliary contacts that will provide the following status information:
   - simple signaling:

- engine start confirmation;
- faulty ventilation motor.

Ventilation groups will be able to receive the following commands:

- double command:
  - start the ventilation group;
  - stop the ventilation group.

6. Technical requirements
   The potential free auxiliary contacts listed above must have the following parameters:
   - operating voltage 250 VDC;
   - continuous current min. 5 A.
     The control elements (coils, electromagnets, etc.) of the actuators mentioned above must have the following parameters:
   - operating voltage 250 VDC;
   - continuous current max. 5 A.

A.2 Command-Control Interface of Process at 110 kV Station

The command-control interface of the process at the 110 kV station requires the purchase of information exchanged with the primary equipment, as well as the acquisition of the information exchanged with the secondary switching and measuring equipment.

The amount of information required for the protection-command-control system is further shown on equipment types.

A.3 Volume of Information about the 110 kV/MT Transformer (110 kV Transformer Cell and Transformer Itself)

The volume of the 110/MT transformer information is detailed in the annexes. Outside the information volume exchanged with 110 kV primary switchgear (switches, separators, CLPs), measuring transformers and power transformer with its accessories (the actual transformer, the load-change switch and the ventilation groups), Next, the amount of information exchanged with the associated secondary switching apparatus is shown.

Corresponding to 110 kV cells the additional information volume is:

- simple signaling:
  - electromagnetic interlock.
    In addition to software interlocking, a numerical input will be provided to signal the condition of electromagnetic interlocking;
- simple commands:
  - voltage measurement selection.
    Determines which of the 110 kV voltage transformers is used for calculating power and energy on the high voltage side of power transformers;

- incident at the cell level, cumulative incidents on the signaling cell at the station level;
- preventive signaling at the cell level, cumulative preventative signaling on the transmitting cell at the station level.

Corresponding to power transformers with their accessories (the actual transformer, the switch on the load, the ventilation groups), the additional information volume is:

- double signaling on how to properly control the transformer:
  - local command (from the control cabinet);
  - remote command (from console in control room or dispatcher);
- simple transformer signatures proper:
  - simple signaling of automatic circuit breakers (representing the sum of the signals on the primary circuit breakers of the primary equipment);
- simple signaling switch under load:
  - on/off, locally. Takes the power supply switch position of the on-load tap changer for local handling;
  - automatic/manual, local. Takes the position of the key to select the automatic voltage adjustment in the power transformer substation;
- simple ventilation fan signaling:
  - automatic/manual, local. Takes the position of the automatic start selection key of each power transformer ventilation group;
- simple transformer commands itself:
  - power transformer protection release;
  - transformer level incident, cumulative preventative signaling for transmission at the station level;
  - preventive signaling at the transformer, cumulative of the preventive signaling for transmission at the station level;
- simple switching controls under load:
  - on/off, remotely;
  - automatic/manual, remote;
- simple ventilation groups orders:
  - automatic/manual, remote.

Transformer control device itself or 110 kV/MT power transformer control—control—control unit (in which all transformer control and control functions are included in a single unit) will have a communication port for retrieving information from the RATT automation device of the load switch.

B. Amount of Information at Level of Internal AC Services of Station

Breakers and automatic internal Power circuit breaker must be equipped with a 220 VDC powered actuator. All other automatic circuit breakers must provide auxiliary contacts. The amount of information on internal AC services is presented in the attached tables.

The following is the amount of information exchanged with the secondary switching device:

- double signaling for internal service supply circuit breakers and coupling switch:
  - automatic circuit breaker SI connected;
  - automatic and disconnected circuit breaker;
- double signaling of how to command internal AC services:
  - local command (next to equipment)
  - remote command (from the control room);
- simple, analog signals:
  - automatic motor circuit breaker triggered drive mechanism;
  - triggered automatic breakers SI loops and so on. This signal is a sum of the signals taken from the auxiliary contacts of the AC circuit breakers;
- measures:
  - the currents on the secondary windings of the current measuring transformers from the supply of internal services;
  - stresses on internal AC services;
- double commands for internal power supply circuit breakers and coupling switch:
  - automatic circuit breaker connection SI;
  - automatic circuit breaker disconnection SI;
- simple signaling:
  - incident at the level of the internal AC services, accumulates the incidents for signaling at the station level;
  - preventive signaling at the level of internal AC services, cumulative preventive signals for transmission at the station level.

    The volume of information can be purchased at the level of the central numeric equipment that acquires the signals in the control room or at the level of individual control-command equipment located on the internal AC cabinet.

C. Volume of Information at Level of Internal DC Services of Station

The automatic DC power supply circuit breakers through the rectifiers and the automatic coupling circuit breaker must be equipped with a remote actuator.

The rectifiers must have a switchable power switch, closing and opening, and provide their status. All other power switches must provide auxiliary contacts.

The amount of information on domestic DC services is detailed in the attached tables. The following is the amount of information exchanged with the secondary switching device:

- double signaling for DC power supply circuit breakers through rectifiers, accumulator battery and coupling switch:
  - connected circuit breaker SI;
  - disconnected SI switch;
- double signaling on how to control internal DC services:
  - local command (next to equipment);
  - remote command (from the control room or from the dispatcher);
- simple signaling for the circuit breakers:
  - connected switch;
- simple signaling:
  - grounding on the bar for each bar;
  - triggered circuit breaker from battery terminals;
  - triggered automatic circuit breakers SI and c.c. This signal is a sum of the signals taken from the auxiliary contacts of the automatic circuit breakers on the DC loops;
  - faulty rectifier for each rectifier;
- measures:
  - the currents from the DC power supply through the battery. For the measurement of these currents, the provider will provide shunts corresponding to the maximum current of the battery of the accumulator;
  - stresses on internal DC services;
  - voltage on the telecommunication battery;
  - the voltages of the rectifier outputs;
- double commands for DC power supplies through rectifiers and accumulator battery, coupling switch and rectifier power supply circuit breakers:
  - breaker connection;
  - disconnect switch;
- simple signaling:
  - incident at the level of internal DC services, accumulates incidents for signaling at the station level;
  - preventive signaling at the level of internal DC services, accumulates the signals for transmission at the station level.

This volume of information can be purchased at the level of the central purchasing numerical equipment in the control room or at the level of individual control-command equipment located on the internal DC cabinet.

D. Volume of General Station Signal Information

The amount of information associated with the secondary switching equipment:

- simple signaling:
  - minimum control room temperature, 1st stage;
  - minimum control room temperature, 2nd stage;
  - maximum control room temperature, 1st stage;
  - maximum control room temperature level, 2nd stage;
  - functioned DRRI, MT-AAR AAR-JT, etc.;
  - frequency indication.
- double signaling on how to control the station:
  - local command (from the control room);
  - remote command (from dispatcher—will be foreseen in the future).
- simple commands:
  - functioned DRRI, ATS, etc.;
  - hupa.

    Numerous inputs and outputs for interfacing with fire detection and fire detection and fire detection and fire alarm systems as well as room air conditioners are provided in this volume of information.

E. Control-Command Control and Equipment

The command-control equipment at the control room level will be built into an architecture set by the supplier according to the type of equipment.

E.1 Central Numerical System

This central system will ensure:

- the acquisition of numerical and analogue sizes related to the control room:
  - local processing of the data acquired for the purpose of performing SCADA functions;
  - communication with hierarchically superior level, zonal dispatcher, two distinct paths with automatic passage from the base path to the reserve;
  - communication with other satellite digital equipment (RTUs, automation, etc.);
  - data storage during the interruption of the power supply of the numerical system and the fall of telecommunications links;
  - running specific programs, in real or off-line (process database management, SCADA, verification and debugging);
  - self-testing and diagnostics, as well as the possibility of system maintenance, from the console or through a specialized port;

- to support protocols with a wide use in the field:
  - IEC 60870-5-101—for communication with the master station at the control point;
  - IEC 61850—for communication with protection and control equipment—subordinate control.

E.2 Automation in Station

The automation at the station level will be done as follows:

- DRRI, AAR-MT, RATT, DAS-MT, AAR-JT, etc. will be made with individual equipment. They must be digital and have communication ports (of the type with which the control equipment is distributed—control) to integrate in the control system—station control by bidirectional data transfer.
  Through these ports the automation equipment must change with the central numerical system all of its own state information and analogue sizes purchased from the process.

E.3 Telecommunications Equipment

The telecommunication links between the control room of the power station and the dispatcher control point must ensure the transfer of information in the form of data and voice.

The SCADA cabinet will include at least the following equipment:

- router;
- switches;
- fiber optic media converter (FO).

Support for handling the types of information is:

- data:
  - optical fiber;
- voice:
  - leased line;
  - optical fiber;
  - radio.

12.4 Smart Grid

12.4.1 General Presentation

The term "Smart Grid" was attributed to M. Amin that define the concept of the intelligent grids (Fig. 12.7) [24].

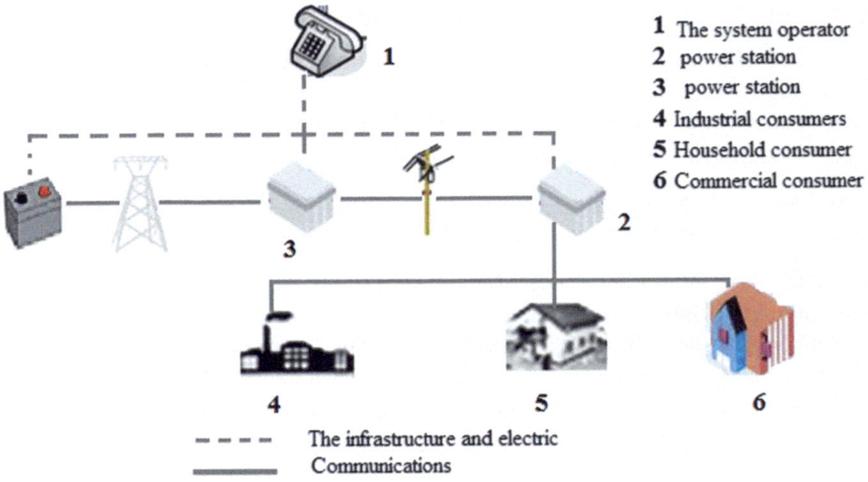

Fig. 12.7 Smart Grid System—the starting point

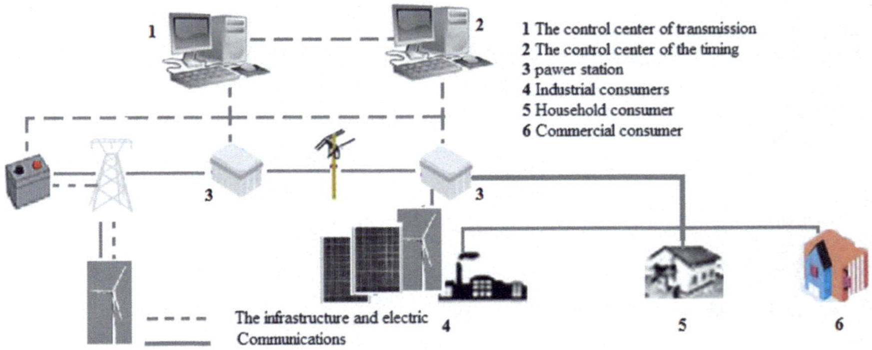

Fig. 12.8 The Smart Grid scheme in the current version

The features of a Smart Grids are:

- autonomy;
- configuration with a high degree of utility;
- security high.

Such infrastructure is characterized by: interdisciplinarity, the component elements are managerial and operational independent, it is distributed over extended surfaces, heterogeneity and high emergence behavior (Fig. 12.8) [24].

Energy efficiency and reliability, optimal management of existing resources and the integration of renewable sources are part of the goals pursued in the development of Smart Grid (Fig. 12.9) [24–26].

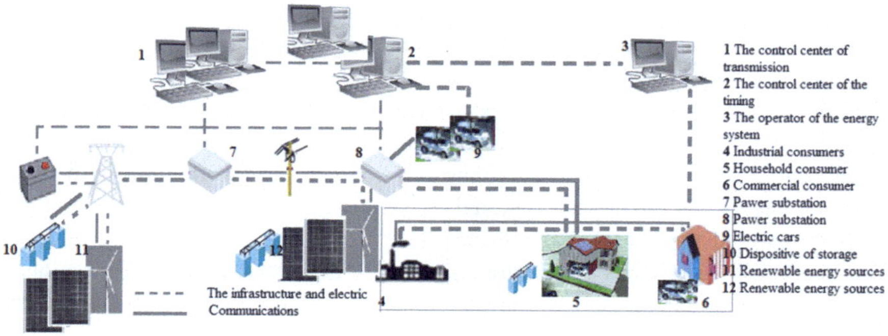

Fig. 12.9 The Smart Grid scheme expected to be achieved

12.4.2 Smart Grid Security Problems

12.4.2.1 Smart Grid Issues

Because it provides real-time information to the consumer the Smart Grid technology is called "the internet of energy". Thus, the Smart Grid does not require the replacement of the existing grid system, while offering the possibility of further upgrading it [27–29].

Regardless of the attempt to define the term SMART GRID, we often meet the three aspects:

- the actual network;
- Intelligent devices either distributed or centralized, representing computer information systems or parts thereof, specific to information technology and controlling the network elements according to various algorithms implemented by numerical programs (firmware or software);
- the communications infrastructure interchanges the two-way information between component parts.

Among the more common technologies that are related to the notion of Smart Grid we can remember:

- at the power supply level, intelligent meters or even complex telegraphy systems can be listed, capable of being programmed to make decisions by hour, depending on consumption or other criteria, and supporting a bilateral communication path;
- at the power distribution level, distribution automation systems that are correlated with SCADA type-specific technologies at the level of power stations based on RTU equipment or other distributed intelligence IEDs (e.g. numerical protections or relay control and protection modules) controlling primary equipment mounted at the station, transformer station or supply points,

respectively reclosers or remote-controlled separators, and performing its functions by running specialized software applications for distribution automation functions;

- at the level of transport or even electricity production, the synchronous phasor measurement systems and the systems that can be developed from them which are usable both for tracking new renewable sources and for monitoring the state of the energy system at least regional or even implementation of protection and automation at the energy system level;
- at the level of electricity generation, a matter of high relevance is the integration of the distributed sources of renewable energy—wind or even solar, which can have a significant impact on the system.

As the energy markets become more and more liberalized and dynamic, the number of stakeholders (stakeholder) increases. All interested parties, starting from governments and corporations and ending with normal users, will help to change the Smart Grid. These microgrids can increase the efficiency of a regional energy system when it faces a high demand for energy, thus avoiding the occurrence of power surges. Microgrid applications can eliminate the need to install additional voltage lines in areas where demand is high [30–33].

In a classic electrical network, electricity has a fixed price for all users. Classical energy systems have no control over tasks, except for emergencies where certain tasks can be "cut" to balance demand and production. Therefore, many network elements are used for a short period, during peak hours, remaining unused for the rest of the day.

The Smart Grid system allows users to prioritize their energy consumption according to daily schedule and needs, taking into account a variable cost of electricity over a day. Integration of smart devices at the consumer level will allow for automatic control of electrical appliances, identifying the right moment for their operation in order to optimize costs. Manufacturers will be able to make far more accurate estimates of consumption, balancing more efficiently between the use of thermal power plants and hydropower plants.

The reconfiguration capability is another important feature of the Smart Grid system. This implies that power streams are automatically adjusted and redirected if a line becomes inoperative. Automatic reconfiguration is achieved through continuous monitoring of system status. With this capability, it will be possible to reduce the frequency and number of power cuts, thus minimizing the economic losses caused by these events [34–37].

12.4.2.2 Contribution of the Smart Grid Security

In the future, the smart grid will reach any household or industrial consumer. Because it incorporates IT subsystems, this system is exposed to many security threats. Given its large size, it is almost impossible to guarantee a high level of security for each subsystem. The large number of Smart Grid components, as well as their diversity and complexity, introduce additional vulnerabilities to those already existing in a traditional power grid [38–42].

A. Adjustment Systems

Because the SCADA network provides communications in these large-scale systems, cyber-attacks may occur. In the case of system malfunctions, the automation equipment operates in perturbations. You cannot use high confidentiality policies by working in real-time control systems because they cause delays. Intrusion Detection Systems (IDS) provide security for control systems.

With data traffic, the control system has a static configuration. For this reason, the model is used. Besides this approach, a digital signature is also used. This signature allows detection of predictable attacks.

Through digital signature, potential security issues can be identified. A model of the detection system is shown in Fig. 12.10. The disadvantage of a model based on the IDS system is that for each control system a model must be built, aspect directly

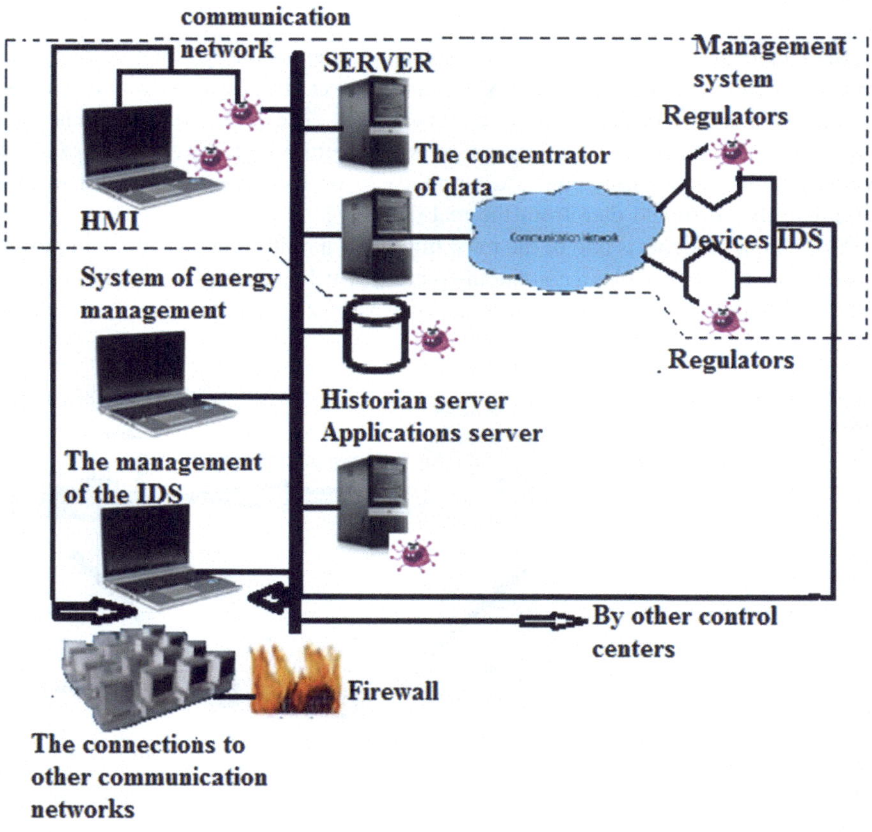

Fig. 12.10 Intrusion detection

proportional to the complexity of the controller. In addition, a static configuration for the regulator must be ensured so that the model will be consistent with actual behavior.

B. Smart Meter's Security

Smart Meter delivers intelligent power measurement. They are installed at the consumer's location and represent a new generation of electric meters. Against the old ones, they can be data concentrators and can communicate. Smart Meters provide the Smart Grid with a feedback mechanism through which to build a realistic estimate of future consumption.

Energy invoice values and erroneous consumption estimates are obtained by affecting the security of Smart Meters. They can easily get physical access to them by attackers for data modification. This way you can get money benefits. Security of Smart Meters can be ensured by implementing an intrusion detection or redundancy system [7, 43, 44] (Fig. 12.11). Integrity of data transmitted by Smart Meter is ensured by the identity of the two measurements [45, 46].

C. Security of State Estimation

Management uses state estimates to maintain system stability. By modifying the data transmitted to the status estimator, an attacker may destabilize the energy system or intervene in the real-time pricing system. The most commonly encountered computer attack on state estimation is the corrupt data injection attack. The security of the state estimation is a complicated problem as it is difficult to differentiate the corrupted data from the real data [47].

Estimation of status is one of the most important monitoring algorithms in power systems because it provides a convincing picture of voltage and phase angles. One of the most effective ways to guarantee a safe state estimate is the strategic placement of

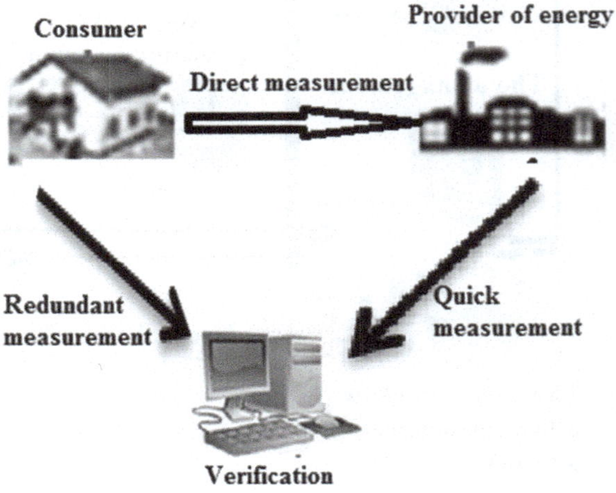

Fig. 12.11 Secure data provided by Smart Meter by redundancy

phasor units and combining the measurements they provide with traditional measurements [48].

D. Security of Communications Network

To effectively control the power system, the intelligent grid is based on the subsystem's ability to communicate. The requirements for a communications system vary from transmission speed and bandwidth to latencies introduced in control commands by regulators.

A major challenge is the safety of the communications networks involved in the smart grid, as it is necessary to integrate a large number of protocols for the requirements of each subsystem. Introducing old systems into configuring new smart grids is causing problems due to low security [49].

12.4.3 Attacks Types in Smart Grid

The main types of computer grid attacks in a Smart Grid system are illustrated in Fig. 12.12 [39, 40].

- *Protocol Attacks*
  IT attacks can affect communication protocols used in the Smart Grid if they are not securely secured [50].
  Examples of protocols used in Smart Grid:

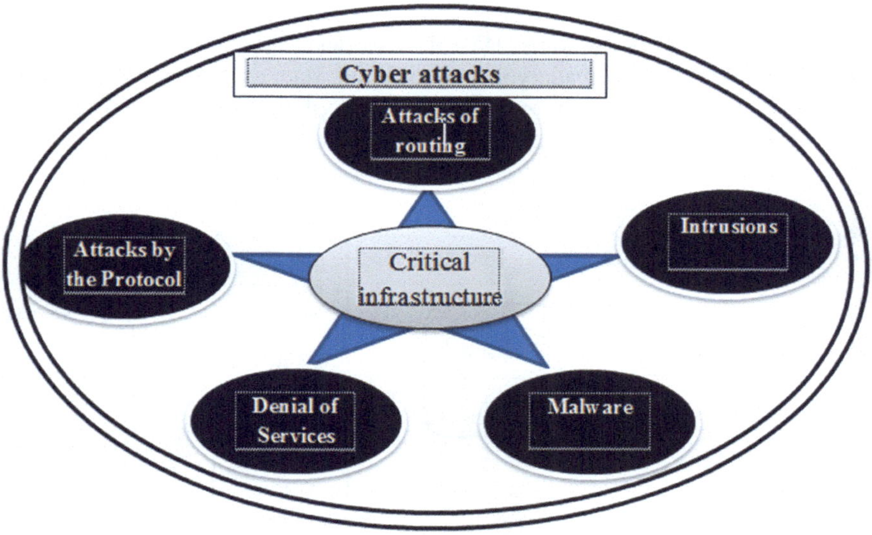

Fig. 12.12 The main types of computer grid attacks in Smart Grid

- ICCP;
- IEC 61850;
- DNP3.

- *Attack Routing*
  The routing attacks are the computer attacks that affect the communication network infrastructure. Such attacks can affect the software applications involved.
- *Intrusions*
  Intrusions are about exploiting vulnerabilities in software and communications infrastructure. They can start either inside or outside the system when the operator abuses its system administration privileges.
  Example:
  Handling data used by human machine interfaces (HMI) by shorting security systems (firewalls, authentication mechanisms, etc.).
- *Malware Attacks*
  Malware Attacks—refers to software applications that may affect software infrastructure, communications, or programmable machines. Malware applications scan the system in search of potential victims, exploit their vulnerabilities, and then propagate to the other computers in the system. One of the most known computer attacks on an energy system is the Stuxnet case.
- *Denial of Service attacks (DoS)*
  Denial of Service attacks (DoS) deprives the user of using a service provided by the system in question. A DoS attack can also involve denial of control. Uploading the communications network with a great deal of unnecessary data leads to these attacks (Fig. 12.13).

On the IT resources (SCADA, HMI devices field, communications protocols) and control systems (Automatic voltage protection, state estimation, detection and isolation of faults, reactive power compensation) of the Smart Grid is realized attacks cybernetics such as [49, 50]:

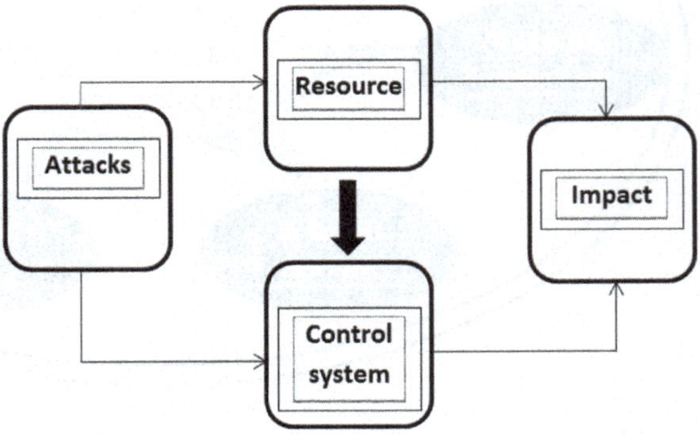

Fig. 12.13 Computer attacks and their impact

- attacks on protocol;
- intrusion;
- malware;
- denial of service.

Regarding the security of a critical system, risk is defined as the product of their threats, vulnerabilities and impact:

$$[\text{Risk}] = [\text{Threat}] \times [\text{Vulnerability}] \times [\text{Impact}] \quad (12.1)$$

The threat is defined by the presence of a potential attack, its motivation and the resources available. In order to succeed in attacking a critical system, the attacker will seek to bypass security software as well as redundancy at the physical level. The impact of these attacks is determined by how far these attacks affect the stability of the system.

12.5 Smart Grid as Critical Infrastructure—Conclusions and Trends in the Evolution

Most of the principles of Critical Infrastructure Protection and their applicative aspects are general and should be considered throughout the lifespan of installations, systems, processes (from the design phase to the decommissioning and decommissioning thereof).

Modern power system operations are heavily dependent on information technology (IT) technology, many of which operate in real time. The introduction of competition and separation has brought many new organizations into the energy sector. Much of the interaction between the participants in the electricity sector is achieved through computer systems. There is a wide variety of mechanisms through which cyber threats—viruses, worms, etc. can spread and affect the integrity of computer systems.

Security is an evolving process and is not static. Continuous work and education are needed to help security processes to keep up with the requirements that will be placed in electrical systems. Security will continue to be a race between the company's security policies and hostile entities. Security processes and systems will continue to evolve in the future. By definition, there are no communication systems that are 100% safe. There will always be residual risks to be considered and managed. Thus, in order to maintain security, vigilance and constant monitoring, as well as adaptation to changes in the global environment, are required.

The main approaches to the security of protection, automation and control systems against cyber-attacks are:

- defense in depth;
- separation of the network;
- electronic perimeter;
- best security practices.

Security practices, such as computer running and network management policies, must be defined in accordance with the guidelines for specific standards and procedures such as the choice of passwords and their expiry date; using a limited number of privileged computer accounts and turning off the rest; closing unwanted communication ports and computers; the implementation of access control mechanisms; frequent updating of anti-virus signature databases.

Critical infrastructures remain an area that is very well investigated, monitored, analyzed, evaluated, predicted and improved. Modern energy technologies have already existed and are perfectly functional—in a very near future, so, at least partially, we will have an SMART GRID operational that can be used for the purpose for which the smart grid will be created: making the energy system safer, more economical and more reliable.

An important role is played by relevant solution providers and service providers, as well as professional and academic environments, which must provide the needs of specialists, inventions, innovations, development and change.

References

1. A. Badea, I. Chiuta, A. Valciu, G. Paun, Infrastructure management critical power electronics systems. Bull. AGIR (2) (2012)
2. G. Alexandrescu, G. Vaduva, Mission-Critical Infrastructures, J. Dangers, Threats to their address, protective systems, Publishing House of the University of National Defense, Carol I, Bucharest, 2006
3. T. Flick, J. Morehouse, in *Securing the Smart Grid—Next Generation Power Grid Security*, Syngress, 2011
4. J.C. Foreman, D. Gurugubelli, Identifying the cyber attack surface of the advanced metering infrastructure. Electr. J. **28**(1), 94–103 (2015)
5. M. Fabro, Conducting cyber threats assessments at nuclear facilities, IAEA Nuclear Energy Series, 2016, https://www-pub.iaea.org/books/iaeabooks/10999/Conducting-Computer-Security-Assessments-at-Nuclear-Facilities
6. A.V. Gheorghe, M. Masera, M. Wiejnen, L. De Vries, in Critical infrastructures at risk, Springer, 2006
7. M. Ficco, M. Chora, R. Kozik, Simulation platform for cyber-security and vulnerability analysis of critical infrastructures. J. Comput. Sci. (2017)
8. C. Alcaraz, S. Zeadally, Critical infrastructure protection: requirements and challenges for the 21st century. Int. J. Crit. Infrastruct. Prot. **8**, 53–66 (2015)
9. F. Birleanu, N. Bizon, Reconfigurable computing in hardware security—a brief review and application. J. Electr. Eng. Electron. Control Comput. Sci. (JEEECCS) **2**(1), 1–12 (2016)
10. N. Bizon, N. Mahdavi Tabatabaei, F. Blaabjerg, E Kurt (Ed.), *Energy Harvesting and Energy Efficiency: Technology, Methods and Applications*, Springer, 2017
11. J. Vijayan, Stuxnet renews power grid security concerns. Computer World, International Energy Agency, Technology Roadmap in Smart Grids, OECD/IEA, Paris, 2011
12. N. Kayastha, Smart grid sensor data collection, communication, and networking: a tutorial. J. Wireless Commun. Mob. Comput. Arch. **14**(11), 1055–1087 (2014)
13. E.D. Knapp, in *Industrial Network Security—Securing Critical Infrastructure Networks for Smart Grid, SCADA, and Other Industrial Control Systems*, Syngress, 2011
14. E. Zio, Challenges in the vulnerability and risk analysis of critical infrastructures. Reliab. Eng. Syst. Saf. **152**, 137–150 (2016)

15. A. Nicholson, S. Webber, S. Dyer, T. Patel, H. Janicke, SCADA Security in the Light of Cyber-Warfare. J. Comput. Sec. **31**, 418–436 (2012)
16. B. Genge, C. Siaterlis, Physical process resilience-aware network design for SCADA systems. Comput. Electr. Eng. **40**, 142–157 (2014)
17. W. Li, L. Xie, Z. Deng, Z. Wang, False sequential logic attack on SCADA system and its physical impact analysis. J. Comput. Sec. **58**, 149–159 (2016)
18. N. Bizon, N. Mahdavi Tabatabaei, Hossein Shayeghi (Ed.), in *Analysis, Control and Optimal Operations in Hybrid Power Systems—Advanced Techniques and Applications for Linear and Nonlinear Systems*, Springer Verlag London Limited, London, UK, 2013
19. N. Bizon and N. Mahdavi Tabatabaei (Ed.), in *Advances in Energy Research: Energy and Power Engineering*, Nova Science Publishers Inc., USA, 2013
20. S. Kunsman, M. Braendle, B. De Wijs, F. Hohlbaum, Replacing Fear With Knowledge—Cyber Security for Substation Automation, Protection and Control Systems, Texas A&M University, in 68th Annual Conference for Protective Relay Engineers, 2015
21. N. Bizon, L. Dascalescu, N. Mahdavi Tabatabaei (Ed.), in *Autonomous Vehicles: Intelligent Transport Systems and Smart Technologies*, Nova Science Publishers Inc., USA, 2014
22. A. Valdes, S. Cheung, Intrusion Monitoring in Process Control Systems, in 47th Hawaii International Conference on System Sciences, 2009
23. A. Ashok, A. Hahn, M. Govindarasu, Cyber-physical security of wide-area monitoring, protection and control in a smart grid environment. J. Adv. Res. Cairo University (2014)
24. C. Tu, X. He, Z. Shuai, F. Jiang, Big data issues in smart grid—a review. Renew. Sustain. Energy Rev. **79**, 1099–1107 (2017)
25. H.R. Nemati, L. Yang, in *Applied Cryptography for Cyber Security and Defense: Information Encryption and Cyphering*, Information Science Reference, 2011
26. Z. Lukszo, G. Deconinck, M.P.C. Weijnen, in *Securing Electricity Supply in the Cyber Age—Exploring the Risks of Information and Communication Technology in Tomorrow's Electricity Infrastructure*, Springer, 2010
27. C.W. Probst, J. Hunker, D. Gollman, M. Bishop, in *Insider Threats in Cyber Security*, Springer, 2010
28. M. Chowdhury, A. Apon, K. Dey, *Data Analytics for Intelligent Transportation Systems* (Elsevier, UK, 2017)
29. J. Graham, R. Howard, R. Olson, in *Cyber Security Essentials*, CRC Press, 2011
30. K. Pipyros, C. Thraskias, L. Mitrou, D. Gritzalis, T. Apostolopoulus, A new strategy for improving cyber-attacks evaluation in the context of Tallinn manual. Comput. Sec. (2017)
31. N. Nezamoddini, S. Mousavian, M. Erol-Kantarci, A risk optimization model for enhanced power grid resilience against physical attacks. Electr. Power Syst. Res. **143**, 329–338 (2017)
32. L. Hughes, M. de Jong, X.Q. Wang, A generic method for analyzing the risks to energy systems. J. Appl. Energy **180**, 895–908 (2016)
33. B. Karabacak, S.O. Yildirim, N. Baykal, Regulatory approaches for cyber security of critical infrastructures: the case of Turkey. J. Comput. Law Sec. Rev. **32**, 526–539 (2016)
34. NIST 2 The Smart Grid Interoperability Panel—Cyber Security Working Group, Guidelines for Smart Grid Cyber Security, NISTIR 7628, pp. 1–597, 2010
35. W. Wang, Z. Lu, Cyber security in the smart grid: survey and challenges. J. Comput. Netw. **57**, 1344–1371 (2013)
36. M. Donohoe, B. Jennings, S. Balasubramaniam, Context-awareness and the smart grid: requirements and challenges. J. Comput. Netw. **79**, 263–282 (2015)
37. N. Nezamoddini, S. Mousavian, M. Erol, Kantarci, a risk optimization model for enhanced power grid resilience against physical attacks. J. Electric Power Syst. Res. **143**, 329–338 (2017)
38. H. Suleiman, I. Alqassem, A. Diabat, E. Arnautovic, D. Svetinovic, Integrated smart grid systems security threat model. J. Inf. Syst. **53**, 147–160 (2015)
39. C. Pursiainen, Critical infrastructure resilience: a Nordic model in the making? Int. J. Disaster Risk Reduction (2017)

40. L. Langer, F. Skopik, P. Smith, M. Kammerstetter, From old to new: assessing cybersecurity risks for an evolving smart grid. J. Comput. Sec. **62** (2016)
41. N. Mahdavi Tabatabaei, N. Bizon, A. Jafari Aghbolaghi, Frede Blaabjerg (Ed.), in *Fundamentals and Contemporary Issues of Reactive Power Control in AC Power Systems*, Springer Verlag London Limited, 2017
42. K.M. Muttaqi, J. Aghaei, V. Ganapathy, A. Esmaeel Nezhad, Technical challenges for electric power industries with implementation of distribution system automation in smart grids. J. Renew. Sustain. Energy, 129–142 (2015)
43. T. Liu, Y. Sun, Y. Liu, Y. Gui, Y. Zhao, D. Wang, C. Shen, Abnormal traffic-indexed state estimation: a cyber-physical fusion approach for smart grid attack detection. J. Future Gener. Comput. Syst. **49**, 94–103 (2015)
44. N. Moreira, E. Molina, J. Lazaro, E. Jacob, A. Astarloa, Cyber-security in substation automation systems. Renew. Sustain. Energy Rev. **54**, 1552–1562 (2016)
45. Y. Xiang, L. Wang, N. Liu, Coordinated attacks on electric power systems in a cyber-physical environment. J. Electric Power Syst. Res. **149**, 156–168 (2017)
46. A. Jacobsson, M. Boldt, B. Carlsson, A risk analysis of a smart home automation system. J. Future Gener. Comput. Syst. **56**, 719–733 (2016)
47. S. Massoud, Amin, smart grid: overview, issues and opportunities, advances and challenges in sensing, modeling, simulation, optimization and control. Eur. J. Control **5**(6), 547–567 (2011)
48. D.P. Varodayan, G.X. Gao, Redundant metering for integrity with information-theoretic confidentiality, in IEEE International Conference on Smart Grid Communications, 2010, pp. 345–349
49. M. Emmanuel, R. Rayudu, Communication technologies for smart grid applications: a survey. J. Netw. Comput. Appl. **74**, 133–148 (2016)
50. N. Nafi, K. Ahmed, M. Gregory, M. Datta, A survey of smart grid architectures, applications, benefits and standardization. J. Netw. Comput. Appl. **76**, 23–36 (2016)

Chapter 13
Continuity of Electricity Supply and Specific Indicators

Doru Ursu and Mariana Iorgulescu

Abstract The power quality in supplying the consumers is very important taking into consideration the plants' diversity. In fact, the quality of electricity includes two components:

- the quality of the voltage curve—symmetrical and sinusoidal of this;
- quality of service—uninterrupted or interrupted short/long term.

The Performance Standard imposes the quality of the distributed power in distribution service and establishes performance indicators in the provision of the distribution service, "the quality of distribution service is measured with respect to the supply continuity to the end users". The Performance Standard sets out the performance indicators for:

- continuity of customers electricity supply;
- technical quality of distributed electricity;
- commercial quality of the power distribution service.

The chapter aims to present an analysis of one of the components of the electricity quality, the continuity in the electricity supply of the consumers, indicating possibilities for improvement of the electricity. The indicated improvement is related to the installation of remote-controlled equipment for the rapid isolation of defects in the medium voltage network, correlated with the identification of the mounting location, which brings maximum benefits in terms of reducing the continuity indicators, especially:

- System Average Interruption Frequency Index (SAIFI) represents the number of customer interruptions divided by the total customers served for one year.

D. Ursu (✉)
Energy Distribution Oltenia, Craiova, Romania
e-mail: doru.ursu@distributieoltenia.ro

M. Iorgulescu
Faculty of Electronics, Communications and Computers,
University of Pitesti, Pitesti, Romania
e-mail: marianaiorgulescu@yahoo.com

© Springer Nature Switzerland AG 2019

N. Mahdavi Tabatabaei et al. (eds.), *Power Systems Resilience*, Power Systems,
https://doi.org/10.1007/978-3-319-94442-5_13

– System Average Interruption Duration Index (SAIDI) represented by the sum of customer-sustained outage minutes per year divided by the total customers served for one year.

Beside the definitions of the continuity indicators, it also presents the formulas based on which they are calculated as required by the regulations in force, ways of continuously decreasing them, how to predict the targets of these indicators, both for planned interruptions and for unplanned interruptions. A case study is presented as a "self-healing" automation to isolate faults on a medium voltage line through reclosers and a General Packet Radio Service (GPRS) as a packet oriented mobile data service on the 2G/3G/4G cellular communication system's global system for mobile communications (GSM), communication using a protocol specific to data transfer.

The contents of the chapter will be structured as follows:

1. General notions about continuity in power supply, one of the components of the electric power quality
2. SAIFI and SAIDI continuity indicators, definitions and calculation mode indicated by the regulation
3. How to determine the mounting location for remote control equipment in order to obtain maximum efficiency in decreasing the continuity indicators
4. Case study on the automation of the medium voltage distribution network.

Keywords Automatic isolation of defects · Continuity in power supply
Performance standard · SAIDI · SAIFI · Way to determine target indicators

13.1 General Notions About the Quality of Electricity

The quality of electric energy, defined as a general concept, is "the manner in which electric receivers are supplied in a way that allows them to function properly." In fact, the term "power quality" is used in a much wider sense, addressing both the problem of harmonic pollution generated by nonlinear loads and other types of electromagnetic disturbances in power systems, so it is possible to define the primary quality indicators [1–3]:

– *variations in the supply voltage frequency (occurring when changing the dynamic balance of the power generated with the power consumed)*
– *slow variations in the voltage supply amplitude (due to line voltage drops, transformers, slow load variations)*
– *voltage drops and short interruptions (sudden drop in voltage amplitude (between 90 and 5% of the contract voltage) for a duration of between 10 ms and 60 s, respectively the voltage drop for a duration between 1 s and 3 min, due to network failures and determines the functioning of the protections as well as the resonant automation systems when defects are passing*

- *long voltage interruptions with a duration >3 min (planned and unplanned = persistent network failures).*
- *transient overvoltages (due to commutations in networks and lightnings) and with a duration of maximum 1 s,*

but also secondary quality indicators:

- *harmonics and interharmonics (non-sinusoidal regimes due to producers, but especially to consumers)*
- *flicker—rapid fluctuations of cyclic or random voltage amplitude (produces by WF (wind farms) and PhVPP (photovoltaic power plants), arc furnaces, electric welding = sudden variations in load)*
- *non-symmetries—temporary and permanent (network or consumer defects, unbalanced loads).*

Therefore, the quality of the electricity depends not only on the distributor, but on the suppliers of other services, but also on all the consumers connected to the same distribution network, some of which can cause disturbing influences in the distribution network's electrical network, the operation of other consumers connected to the same network. In such situations, consumers who contribute to altering the quality of the electricity beyond the permissible values must take measures to accommodate the disturbances produced within the limits provided by the country's performance standard.

The traceability of electricity quality indicators and the adoption of measures to keep them within acceptable limits, as an obligation of the electricity distribution operator, can only be correlated with the observation of disturbances introduced into the electricity supply network of certain consumers and sometimes even by PhVPP (photovoltaic plants) that convert continuous voltage into alternating voltage through inverters for which the distribution operator imposes limits even from the commissioning of the power plants [1, 2] (Fig. 13.1).

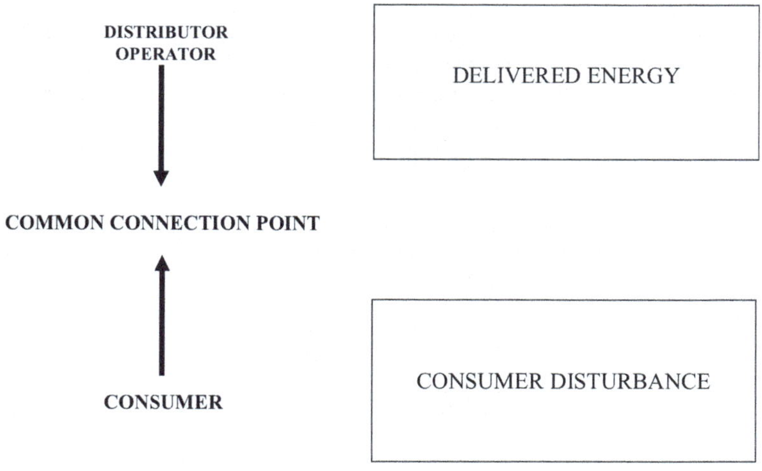

Fig. 13.1 Relief of the disturbance point in electrical networks

The quality of the electricity supplied in distribution networks and provided to consumers is one of the important factors that determine the economic efficiency of both power grids and consumers.

The components of the electricity quality, taking into account the primary and secondary indicators, as well as the ability of the electricity distribution service to meet the consumer's needs, as stated in the Performance Standards, are the following:

- The quality of the voltage curve—the amplitude and frequency;
- Continuity in power supply; Commercial quality—the commercial relationship between the electricity distributor and the consumer through the insured services.

Out of the three components, in the next chapter are presented aspects related to the second component, the safety of the power supply, the continuity in the electricity supply of the consumers. It also defines the continuity indicators, indicating the predictive mode and some of the possibilities for improvement.

13.2 Continuity in Power Supply, Mode of Calculation of the Continuity Indicators Indicated by the Regulator

For the calculation of continuity indicators, long-term interruptions as well as short-circuit breaks of the power/evacuation path of the power and/or production sites connected to the electrical networks (regardless of their voltage) must be recorded during a calendar year.

For each *long-term* supply and evacuation of electric energy, it is necessary to record the following data, according to [1]:

- the voltage at which the interruption occurs (the origin of the interruption), for calculating the indicators for each voltage level;
- the scheduled or unplanned nature of the *interruption*, for calculating the continuity indicators by categories of interruptions;
- the date, hour and minute of the interruption;
- the number of reconnection stages, if applicable;
- the number of users supplied at each reconnection stage, as well as the date, hour and minute of the end of the interruption for them;
- the date, hour and minute of the end of the interruption for all users affected by the interruption;
- total time (from the moment of voltage dropping to reconnection) in minutes of the interruption or refueling stage;
- the number of users, for each voltage level, affected by the interruption, respectively the stage;

- the number of phases affected by the interruption if it occurs in the low voltage network;
- the interrupted electric power (last power measured before *interruption*) at HV.

Thus, based on the stipulated records, on a yearly basis, calculation for a distribution operator can be made concerning the data related to continuity of supply/evaluation of energy for the users in their area of activity.

13.2.1 Mode of Calculation of the Continuity Indicators Indicated According to the Romanian Regulation

a. the number of long interruptions (duration > 3 min);
b. *System Average Interruption Frequency Index* (*SAIFI*)—This index for a user is the average number of interruptions supported by users connected to the grid. It is calculated by dividing the total number of users who have experienced an interruption of more than 3 min to the total number of users served, according to [1]:

$$SAIFI = \frac{\sum_{i=1}^{n} N_i}{N_t} \quad (13.1)$$

c. *System Average Interruption Duration Index* (*SAIDI*)—This index for a user is the average user interruption time at the operator level (a weighted average). The second formula applies if users are reconnected gradually, in several steps, not simultaneously for all users. The indicator is calculated by dividing the cumulative duration of long interruptions in the total number of users served by the electricity distributor, according to [1]:

$$SAIDI = \frac{\sum_{i=1}^{n} N_i D_i}{N_t} \quad \text{or} \quad SAIDI = \frac{\sum_{i=1}^{n} \sum_{j=1}^{k_i} N_{ij} D_{ij}}{N_t} \quad [\text{min/year}] \quad (13.2)$$

d. *Energy Not Supplied* (*ENS*)—Undelivered energy, defined as total energy not supplied to the places of consumption connected to the network of the electric energy distributor due to *interruptions*, according to [1]:

$$ENS = \sum_{i=1}^{n} P_i D_i \quad [\text{kWh, MWh or GWh}] \quad (13.3)$$

e. *Average Interruption Time* (*AIT*)—This represents the average equivalent time period in which the electricity supply was interrupted at the power distributor level, according to [1]:

$$AIT = 8760 \times 60 \times \frac{ENS}{AD} \quad [\text{min/year}] \tag{13.4}$$

where, in the formulas above, the notations represent, according to [1]:

n the total number of long interruptions;
k_i the number of reconnection stages corresponding to the *interruptioni*;
N_i the number of users who suffered an interruption of more than 3 min after the *interruption i*;
N_{ij} the number of users who suffered an interruption of more than 3 min in phase j of the *interruption i*;
P_i electrical power interrupted at *interruption i*, only at IT;
D_i the user interruption time (time) (from the moment of the outage to the reconnection) to *interruption*;
D_{ij} the user interruption time (time) (from the moment of the outage to the reconnection) for stage j of the *interruption i*;
N_t total number of users served;
AD Annual Demand—annual electricity consumption (without grid losses) at the distributor's electricity.

For short interruptions, the following data shall be recorded and calculated annually to provide information on network reliability and the performance of automation equipment:

a. Number of short-term interruptions (time < 3 min);
b. *Momentary Average Interruption Frequency* (*MAIFI*), as a ratio between the total number of users interrupted for short-term and the total number of N_t users served in the analysed system, according to [1]:

$$MAIFI = \frac{\sum_{m=1}^{M} N_m}{N_t} \tag{13.5}$$

where:

M the total number of short-term interruptions;
N_m the number of users who suffered a short-term interruption (less than 3 min) at each interruption *m;*

As specified in the performance standards, the SAIFI, SAIDI, and MAIFI indicators are typically determined on the basis of automatic recordings of MV and HV interruptions, and LV is estimated from the calculations [1]. The ENS and AIT indicators are calculated only for users connected to the HV grid [1].

13.2.2 *Forecast of SAIFI and SAIDI Indicators Targets for Planned and Unplanned Interruptions*

All of the indicators defined according to the performance standards, the SAIFI and SAIDI indicators are those monitored by the electricity distribution operators with a view to continuously reduce them, thus aiming at performances in terms of isolation and elimination of deficiencies in the electrical networks, in particular in the medium voltage ones that have the greatest impact in the continuity of the electricity supply to consumers, due to the fact that over 95% of consumers are supplied from the medium and low voltage. It is also possible to draw conclusions on the performance of the network and to direct investments in medium and low voltage networks taking into account these indicators.

Within Distributie Energie Oltenia—CEZ Group in Romania, a way of predicting the continuity indicators SAIFI, SAIDI for unplanned interruptions has been established based on the indicators achieved in the previous years, which is based on an identified exponential function, passing through the points of achievement of previous years.

Also, a forecasting mode has been established for SAIFI and SAIDI indicators for planned interruptions on the basis of the maintenance and investment plans of the forecasting year.

For the calculation of the target values of SAIFI and SAIDI for unplanned interruptions a non-linear approximation method was used, the approximation of numerical functions being useful when the function does not have a known analytical expression, as it is given in tabular form through points as in our case (knowing the achievements in previous years of SAIFI and SAIDI indicators).

In this sense, there are several methods of solving, but the one recommended to achieve results close to reality is "Approximation by the least squares method"—in this case the function of approximation is determined by imposing the condition that the sum of the squares of the distances between the original function and the approximation in some points is minimal [4].

13.2.2.1 Least Squares Method

The functional dependence of a random variable y (dependence-effect) on another variable x (independence-cause) can be empirically studied experimentally by making a series of measurements on the variable y for different values of the variable x. The problem that arises in this case is to find the analytical representation of the desired functional dependence (fitting process), is to choose an expression (mathematical formula or model) that describes the results of the experiment through a mathematical model.

The formula (mathematical model—analytical expression) is chosen from a set of determined formulas (nonlinear approximation models), for example [4]:

$$y = ax^2 + bx + c \quad \text{(parable)}$$
$$y = a + b\ln x \quad \text{(logarithm)}$$
$$y = a\,\mathrm{e}\,bx \quad \text{(exponentially)}$$

Therefore, the problem is to determine the parameters a, b, c, as the case may be, while the formula (*analytical expression*) is known in advance, as a result of some theoretical considerations or the graphical presentation of the data.

Let us consider the general case when we have p parameters, and thus we will note the functional dependence by $y = f(x;\ a_1,\ a_2,\ \ldots\ ,\ a_p)$ [4]. The parameters a_1, a_2, … , a_p can not be determined exactly based on known values from previous historical, y_1, y_2, … , y_n data of the function, the latter containing random errors [4].

13.2.2.2 Question to Get a "*Good Enough*" Estimate

Therefore, if all the measurements of the variable y values are y_1, y_2, … , y_n, then the estimates of the parameters a_1, a_2, … , a_p are determined by the condition that the sum of the squares of the deviations of the measured values y_k from the calculated $f(x_k;\ a_1,\ a_2,\ \ldots\ ,\ a_n)$, take the minimum value, that is, the minimum expression:

$$S(a_1, a_2, \ldots, a_p) = \sum_{k=1}^{n} [y_k - f(x_k; a_1, a_2, \ldots, a_p)]^2 \tag{13.6}$$

The determination of the values of the parameters a_1, a_2, … , a_p is done by applying the conditions for obtaining the minimum value in the partial derivatives of the function S considered in the variables a_1, a_2, … , a_p i.e. the function with p variables $S(a_1,\ a_2,\ \ldots\ ,\ a_p)$. Obtaining these values means solving the p unknown system [4]:

$$\begin{cases} \dfrac{\partial S}{a_1} = 0 \\ \dfrac{\partial S}{a_2} = 0 \\ \quad \ldots \\ \dfrac{\partial S}{a_p} = 0 \end{cases} \tag{13.7}$$

13.2.2.3 Calculation of Continuity Indicators Using Exponential Function $f(x) = ae^{bx}$

In the case of the exponential model we study only two variables x (cause), y (effect) and we want to find the dependence $y = f(x)$, where $f(x) = ae^{bx}$ is a nonlinear dependence (exponential function) a and b.

If the variables x (cause), y (effect) are known in the measurements or values obtained for SAIFI/SAIDI by the data (x_i= years, y_i= values obtained for SAIFI/SAIDI), $i = 1, \ldots$, the exponential model $f(x) = ae^{bx}$ is determined by the coefficients a and b having the following expressions, according to [4]:

$$b = \frac{\sum_{i=1}^{n} x_i \sum_{i=1}^{n} \ln y_i - \sum_{i=1}^{n} x_i \ln y_i}{\left(\sum_{i=1}^{n} x_i\right)^2 - n \sum_{i=1}^{n} x_i^2}, \quad a = e^p$$
$$p = \frac{\sum_{i=1}^{n} x_i \ln y_i - b \sum_{i=1}^{n} x_i^2}{\sum_{i=1}^{n} x_i} \text{ or } p = \frac{\sum_{i=1}^{n} \ln y_i - b \sum_{i=1}^{n} x_i}{n} \tag{13.8}$$

Applying the presented calculation method for the use of the exponential function and taking into account the history of SAIFI/SAIDI continuity index achievements, we can obtain with good approximation the curves (with blue) of the decreases of SAIFI and SAIDI indicators for predictions of future years (Figs. 13.2 and 13.3).

Of course, all these predictions are based on the maintenance activity and the investments made in the electric grid, so it is possible to make corrections in the forecasts if in the respective year for which the determinations are done additional

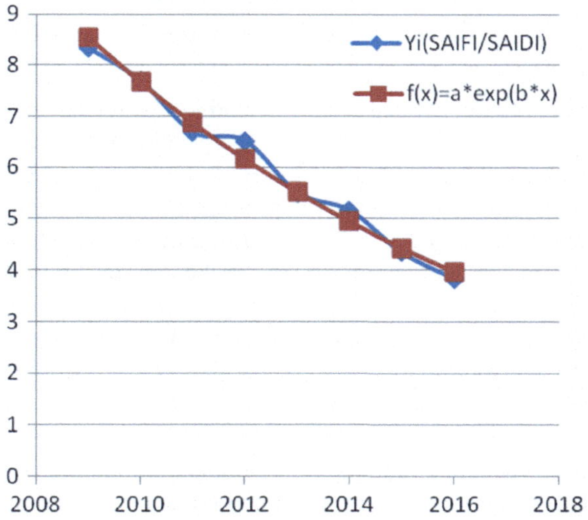

Fig. 13.2 The approximate downward trend of the indicator SAIFI

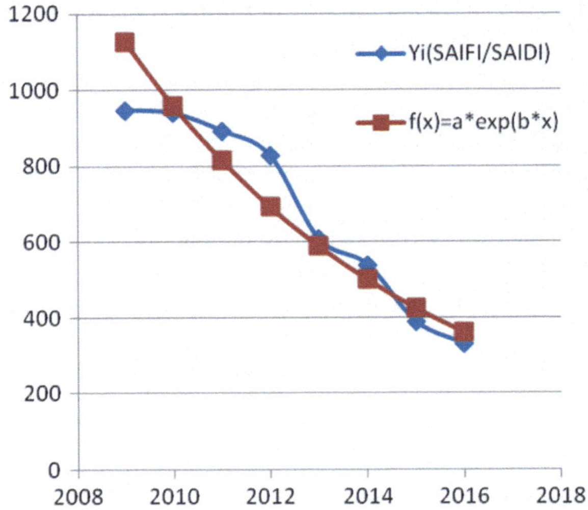

Fig. 13.3 The approximate downward trend of the indicator SAIDI

works are carried out or in the network are mounted reclosers or remotely controlled separators for remote operation and limitation of deficient areas after network incidents.

13.3 Determination the Mounting Location for Remote Control Equipment in Order to Obtain Maximum Efficiency in Decreasing the Continuity Indicators

It is well known that re-closure and load-carrying separators must be fitted to reduce network continuity indicators, and positioning is very important to determine where they can bring maximum influence in decreasing SAIFI/SAIDI indicators [5].

The Remote Control Separators (RCS) are switchgears consisting of a load separator equipped with arc extinguishers, possibly a fault detector, disconnection automation and command unit, and communication with a Central Point (PC). Due to the fact that separators can not switch off fault currents, all maneuvers needed to detect and eliminate the fault can only be performed by remote control reclosers (RCR) from the overhead power line axis or medium voltage cell breaker from the source transformer station.

Remote control reclosers are switchgear devices consisting of a circuit breaker with vacuum-extinguishing chambers, a numeric control-protection control terminal, a measure and a control and communication unit with a central point. Due to the fact that the reclosers can switch faults currents (12–16 kA), all the maneuvers needed to detect and eliminate the fault can be performed with them or with the medium voltage cell breaker in the source transformer station [6].

Communication between equipment and central dispatcher points is generally done through GPRS, as a cheaper communication environment, while fiber optics is a much safer environment, the execution of such a communication path is very expensive.

In order to determine the optimum location of the telecommunication equipment, several basic criteria should be considered [5]:

- Determination of the classification of the medium voltage overhead lines where new sectioning equipment will be installed and for which SAIDI (SAIFI) annual average should weigh 80%, average calculated for a relevant statistic period available in the database (minimum the last three years), the remaining 20% taking into account the number of consumers and the number of complaints from consumers.
- The so-called "analysis areas" set between the existing equipment are established on the lines chosen by the classification for mounting the remote-controlled equipment. For these areas of analysis, we determine the number of consumers and the annual average of the SAIDI in part, corresponding to the respective area (part of the annual average SAIDI calculated for the whole line).
- In order to achieve a balance between medium voltage lines, in terms of the density of the sectioning equipment, the maximum number of such equipment to be mounted on a medium-voltage airline will be 7, including the equipment mounted for taking over from other sources.
- In the event of incidents, in order to ensure the reinstatement of the supply for as many consumers as possible, the possibility of supplying the line and the derivations from other sources of supply than the source transformation station must be insured.
- The areas initially established for the location of remote-controlled equipment will be subdivided into sub-zones so as to include groups of less than 600–800 consumers as far as possible (or in the vicinity of this range).
- Preferably, up to three (13.3) remote controlled equipment connected in series to the line axis, with the exception of very long lines (≥ 50 km), shall be installed, taking into account that their protection shall be selective between them and the source transformer station.
- Major derivations (serving a number of consumers $\geq 15\%$ of line consumers) must be quickly isolated by introducing remotely controlled separation equipment.

13.3.1 Analysis Areas and Sharing Example

After selecting the medium voltage overhead lines on which to install new sectioning equipment (reclosers or remote-controlled separators), the next essential step is to identify the areas. The selection criteria should lead to the creation of similar areas for the sectioning equipment to be allocated. For a better representation of weights in the evaluation formula, the coefficients assigned to the criteria are adjusted so that their sum is 100% [7].

At this stage, it will be taken into account the topology of the network and the objective pursued:

- the length of the medium-voltage overhead line
- the maximum number of consumers supplied from that line
- similar sectioning equipment available on the line
- available pillars existing in the delimited areas and GPRS signal level to ensure good communication
- the basic criteria for determining the optimal mounting locations of the remote-controlled sectional equipment
- the objective of reducing the SAIFI/SAIDI indicators for the medium voltage electric power line.

Figure 13.4 exemplifies a part of a distribution grid split into relevant areas according to all the specified core criteria and taking into account both the network topology and the SAIFI/SAIDI continuity indicators reduction goal.

where:

$n_1 - n_8$ number of relevant areas for sectioning;
$DG_1 - DG_4$ derivatives with and without looping possibilities;

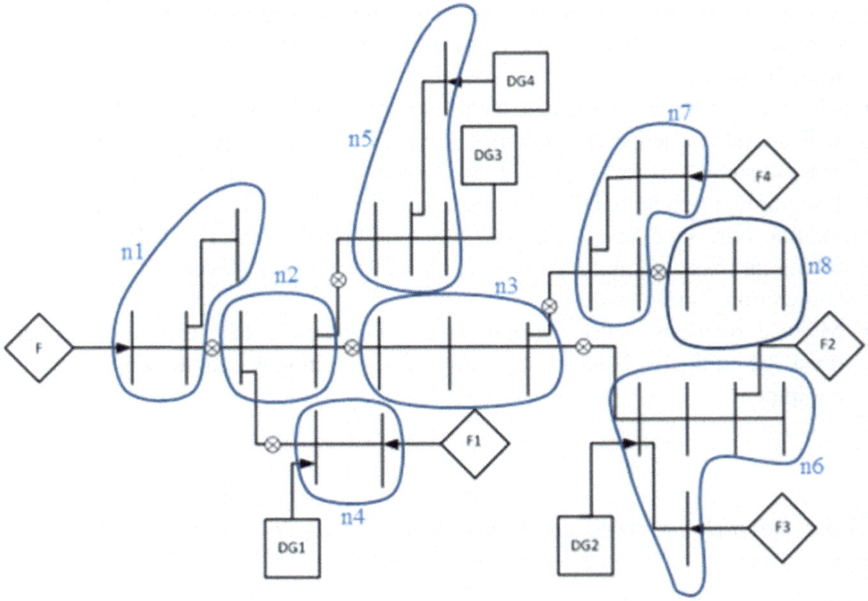

Fig. 13.4 Distribution network split in relevant areas for determining the locations of installation of sectioning equipment

F	the main power source of the overhead power line in the normal scheme (direct power from the transformer station);
$F_1 - F_4$	possibilities of looping through remote reclosers or separators.

13.3.2 Algorithm to Determine the SAIFI/SAIDI Continuity Indicator Reduction

Line analysis and line ranking with high SAIDI risk:

(1) In order to determine the location of the installation of remote-controlled separation equipment, the classification of the medium voltage overhead lines is prepared, using the statistics of the available events, using both the mean contribution of SAIDI on the respective lines and the set of factors defined for these determinations.
(2) In order to determine the behavioral nature of interruptions on medium-voltage overhead lines, the statistics of the last years, both in terms of the number of events (interruptions) and in terms of the duration of these events (interruptions), should be used.
(3) Below, according to [5] is an example of a fragment of the classification of medium voltage overhead lines, based on existing statistics, which includes data from the last 3 years of a distribution operator. According to the maximum density of separation equipment established on the medium voltage lines, the last column specifies the maximum number of new equipment proposed to be installed on each such line, as well as any proposals for exceptions to the defined criteria.

13.3.2.1 Analysis of a Line Susceptible to the Installation of Sectioning Equipment—Establishment of Optimal Areas for Installation of Sectioning Equipment

Through this case study it is done an actual analysis on a real line, with real calculated indicators, for which using the area analysis and using the defined criteria, it is determined the required number of reclosers and the approximate location of installation for these reclosers in order to obtain maximum reductions of SAIFI/SAIDI continuity indicators.

(1) First, it is prepared the simplified single phase diagram of the line taken from a classification as outlined in Table 13.1. Next, the diagram of this line is analysed, from a topological point of view (analysis of the normal scheme), and define the "delimited areas of analysis" either by the existing equipment

Table 13.1 Top SAIDI for 5 medium voltage power lines, an example from [5]

LEA (Medium voltage electric air line—20 kV)	SAIDI/year [min]	Number of consumers	Existing equipment	New equipment (Exceptions)
L_1	14.28	8.014	7	0 (4)
L_2	11.38	4.550	2	5
L_3	10.08	9.476	7	0 (3)
L_4	7.04	6.676	7	0 (2)
L_5	6.87	5.276	7	0

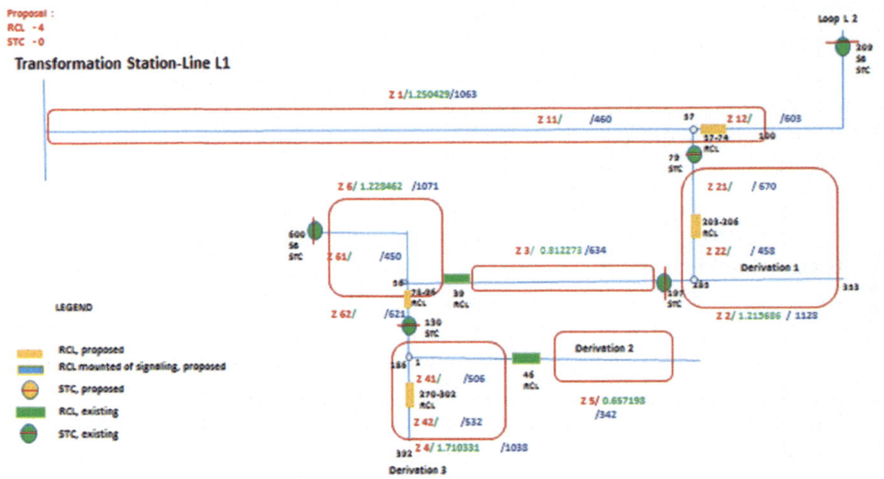

Fig. 13.5 Electric overhead Line split in 6 relevant areas for determining 4 recloser mounting locations

(remote separators, reclosers, manual section separators) or the topological characteristics of the line (derivatives, loops)—as in Fig. 13.5.

(2) For each analysis area established under the above rule the main features will be calculated from which the analysis of priorities and the need for new section equipment will be started. These include the number of consumers (eventually the power corresponding to the analysis areas), partial SAIDI corresponding to the respective areas calculated according to the average number of failures (events) per year and the average duration of the failures (events) on the respective part of the network for the events in a given area in one year—as in example from Table 13.2.

In the following, processing steps will be exemplified for a typical case of a line that has remote reclosers and separators installed, as shown in the legend of Fig. 13.5, steps for identifying the installation locations for reclosers or remote controlled separators and their prioritization [5]:

Table 13.2 Average SAIDI values per year per areas of an overhead power line

Line L_1	$N_i\ D_i$	SAIDI cumulated for 3 years	SAIDI/year	Prioritization
Z_1	5,293,064	3.751286	1.250429	2
Z_2	5,146,000	3.647059	1.215686	2
Z_3	3,438,353	2.43682	0.812273	
Z_4	7,239,829	5.130992	1.710331	1
Z_5	2,781,919	1.971594	0.657198	
Z_6	5,200,081	3.685387	1.228462	2
Total	29,099,246	20.62314	6.874379	

(1) In the defined areas, taking into account the number of consumers and the existing sectioning equipment, the values for SAIDI/year and the number of consumers delimited by the area.
(2) In the split scheme, the derivations Der 1–3 on the line and the possibility of looping of L_1 line analyzed from the L_2 loop line are identified, thus identifying the proposed positions for the equipment and determining the benefit brought about by the reduction of the continuity indicators by mounting the new equipment.

It can be noticed that for a line, in the areas where SAIDI is identified as the maximum of the statistical determinations, area 4, according to Table 13.2, also results in an immediate prioritization of recloser mounting. For the calculations to reduce the indicators after reclosers installation, determinations that take into account proposed locations, split areas are subdivided into computational subzones, such as zone 4 being divided into Z_{41} and Z_{42}. Of course, the order of priority 2 for recloser installation may be a future step or one to be made with the priority order 1, taking into account in this choice of all the other criteria defined for that purpose.

13.3.2.2 Calculation of the Estimated Benefits of Fitting Reclosers on Top SAIFI/SAIDI Overhead Power Lines

The benefits calculation is one that takes into account the number of reclosers or remote-controlled separators that will be installed as in the medium voltage network. Because in the future it is also necessary to consider the automation of the network, the switching equipments should be chosen taking into account this aspect, choosing to install mainly or only reclosers.

Going on the example of a Distributie Energie Oltenia—CEZ Group in Romania—who mounted such reclosers on average about 100 reclosers/year for 5 years (2015–2019), through a project in which there were analyzed about 300 medium voltage lines for which the installation of reclosers would bring the maximum benefits to SAIFI/SAIDI indicators for unplanned interruptions, considering that at

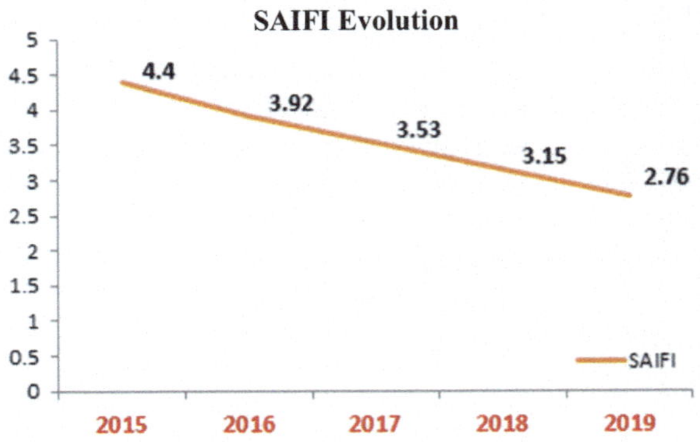

Fig. 13.6 The evolution curve of SAIFI unplanned interruptions after the installation of about 500 reclosers in a medium voltage network

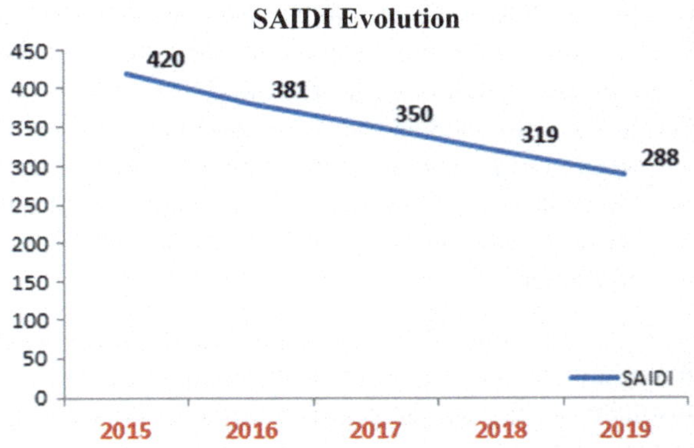

Fig. 13.7 The evolution curve of SAIDI unplanned interruptions after the installation of about 500 reclosers in a medium voltage network

2015 level it had already installed in the medium voltage network another 700 remote controlled sectioning equipment, were obtained from the calculations of the decreasing of the indicators with the annual average determined in the paragraph "Analysis of a line susceptible to the installation of sectioning equipment. Establishment of optimal areas for the fitting of sectional equipment", the downward curves of Figs. 13.6 and 13.7.

13.3.2.3 Proposals Based on the Evolution of Performance Indicators

(1) After the implementation of the sectioning equipment, the operating phase will follow the performance attributed to the respective equipment and the expected benefits. Within this operation, at the end of each operational year, the SAIFI/SAIDI indicators will be assessed and their framing will be checked within the estimated performance limits. Performance graphs for each line on which reclosers have been installed over several operating years will be made and the trends of these performances (SAIFI/SAIDI) will be followed. Depending on these trends, complementary measures for future implementation can be proposed.

(2) The second major operation, to be performed in the operational phase, is to review the performance targets based on the results obtained in the previous operational year. Based on the results of unplanned interruptions, the SAIFI/SAIDI savings on that line will be recalculated and performance estimates graphs will be recalculated. The overall trend should be to improve the limits, in the sense of maintaining the majority of future results in the area designated as the one with highest probability of production/evolution.
(3) Of course that besides these actions related to specific investments in the network, maintenance works can be done in the network, investments to improve the state of the network (replacement of insulation, poles, replacement of parts of overhead lines with underground ones) or switch to network automation so that many of the long-term interruptions (>3 min) turn into short interruptions (<3 min).

Depending on the evolution of the indicators, taking into account the deviations from estimation to achieved results, individual measures are taken on the lines with big deviations regarding the deviations, coming with investment proposals for the improvement of the state of the network (punctual for those medium voltage distribution lines), thus maintaining the trend of decreasing the continuity indicators in the electricity supply of consumers as an ongoing stage, and in the next stage to implement step by step automation of the medium voltage grid for quick isolation of defect areas.

13.4 Reducing the Continuity Indicators Using Automation Methods of the Medium Voltage Distribution Network

In order to ensure continuity in the electricity supply to consumers, it is possible to use the automatic resupply of the consumers with isolation of the fault zone without the intervention of the dispatcher by remote manipulation of the remote control equipment.

13.4.1 Automatic Restoration

For example, a medium voltage line (20 kV LEA) is being considered as an example, which is powered by the PA—Power Point or Transformer Station, as base source, with the possibility of looping with the 20 kV LEA loop no. 1 and LEA 20 kV Loop no. 2 and which are reserve sources for the takeover of consumers L 20 kV no. 1.

Automation on the overhead line LEA 20 kV no. 1 is a system for automatic reconfiguration of the distribution network, which manages a number of

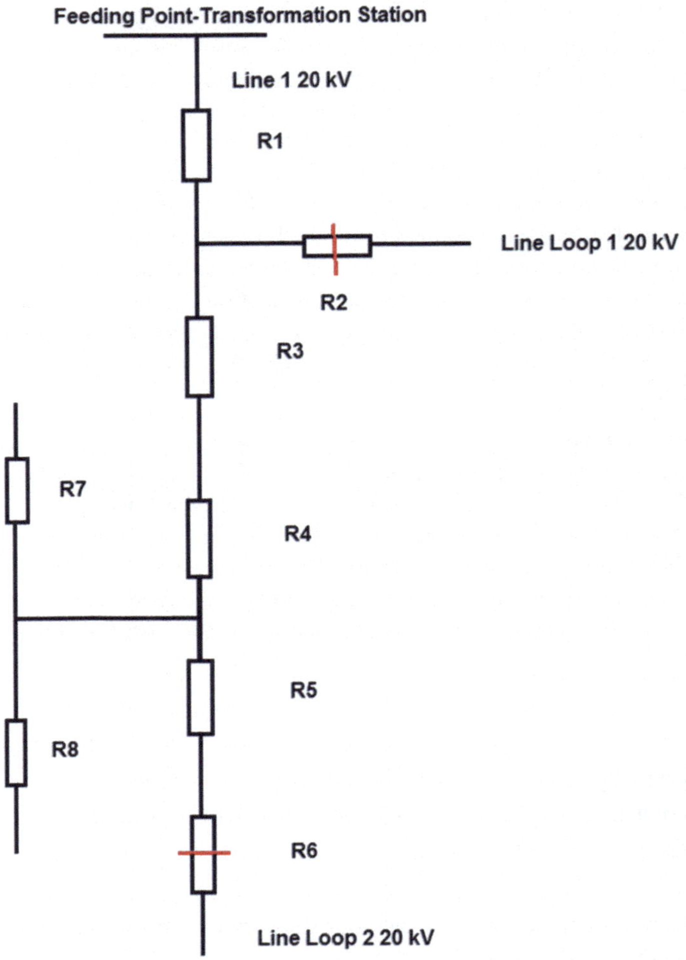

Fig. 13.8 Scheme of positioning of the reclosers included in the automation of the LEA 20 kV no. 1 [8]

8 reclosers, of which 4 in the axis of the line, 2 on the derivations and 2 in the loop points with the two LEA 20 kV loop no. 1 and 2, as specified in the supply scheme of Fig. 13.8 [8].

Automation has as a central element a Restauration controller that has implemented a set of algorithms that can handle multiple field equipment (reclosers, load separators, or circuit breakers) according to predefined scenarios.

Sequences for isolating and reconfiguring the distribution network in the event of network failures are created using the controller-specific software and taking into account the recloser protection functions. From the point of view of communication, GPRS or optical fiber is used as the physical environment, and the data traffic can be done through the IEC protocol.

13.4.1.1 Advantages of Automation

- Limits the number of consumers affected by long-term interruptions (SAIFI)
- Reduces the time to isolate the fault, which leads to a decrease in the consumer interruption time and implicitly to the SAIDI indicator
- Reduces the time for fault detection by controlling the limited fault area, considering the density of reclosers installed for the automation of an overhead medium voltage power line.

13.4.1.2 Self-healing Automation Includes 3 Stages

- In the first stage the protection will initiate triggering and therefore the set of adjustments and the protection coordination will be designed so that it will trigger the recloser closer to the defect area. The protection relay will have available and enabled the specific protection functions as well as the automatic re-start function which will restore the power supply in the case of passing faults [8].
- The second stage is the isolation of the defective network area, this being done automatically after the controlled triggering and will be monitored by the self-healing controller from the central point [8].
- The third stage is the energy resupply of the unaffected areas, which will be automatically monitored and controlled by the central self-healing system controller, after this stage, depending on the restoration of the power supply, the local dispatcher intervenes for additional commands and defect identification with the operational personnel of the field [8].

13.4.1.3 Logic of Isolation of the Fault Area and Reconfiguration of the Network

Figure 13.9 shows the position of the reclosers in the normal operating situation, all reclosers connected, as well as the supply point switch, and disconnected only the L 20 kV loop reclosers 1 and 2 loop [8]. With P01 and P03 were marked the switches from the transformation station.

In the event of a permanent fault (short-circuit or grounding) on one of the network sections in Zone 3 marked with C03 in Fig. 13.10, the fault will be

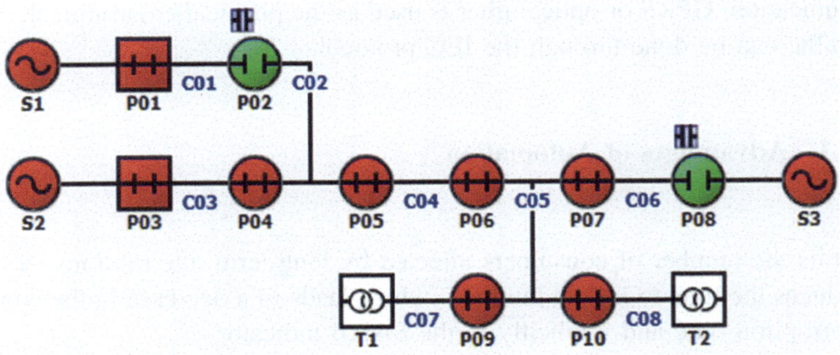

Fig. 13.9 Normal operation of reclosers, red = closed, green = open [8]

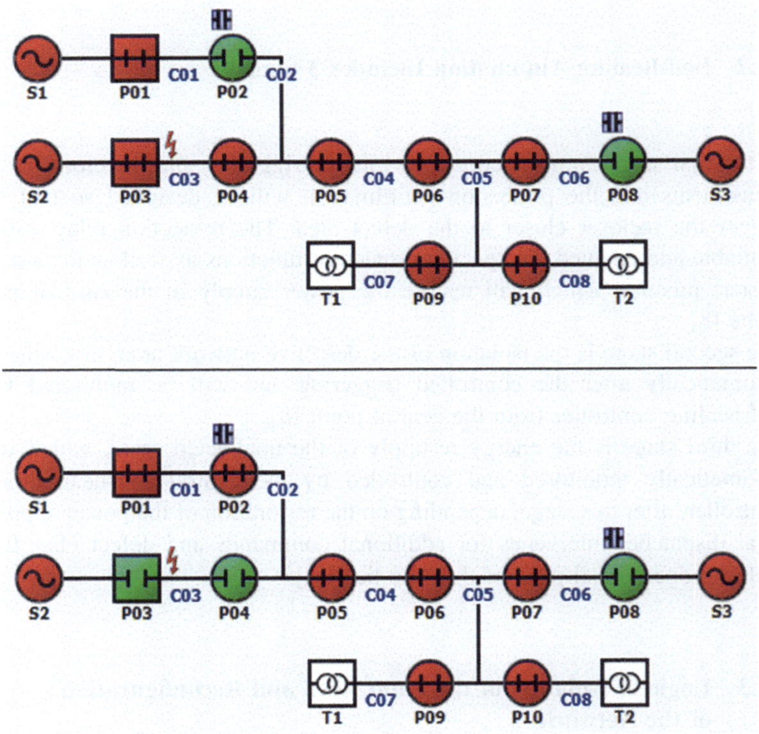

Fig. 13.10 Defect in ZONE 3 and the status after restoration, red = closed, green = open [8]

detected and removed by the self-protection of the reclosers (after performing RAR cycles), and via the programmable logic controller that communicates (via GPRS) with each recloser the fault isolation command will be given and the power supply from one of the two sources to the healthy network areas will be restored. In the case illustrated in Fig. 13.2, the switch in S2 (P03) and the first recloser after PA (P04) will be disconnected, thus bypassing the fault, then connecting the recloser P02 to resupply the consumers from the 20 kV loop LEA. With S were noted the sources.

The simplest case of defect isolation is when the defect occurs on one of the derivations supplied by the P09 or P010 reclosers, because after the automatic reclosing cycle, if not successful, the reclosers remain triggered, without the automation having to restore the supply, these derivations being supplied radially.

For all network defects/triggering events, an automatic network reconfiguration will be initiated if the system is set to automatic state. Sequences will automatically run, controlled and monitored by the self-healing controller. Automation will be fully controllable, both in terms of working arrangements, both by the dispatcher and by the system administrator.

Any manual command (locally given by the operational staff or remote from the dispatcher) will have priority and will inhibit the operation of the automation until the dispatcher reactivates the operation of the automation. All operating conditions are monitored and events are generated in the SCADA Event Log (Supervisory Control and Data Acquisition) from dispatchers, thus alerting the dispatcher to possible problems. After an incident has been remedied, the return to normal operation must be done only manually through dispatch center [9].

In order to increase the safety of the consumers' power supply and to improve the performance indicators, this automation is very suitable, with the mention that it requires the need for a controller at the central control point [10].

Extending the already existing self-healing automatic system to a line can only be done for a limited number of other lines, requiring an up-grade system for an unlimited number of lines.

13.4.2 Automation for Decreasing of Maneuver Time

Given that the desire to achieve an automation is to use only the numerical terminals of the reclosers, without the need for communication between them or with a central point, below is presented such an automation that uses only the conditions given in the field, without interfering with other equipment.

Automation consists of a logic that will be implemented in Remote Terminal Unit (RTU) equipment and can be used for radial or rectangular medium voltage lines that are fitted with reclosers equipped with numerical protection terminals and RTU for GPRS communication, fibres optics or radio and integration into the SCADA system of these segmentation equipment [11, 12]. No additional physical equipment or equipment is required.

Automation reduces the total time for at the maneuveurs done by the dispatcher for a resupply time $t < 1$ min from voltage failure.

13.4.2.1 Primary Conditions for Implementing Such Quasi-automation

The voltage transformer (VT) for charging the RTU power supply battery and the radio station (if radio communication is made) must be mounted upstream of the recloser. This VT also controls the presence of upstream voltage to the source (transformer station or feed point).

13.4.2.2 Principle of Automation Operation

It is considered a medium-voltage overhead power line that is supplied from a transformation station, a line that we considered to be supplied radially, but which can also be looped, the principle of automation being the same, as represented in Fig. 13.12, similar to the line in Fig. 13.11, but with fewer derivations.

If the Station switch or any Rn reclosers are triggered by protection, all non-voltage downstream equipment is self-disconnected by automation at a short time, until the RAR automations cycle. The self-logic logic is implemented in the

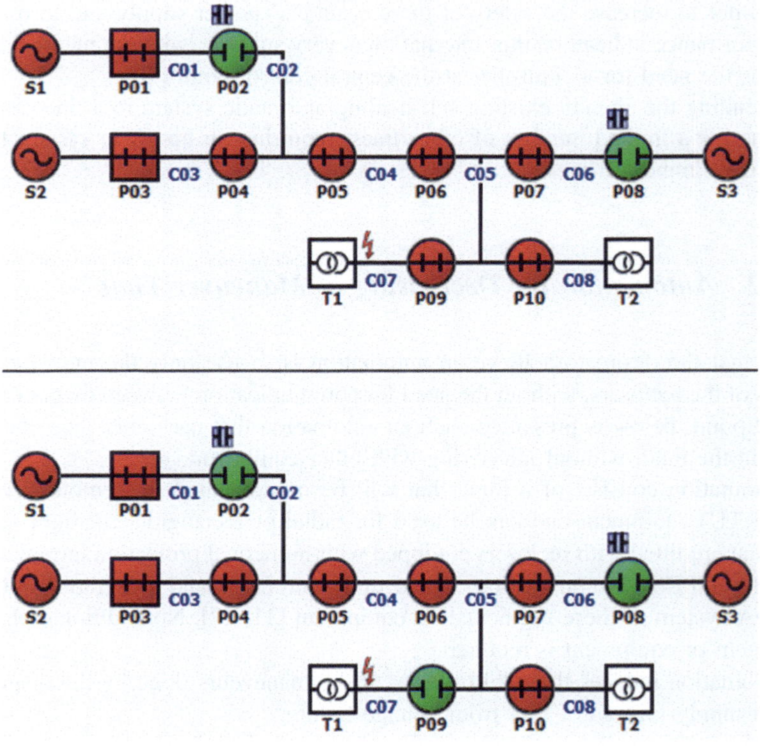

Fig. 13.11 Defect in Zone 7 and the status after restoration, red = closed, green = open [8]

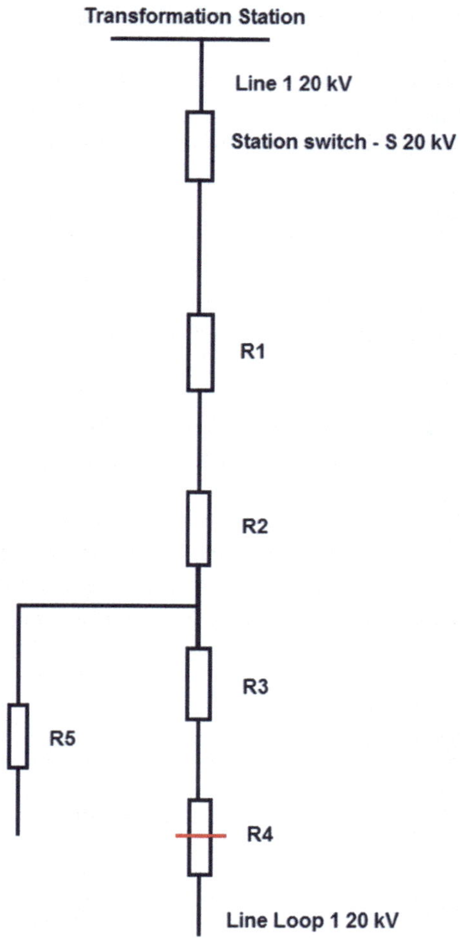

Fig. 13.12 A medium-voltage overhead power line radially supplied from a transformer station

RTU associated to each recloser, here also taking into account the RAR cycle times and the self-disconnection time.

When the voltage comes back on by connecting the triggered equipment (via SCADA, manually, or after a successful RAR (+) automatic re-start), all downstream equipment is connected by automation. In order for the automation to function in the sense of the logic described, the re-switching time at the upstream voltage must be chosen higher than the RAR pause.

If the upstream voltage does not appear, the self-connected reclosers remain in this state, the dispatcher intervening and doing the power recovery maneuvers shortly sealing the defective zone between the first two reclosers, the first one triggering or making the RAR (−), the latter being self-connected, thus reassigning the task of reconnecting the other auto-connected downstream reclosers and connecting a loop recloser to replenish the area from the downstream to the defective area.

13.4.2.3 Automation Lock Conditions

For the correct functioning of the automation and without changing the operating mode of the dispatcher and the operating personnel in operation, the automation still has to comply with some conditions presented in Table 13.3.

13.4.2.4 Returning from the Locked State in Operation

Returning from lockout depends on the reason of the lock and can only be done in the situations shown in Table 13.4.

13.4.2.5 Integration into SCADA and Transmission to the Dispatcher of Automated Signaling

Supervision of operation and automation control can be made from SCADA, in the SCADA system the signals indicated in Table 13.5 are available.

The automations presented are functional in the distribution networks of a Distribution Operator from Romania. This is a first step towards automating the whole network to achieve it in an Advanced Distribution Management System

Table 13.3 Automation lock situations

Ways of locking	Detail
Auto locking	The automation self-locks after 10 min after the voltage drops. The logic of this self-blocking is that each voltage drop is part of a sequence (eventual event situation) that needs to be resolved in max. 10 min automatically
The localkey L/D	Operational personnel are assumed to be working or maneuvering equipment
External disconnect command (SCADA command, local command, protection command transmitted by staff)	In the three situations, there was a reason for staff to intervene to cause a protection, a local disconnection, or SCADA, with the authority returning to these situations only to operational personnel or dispatchers, with automation blocked
SCADA cancellation command of automation	SCADA cancellation command of automation

Table 13.4 Situations of return from the locked state in the operating state of the automation

Ways to unlock	Details
Exit voltage upstream	If the lack of previous voltage lasted more than 10 min, a new occurrence/disappearance sequence unlocks the automation
Remote L/D key	When crossing the L/D key to "Distance", it is assumed that the operative staff completed the work or maneuvers on the spot
Command commissioning from SCADA	If the automation has been canceled from the SCADA, the commissioning command activates the automation
Switch switch command	A connection command (as a sequence of a RAR (+) or SCADA switch connection command) unlocks automation

Table 13.5 Signals in SCADA of automation

Signal type	Meaning	States
Signaling	Automation status	Canceled/in function
	Locking automation	Unlocked/locked
	Disconnect via automation	Act
	Connect via automation	Act
Commands	Automation status	Canceled/in function

(ADMS) where, using special algorithms and a communications environment with reclosers through a particular protocol, it is possible to control the entire network in regarding the localization of defects, their isolation and the restoration of the consumers supply.

References

1. Romanian Standard of Performance for Power Distribution Service, Order no. 11, pp. 11–13, 18.04.2016
2. C. Giron, F. Javier Rodriguez, L. Gimenez de Urtasum, S. Borroy, Assessing the contribution of automation to the electric distribution network reliability. Int. J. Electr. Power Energy Syst. **97**, 120–126 (2018)
3. T. Adefarati, R.C. Bansal, Reliability assessment of distribution system with the integration of renewable distributed generation. J. Appl. Energy **185**(1), 158–171, 1 January 2017
4. M. Vlada, in *Nonlinear Approximation Models: Mathematical Calculations and Applications*, Bucharest University, pp. 17–29, 2012
5. Internal Romanian OD Study, in *Methodology for Prioritizing the Installation of Remote Separators and Reclosers—Part I and Part II*, pp. 3–14, respectively 3–27
6. F. Vatra, A. Poida, A.C. Vatra, Home Area Network and Smart Home Concepts in the Smart Grids Structures, CNEE, pp. 160–180, 2017
7. C. Winzer, Conceptualizing energy security. J. Energy Policy **46**, 36–48 (2012)
8. I. Dobrescu, M. Albu, S. Zamfirache, M. Pasol, Improving the Performance Indicators in the Distribution of Electricity by Implementing a System for Automatic Reconfiguration of the Distribution Network, CNEE, pp. 160–166, 2017

9. H. Albert, A Priority Issue: The Quality of the Electricity to the Consumer, www.sier.ro/Articol_Albert_Hermina.pdf
10. R. Arghandeh, A. Von Meier, L. Mehrmanesh, L. Mili, On the definition of cyber-physical resilience in power systems. Renew. Sustain. Energy Rev. **58**, 1060–1069 (2016)
11. E.R. Larsena, S. Osoriob, A. Van Ackere, A framework to evaluate security of supply in the electricity sector. Renew. Sustain. Energy Rev. **79**, 646–655 (2017)
12. M. Castroa, A. Moona, L. Elner, D. Roberts, B. Marshall, The value of conservation voltage reduction to electricity security of supply. J. Electric Power Syst. Res. **142**, 96–111 (2017)
13. Z. Popovic, B. Brbaklic, S. Knezevi, A mixed integer linear programming based approach for optimal placement of different types of automation devices in distribution networks. Electr. Power Syst. Res. **148**, 136–146 (2017)
14. P.H. Larsen, K.H. La Commare, J.H. Eto, J.L. Sweeney, J. Energy **117**, 29–46 (2016)

Index

A
Artificial Intelligence (AI), 175, 176
Attack, 6, 66, 122, 223–236, 238, 239, 241, 249, 256, 257, 270, 275, 276, 278, 279, 289, 318, 320, 321

B
Big Data, 45, 46

C
Cluster centroid, 126
Communication, 10, 52–55, 58, 59, 64, 65, 81, 83, 98, 223, 224, 227, 230, 231, 234, 238, 242–244, 276, 278, 289, 290, 294, 296, 299, 301, 305, 309, 312, 313, 315, 319, 320, 334, 345
Constraint, 3, 16, 17, 23, 57, 122–125, 132, 135, 136, 147, 163, 164, 170, 184, 197, 199, 239
Continuity in power supply, 326, 328
Control and Communication, 290
Cost function, 17, 123, 126, 130, 149
Critical infrastructure, 7, 66–68, 114, 115, 238, 242, 262, 269–272, 290–293, 321, 322
Critical load, 89, 110, 122, 127
Critical processes, 289
Cryptographic algorithm, 229, 232
Cyber, 46, 65, 66, 81, 271, 276, 280, 289, 294
Cyber-attacks, 46, 64, 83, 97, 225, 228, 231, 238, 248, 256–259, 289, 291–293, 317, 321
Cyber criminal, 223
Cyber-Physical System (CPS), 83–86, 97, 98, 270–275, 277–279, 281–284
Cyber warfare, 229

D
Deliberate attack, 45, 46, 105, 223, 225–228, 230, 243, 247–249, 257, 258
Demand Side Management (DSM), 10, 11, 86, 87
Diagnostic agent, 49
Distributed Energy Resources (DER), 65, 115, 123, 139, 142, 143, 146
Distributed Generation (DG), 11, 50, 68, 70, 81, 82, 122, 123, 198
Distribution network, 3, 6, 9, 12, 14–19, 21, 34, 35, 49, 51, 68–70, 75, 119, 120, 122, 123, 127, 130, 136, 139, 141–143, 145, 152, 159, 195, 205, 263–266, 297, 326–328, 336, 341, 342, 348

E
Electric Vehicle (EV), 48, 64, 81, 85, 122, 143
Electro-energetic, 290
Embedded system, 223–226, 229–232, 234, 235, 238, 243, 244
Emergency condition, 142
Energy Management System (EMS), 48, 51, 86, 139, 142, 144, 146
Energy storage, 11, 47, 58, 88, 94, 108, 123, 139, 150, 152, 157, 159, 250
Evolutionary Computation (EC), 163, 177–181, 183
External shock, 193–197, 199, 205, 210
Extreme event, 105, 107, 111–113, 122, 129, 130, 139

F
Fault Tolerance, 269

© Springer Nature Switzerland AG 2019

N. Mahdavi Tabatabaei et al. (eds.), *Power Systems Resilience*, Power Systems,
https://doi.org/10.1007/978-3-319-94442-5

Field Programmable Gate Array (FPGA), 223–226, 230, 235
File Transfer Protocol (FTP), 54
Fitness function, 3, 17, 21, 29, 32, 35, 178
Flexibility, 9, 56, 68, 81–88, 90–94, 98, 99
Fragility, 3, 12–14, 19, 21, 24, 25, 27, 29, 31, 32, 130–132

G
Geographical location, 3, 25, 27, 31, 32
Greenfield, 119, 124, 132
Grid security, 139

H
Hacker, 98, 223, 224, 226–229, 232, 233, 238, 239, 243, 292
Harmonic, 326

I
Industrial control systems, 283, 290, 292
Information and Communication Technologies (ICT), 83–86
Infrastructure, 6, 7, 9, 11, 18, 46, 57, 59, 65, 67, 68, 83, 84, 86, 87, 95, 96, 98, 101, 104, 106, 108, 112, 132, 164, 224, 231, 238, 248, 249, 256, 258, 262, 314, 320
Intellectual Property (IP), 223, 225, 227, 232
Internal shock, 195, 197, 205, 209
Internet of Things (IoT), 45, 46, 59, 224, 239, 243, 271
Intrusion, 83, 97, 235, 238, 318, 320, 321
Isolation of defects, 325

K
K-Means, 124, 125

L
Levelized cost of energy, 132
Linear Programming (LP), 93, 193
Load loss, 3, 92
Loss factor, 15

M
Malicious, 83, 103, 114, 115, 223–230, 232–235, 239, 242, 243, 276, 277, 279, 283
Malware, 230, 231, 233, 234, 238, 239, 243, 278, 290, 293, 320, 321
Microgrid, 11, 81, 85, 93, 110, 119, 123, 127, 132, 135, 139, 142–144, 146, 147, 149, 152, 153, 157–159

N
Natural disaster, 3, 6, 24, 35, 45, 46, 66, 119, 120, 122, 123, 129, 130, 132, 250, 252, 257
Networked-microgrids, 141, 143, 144, 146, 159

O
Optimal configuration, 21
Optimal Power Flow (OPF), 112, 143, 163–165, 170–172, 179, 181
Optimization, 3, 21, 24, 27, 35, 163, 164, 171, 173, 175, 177–181, 184–186, 201, 213, 236

P
Performance standard, 325, 327, 328, 330, 331
Power quality, 51, 69, 79, 123, 325, 326
Power System (PS), 223, 234, 236, 238, 242, 243, 250
Power system restoration, 96–98, 112
Protection system, 45, 46, 76

Q
Quality of Service (QoS), 53, 54, 279, 325
Quantitative resilience, 101, 103, 113, 115

R
Recovery, 6, 8–10, 46, 64, 69, 81, 83, 90, 95, 96, 99, 103, 104, 106, 112–115, 122, 130, 197, 205, 234, 257, 262, 279, 283, 347
Redundancy, 8, 9, 89, 234, 262, 269, 278, 296, 318
Reliability, 3, 6–8, 12, 15, 19, 57, 83, 86, 87, 89, 94, 95, 97, 122, 132, 144, 164, 166, 195–197, 225, 230, 236, 258, 290, 296, 300, 304, 314, 330
Reliability index, 101, 103, 112, 115, 201
Remote Terminal Unit (RTU), 48, 52, 271, 295, 345–347
Resilience, 6–11, 65, 96, 104, 105, 108, 113–115, 122, 130, 144, 152, 156, 157, 193, 197, 223, 224, 234–238, 262, 269, 270, 273, 279, 281–284
Resilience enhancement, 269, 279, 283
Resiliency, 9, 10, 20, 35, 45, 46, 76, 81, 83–90, 95–99, 122, 123, 141, 144, 150, 152, 154, 159, 195, 197, 243, 248
Resiliency constraint, 119, 132, 135
Resiliency enhancement, 3, 35, 142

Resiliency improvement, 32, 34, 35, 262
Resiliency index, 20, 21, 24–30, 32, 34, 35, 134, 136
Resiliency metrics, 7, 103, 110
Resiliency operation, 139
Resilient, 3, 6, 9–11, 15, 16, 19, 21, 26, 28, 31, 32, 82, 101, 120, 142, 150, 165, 248, 280
Resilient distribution system, 193, 213
Resilient system expansion planning, 195
Restoration, 10, 13, 89, 90, 95–97, 112, 122, 123, 234, 259, 343
Risk, 10, 27, 31, 67, 95, 223–226, 236, 239, 242, 243, 249, 250, 254, 256–258, 277, 279, 321
Risk management, 11, 65, 248, 262

S
Search algorithm, 50, 122, 178, 180, 184, 185
Security, 6, 46, 52, 58, 65–68, 75, 81, 83, 85, 86, 94, 98, 163, 173, 226, 228, 229, 233–239, 242–244, 250, 257, 258, 269–271, 274–281, 283, 290, 293, 299, 316, 318, 321
Security situations, 290
Smart, 239, 242
Smart grid, 6, 45, 48, 50, 84, 103, 122, 132, 143, 144, 224, 236, 242, 294, 297, 299, 313–315, 318, 319, 322
Smart meter, 84, 242, 290, 318
Smart street lighting, 239–241
Spanning tree, 16, 23, 122
Spatial risk index, 3, 120
Stability, 50, 68, 83, 85, 94, 96, 167, 279, 280, 290, 318, 321
Stochastic, 97, 139, 142, 145, 171, 185
Supervisory Control and Data Acquisition (SCADA), 46, 48, 51, 292, 294–296, 299, 305, 312, 317, 320, 345
System Average Interruption Duration Index (SAIDI), 326, 329–331, 333, 335–339, 341, 343
System Average Interruption Frequency Index (SAIFI), 106, 325, 326, 329–331, 333, 335, 336, 339, 343

T
Target indicators, 331
Threat, 6, 7, 65, 66, 68, 70, 101, 104, 105, 115, 223, 225, 228, 238, 239, 242, 243, 256, 258, 259, 274, 276, 279, 290, 292, 321

U
Uncertainty, 90, 95, 98, 127, 132, 144, 145, 196

V
Variable Generation (VG), 82, 92, 98
Viruses, 230, 234
Vulnerability, 6, 12, 45, 46, 66, 67, 89, 225, 239, 240, 242, 249, 263, 289, 291, 293

W
Wind Farm (WF), 74, 263, 327
Wireless Application Protocol (WAP), 233, 235

The manufacturer's authorised representative in the EU is Springer Nature Customer Service Centre GmbH, Europaplatz 3, 69115 Heidelberg, Germany. If you have any concerns regarding our products, please contact ProductSafety@springernature.com

Printed and bound by CPI Group (UK) Ltd, Croydon, CR0 4YY
05/10/2026
02228144-0011